Lecture Notes on
Coastal and Estuarine
Studies

Managing Editors:
Richard T. Barber Christopher N. K. Mooers
Malcolm J. Bowman Bernt Zeitzschel

12

Offshore and Coastal Modelling

Edited by
P. P. G. Dyke, A. O. Moscardini, and E. H. Robson

Springer-Verlag
Berlin Heidelberg New York Tokyo

Managing Editors

Richard T. Barber
Coastal Upwelling Ecosystems Analysis
Duke University, Marine Laboratory
Beaufort, N.C. 28516, USA

Malcolm J. Bowman
Marine Sciences Research Center, State University of New York
Stony Brook, N.Y. 11794, USA

Christopher N. K. Mooers
Dept. of Oceanography, Naval Postgraduate School
Monterey, CA 93940, USA

Bernt Zeitzschel
Institut für Meereskunde der Universität Kiel
Düsternbrooker Weg 20, D-2300 Kiel, FRG

Contributing Editors

Ain Aitsam (Tallinn, USSR) · Larry Atkinson (Savannah, USA)
Robert C. Beardsley (Woods Hole, USA) · Tseng Cheng-Ken (Qingdao, PRC)
Keith R. Dyer (Taunton, UK) · Jon B. Hinwood (Melbourne, AUS)
Jörg Imberger (Western Australia, AUS) · Hideo Kawai (Kyoto, Japan)
Paul H. Le Blond (Vancouver, Canada) · Akira Okubo (Stony Brook, USA)
William S. Reebourgh (Fairbanks, USA) · David A. Ross (Woods Hole, USA)
S. Sethuraman (Raleigh, USA) · John H. Simpson (Gwynedd, UK)
Robert L. Smith (Corvallis, USA) · Mathias Tomczak (Cronulla, AUS)
Paul Tyler (Swansea, UK)

Editors

P. P. G. Dyke
Plymouth Polytechnic, Department of Mathematics, Statistics and Computing
Drake Circus, Plymouth, Devon, PL4 8AA, United Kingdom

A. O. Moscardini and E. H. Robson
Department of Mathematics and Computer Studies
The Polytechnic, Sunderland, Tyne and Wear SR1 3SD, United Kingdom

ISBN 3-540-96054-6 Springer-Verlag Berlin Heidelberg New York Tokyo
ISBN 0-387-96054-6 Springer-Verlag New York Heidelberg Berlin Tokyo

Library of Congress Cataloging in Publication Data. Main entry under title. Offshore and coastal modelling. (Lecture notes on coastal and estuarine studies; 12) Papers presented orally at the seventh POLYMODEL conference held at Sunderland Polytechnic in the U. K. in May 1984 and organized by the North East of England Polytechnic's Mathematical Modelling and Computer Simulation Group-POLYMODEL. Bibliography: p. Includes Index. 1. Ocean engineering–Mathematical models–Congresses. 2. Offshore structures–Mathematical models–Congresses. 3. Storm surges–Mathematical models–Congresses. 4. Tidal currents–Mathematical models–Congresses. I. Dyke, P. P. G. II. Moscardini, A. O. III. Robson, E. H. IV. North East Polytechnics Mathematical Modelling and Computer Simulation Group. Conference (7th: 1984: Sunderland Polytechnic) V. Series. TC1505.Q344 1985 627'.042 84-26860
ISBN 0-387-96054-6 (New York)

This work is subject to copyright. All rights are reserved, whether the whole or part of the material is concerned, specifically those of translation, reprinting, re-use of illustrations, broadcasting, reproduction by photocopying machine or similar means, and storage in data banks. Under § 54 of the German Copyright Law where copies are made for other than private use, a fee is payable to "Verwertungsgesellschaft Wort", Munich.

© 1985 by Springer-Verlag New York, Inc.
Printed in Germany

Printing and binding: Beltz Offsetdruck, Hemsbach/Bergstr.
2131/3140-543210

PREFACE

The papers contained in this volume were presented orally at the seventh POLYMODEL conference, held at Sunderland Polytechnic in the United Kingdom in May 1984 and sponsored by Barclays Bank PLC and Imperial Chemical Industries Ltd. The conferences are organised annually by the North East of England Polytechnic's Mathematical Modelling and Computer Simulation Group - POLYMODEL. The Group is a non-profit making organisation based on the mathematics department of the three polytechnics in the region and has membership drawn from those educational institutions and from regional industry. Its objective is to promote research and collaboration in mathematical and computer-based modelling.

After a short introductory chapter, the volume may be considered as dividing naturally into four parts. Chapters 2 to 5 constitute the first part on Tides, Storm Surges and Coastal Circulations which deals with the hydrodynamics of coastal seas. Chapters 6 to 11 concern Coastal Engineering Modelling and discuss such coastal phenomena as beach erosion, sediment transport, and non-linear waves. The third part (Chapters 12 to 16) on Offshore Structures considers sea structures in general and the connections between the structures (hoses, moorings, pipelines) in particular. The last two chapters focus on Offshore Corrosion problems.

The editors would like to take this opportunity to thank all of the participants of POLYMODEL 7 for making the conference so successful and particularly those who presented the papers. The editors are especially indebted to Avril Barton and Barbara Foulds who not only typed this manuscript but also gave valuable secretarial assistance during the organising of the conference. The help of Neil Purves who re-drew some of the diagrams is also acknowledged.

P.P.G. Dyke,
A. O. Moscardini,
E. H. Robson.

Sunderland Polytechnic, October 1984.

… C O N T E N T S

PAGE

CHAPTER 1	MODELLING IN OFFSHORE AND COASTAL ENGINEERING: P. P. G. Dyke	1
CHAPTER 2	TIDES, STORM SURGES AND COASTAL CIRCULATIONS: N. S. Heaps	3
2.1	Bathymetry	3
2.2	Tides and Tidal Currents	3
2.3	North Sea Storm Surges	10
2.4	Two-dimensional Numerical Storm-surge Models	15
2.5	Surge Forecasting	26
2.6	Three-dimensional Models	40
2.7	Vertical Structure of Current	42
CHAPTER 3	MODELLING STORM SURGE CURRENT STRUCTURE: A. M. Davies	55
3.1	Introduction	55
3.2	Spectral Model Formulation	55
3.3	Form of Vertical Eddy Viscosity	58
3.4	A 3-D Simulation Model of Surge Currents on the North-West European Shelf	65
3.5	A Mechanistic Model of Wind Induced Current Profiles	71
3.6	Concluding Remarks	77
CHAPTER 4	Optimally Controlled Hydrodynamics for Tidal Power from the Severn Estuary: S. C. Ryrie	83
4.1	Introduction	83
4.2	Hydrodynamics	83
4.3	Optimal Control	85

			PAGE
CHAPTER 5		NUMERICAL MODELLING OF STORM SURGES IN RIVER ESTUARIES: E. H. Twizell	93
	5.1	Introduction	93
	5.2	Points to note in Modelling	93
	5.3	The Aims of a Mathematical Model of Storm Surges	95
	5.4	The Differential Equations of the Model	98
	5.5	Computational Aspects	102
	5.6	Numerical Results : The Storm of 1953	105
	5.7	Summary	107
CHAPTER 6		COASTAL SEDIMENT MODELLING: B. A. O'Connor	109
	6.1	Introduction	109
	6.2	Need for Computer Models	109
	6.3	Model Types	111
	6.4	Conclusions	132
CHAPTER 7		THE APPLICATION OF RAY METHODS TO WAVE REFRACTION STUDIES: I. M. Townend and I. A. Savell	137
	7.1	Introduction	137
	7.2	Ray Models	139
	7.3	Application of the Ray Model, a Simple Case	146
	7.4	A Study including Wave Breaking	154
	7.5	A Study including Diffraction and Reflection	161
	7.6	Future Developments	163
CHAPTER 8		A MODEL FOR SURFACE WAVE GROWTH: A. J. Croft	165
	8.1	Introduction	165
	8.2	Formulation of the Problem	166
	8.3	First Order Solution	170
	8.4	Second Order Solution	172
	8.5	Third Order Solution	173
	8.6	Wind Shear Stress	179
	8.7	Conclusion	184

PAGE

CHAPTER 9		POWER TAKE-OFF AND OUTPUT FROM THE SEA-LANCHESTER CLAM WAVE ENERGY DEVICE: F. P. Lockett	187
	9.1	Introduction	187
	9.2	Experimental Tests	187
	9.3	Power Take-off Simulation	188
CHAPTER 10		NUMERICAL MODELLING OF ILFRACOMBE SEAWALL: P. Hewson and P. Blackmore	201
	10.1	Introduction	201
	10.2	Finite Element Modelling of Ilfracombe Seawall	201
	10.3	Need for a Finite Element Model Approach	202
	10.4	Conclusions	215
CHAPTER 11		MODELLING THE PLAN SHAPE OF SHINGLE BEACHES: A. H. Brampton and J. M. Motyka	219
	11.1	Introduction	219
	11.2	General Considerations when Modelling Beach Changes	221
	11.3	Derivation of an Alongshore Transport Formula	225
	11.4	Incipient Motion of Shingle	228
	11.5	Discussion	231
	11.6	Conclusions	232
CHAPTER 12		MATHEMATICAL MODELLING APPLICATIONS FOR OFFSHORE STRUCTURES: P. J. Cookson	235
	12.1	Introduction	235
	12.2	Operational, Environmental and Foundation Condition	236
	12.3	Structural Concepts	237
	12.4	Fabrication	242
	12.5	Construction	243
	12.6	Load Out	244
	12.7	Tow Out	246
	12.8	Installation	247
	12.9	Mathematical Modelling in Platform Design	247
	12.10	Conclusions	250

			PAGE
CHAPTER 13		MATHEMATICAL MODEL OF A MARINE HOSE-STRING AT A BUOY : PART 1, STATIC PROBLEM: M. J. Brown	251
	13.1	Introduction	251
	13.2	Assumptions	254
	13.3	Equations	255
	13.4	Boundary Conditions	256
	13.5	Hose Radius	257
	13.6	The Load	258
	13.7	Method of Solution	259
	13.8	Analytical Solutions for Simplified Models	261
	13.9	Results	262
	13.10	Applications	268
	13.11	Conclusions	274
		Appendix 1	277
CHAPTER 14		MATHEMATICAL MODEL OF A MARINE HOSE-STRING AT A BUOY : PART 2, DYNAMIC PROBLEM: M. J. Brown	279
	14.1	Introduction	279
	14.2	Equation of Motion	279
	14.3	Boundary Conditions	281
	14.4	Method of Solution	282
	14.5	Flanges	284
	14.6	Comparison of Analytical and Numerical Results	288
	14.7	Numerical Results	289
	14.8	Conclusions	298
		Appendix 1	301
CHAPTER 15		THE DESIGN OF CATENARY MOORING SYSTEMS FOR OFFSHORE VESSELS: A. K. Brook	303
	15.1	Introduction	303
	15.2	Representation of the Environment	305
	15.3	Mathematical Model of Moored Vessel	307
	15.4	Calculations of Environmental Forces and Moments	308
	15.5	Calculation of Mooring Forces and Moments	314
	15.6	Static Analysis	319
	15.7	Response of Vessel to Wind Gusting and Wave Drift Action	320
	15.8	Conclusions	320

			PAGE
CHAPTER 16		SOME PROBLEMS INVOLVING UMBILICALS, CABLES AND PIPES: D. G. Simmonds	323
	16.1	Introduction	323
	16.2	The Statics of Cables and Pipes	323
	16.3	Hydrodynamic Forces	329
	16.4	Analytical Solutions	333
	16.5	Typical Problems and Numerical Solutions	338
	16.6	Final Comments	348
CHAPTER 17		MATHEMATICAL MODELLING IN OFFSHORE CORROSION: A. Turnbull	353
	17.1	Introduction	353
	17.2	General Mass Transport Theory	355
	17.3	Mathematical Modelling of the Electrochemistry in Cracks	356
	17.4	Mathematical Modelling in Cathodic Protection Offshore	371
	17.5	Mathematical Modelling of Crevice Corrosion and Protection	372
	17.6	Mathematical Modelling of Corrosion in Concrete	372
	17.7	Conclusions	374
CHAPTER 18		FATIGUE CRACK GROWTH PREDICTIONS IN TUBULAR WELDED JOINTS: S. Dharmavasan	377
	18.1	Introduction	377
	18.2	Fatigue Crack Growth Behaviour of Tubular Joints	377
	18.3	Theoretical Analysis of Crack Growth	385
	18.4	Conclusions	392
Subject Index			394

1. MODELLING IN OFFSHORE AND COASTAL ENGINEERING

P.P.G. Dyke,
Department of Mathematics and Computer Studies,
Sunderland Polytechnic,
Green Terrace,
Sunderland.

As a branch of engineering, offshore engineering is probably one of the newest. On the other hand, coastal engineering is one of the oldest, yet each has benefited from the other. Offshore engineering has had to grow up very quickly and has relied on the expertise of coastal engineers when data and knowhow have been sadly lacking. Coastal engineering, an altogether less glamorous and one can almost say overlooked, branch of engineering, has received a welcome shot in the arm from all the exploration and production now occurring in coastal waters throughout the world. The major difference between them, besides their pedigree, is that Offshore Engineering tends to be related to the extraction of hydrocarbons from continental shelf seas and hence can be seen as a net contributor to the Gross National Product. Coastal Engineering, however, is an expense justified only by the promise of an incursion of much greater expense if the effects of waves, wind and currents on coastal structures and the coast itself are ignored. This difference is a very important one to industry and research grant awarding bodies, especially when they are spoilt for choice on how to distribute monies amongst researchers, and has lead many erstwhile coastal engineers to conduct research in the field of offshore engineering.

Parallel with the rapid rise of offshore engineering has been the well publicised increase in accessibility of computers. The important aspect of this for the engineer is the improvement in the data handling speed and the ease with which very large programs can now be processed by even modest computers. The computer is now an essential tool for the engineer to use in design, and the mathematical model has become as commonplace in the everyday work of the engineer as the laboratory model. In fact the mathematical model, if well enough constructed, has the advantage over the laboratory model of relative cheapness and greater flexibility. In the last twenty years, offshore engineering has given birth to many interesting and challenging

P. P. G. Dyke

problems. At the same time, the problems of coastal engineering, which have hiterto been solved by rule of thumb procedures, are at last becoming tractable thanks to advances in computing. In this volume, some of the latest mathematical models in offshore and coastal engineering are presented.

2. TIDES, STORM SURGES AND COASTAL CIRCULATIONS

N.S. Heaps,
Institute of Oceanographic Sciences, Bidston Observatory,
Birkenhead, Merseyside

2.1 Bathymetry

This account is concerned with tides, storm surges and circulation patterns on the continental shelf surrounding the British Isles. The bathymetry of the area is shown in figure 2.1. The 200m depth contour delineates the shelf break beyond which depths increase rapidly to oceanic values. The North Sea shallows from depths of 200 to 100m in the north to less than 50m in the south. On the West Coast: the Irish Sea, the Bristol Channel and the English Channel constitute a system of inner channels connecting to the North Atlantic Ocean through the outer shelf areas of the Celtic Sea in the south and the Hebridean Sea in the north (figure 2.2).

2.2 Tides and Tidal Currents

Figure 2.3 presents the M_2 cotidal chart for the North Sea derived by Proudman and Doodson (1924) using coastal observations of tidal amplitude, offshore observations of tidal current, and the equations of tidal motion. Along the east coast of England it is evident that the mean tides attain amplitudes exceeding 2m in places with a minimum of around 0.9m near Lowestoft. Large semidiurnal tides develop in the Irish Sea and in the Bristol Channel due to resonances with the incident North Atlantic tides. Thus the principal harmonic constituent, M_2, attains an amplitude exceeding 3m at Liverpool and 4m at Avonmouth (figure 2.4).

Two-dimensional, depth-averaged numerical tidal models have been successfully formulated for the North Sea, the Irish Sea, the English Channel and the entire continental shelf. A shelf model due to Flather (1976), with a grid 1/2° longitude by 1/3° latitude, gave tidal amplitudes and phases generally coming to within 10% and 10°, respectively, of observed values. An extension of that shelf model, covering the North East Atlantic, is shown in figure 2.5. The extended model produced the M_2 cotidal chart shown in figure 2.6, in good agreement with charts based on observations (Flather 1981). The tidal regime of figure 2.6 represents a co-oscillation with boundary tides obtained from tidal measurement (Cartwright et al 1980). A radiation condition is applied along

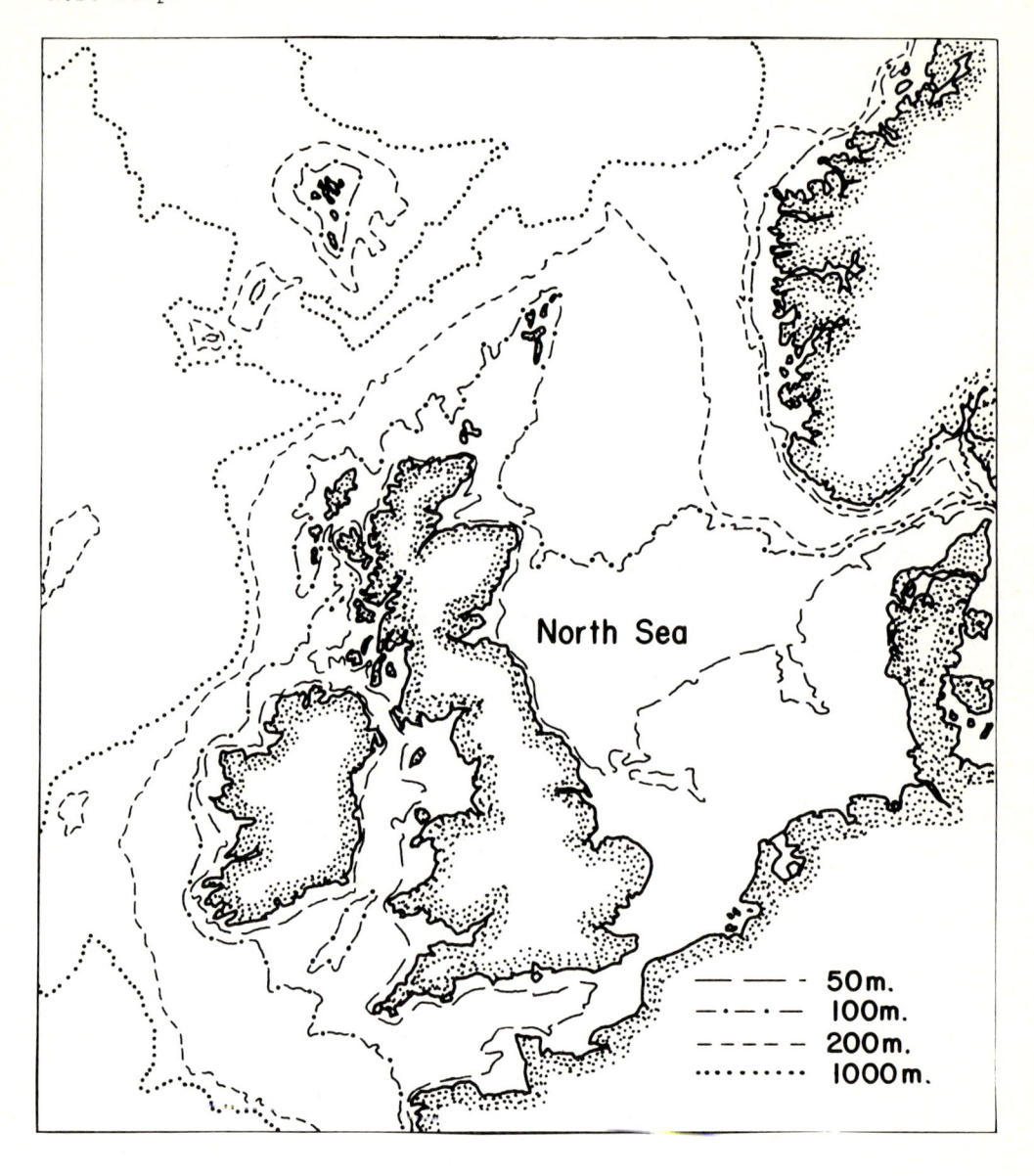

Figure 2.1   Bathymetry of the sea areas around the British Isles

Tides, Storm Surges and Coastal Circulations

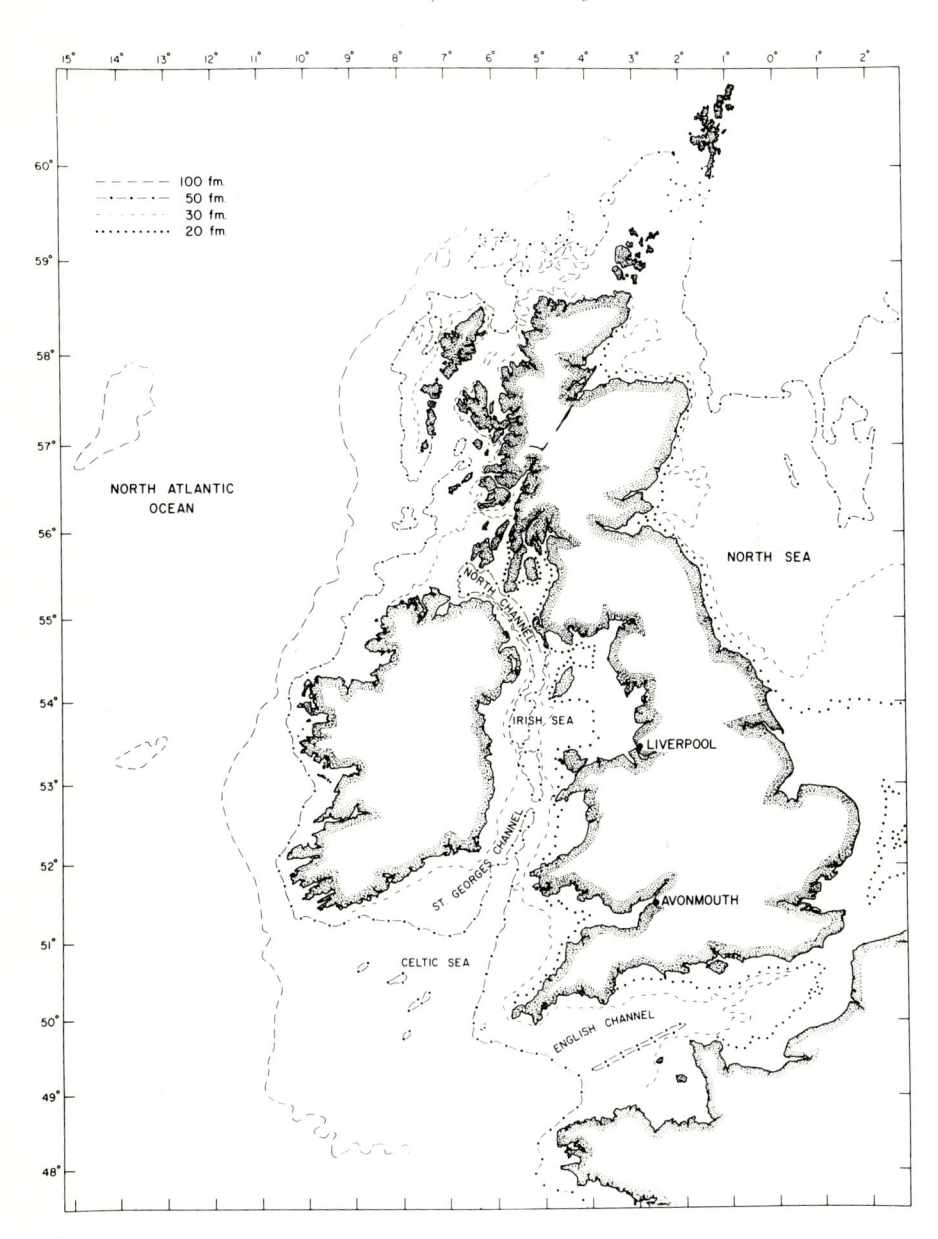

Figure 2.2   Sea areas on the west coast of the British Isles

N.S. Heaps

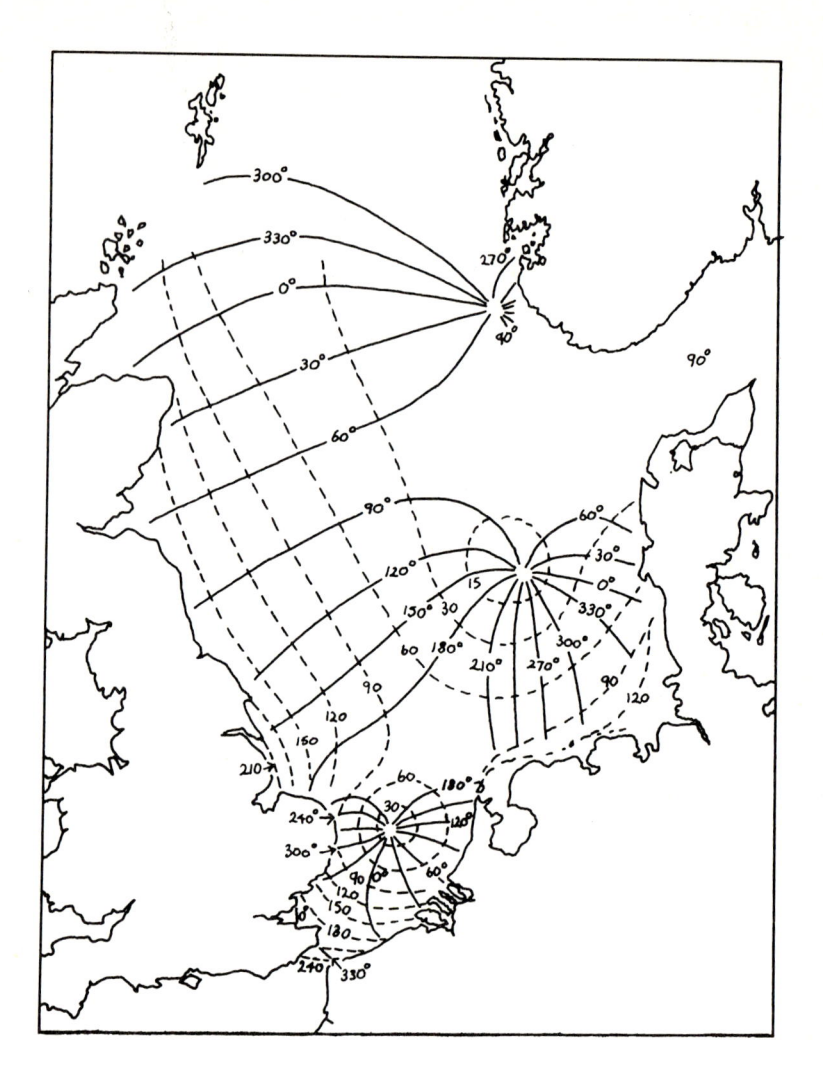

Figure 2.3 M₂ cotidal chart for the North Sea; ─────── phase in degrees, ──────── amplitude in centimetres. From Proudman and Doodson (1924).

Tides, Storm Surges and Coastal Circulations

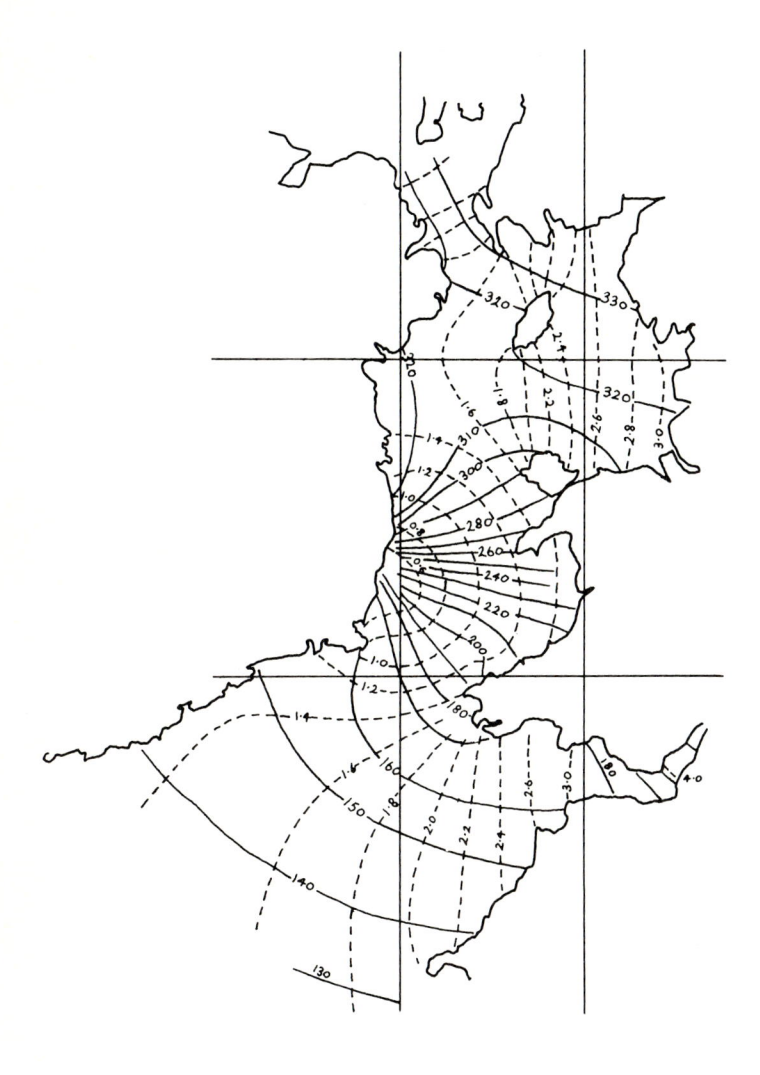

Figure 2.4 M_2 cotidal chart for the west coast of the British Isles;
——— phase in degrees, ------- amplitude in metres.
From Robinson (1979).

N.S. Heaps

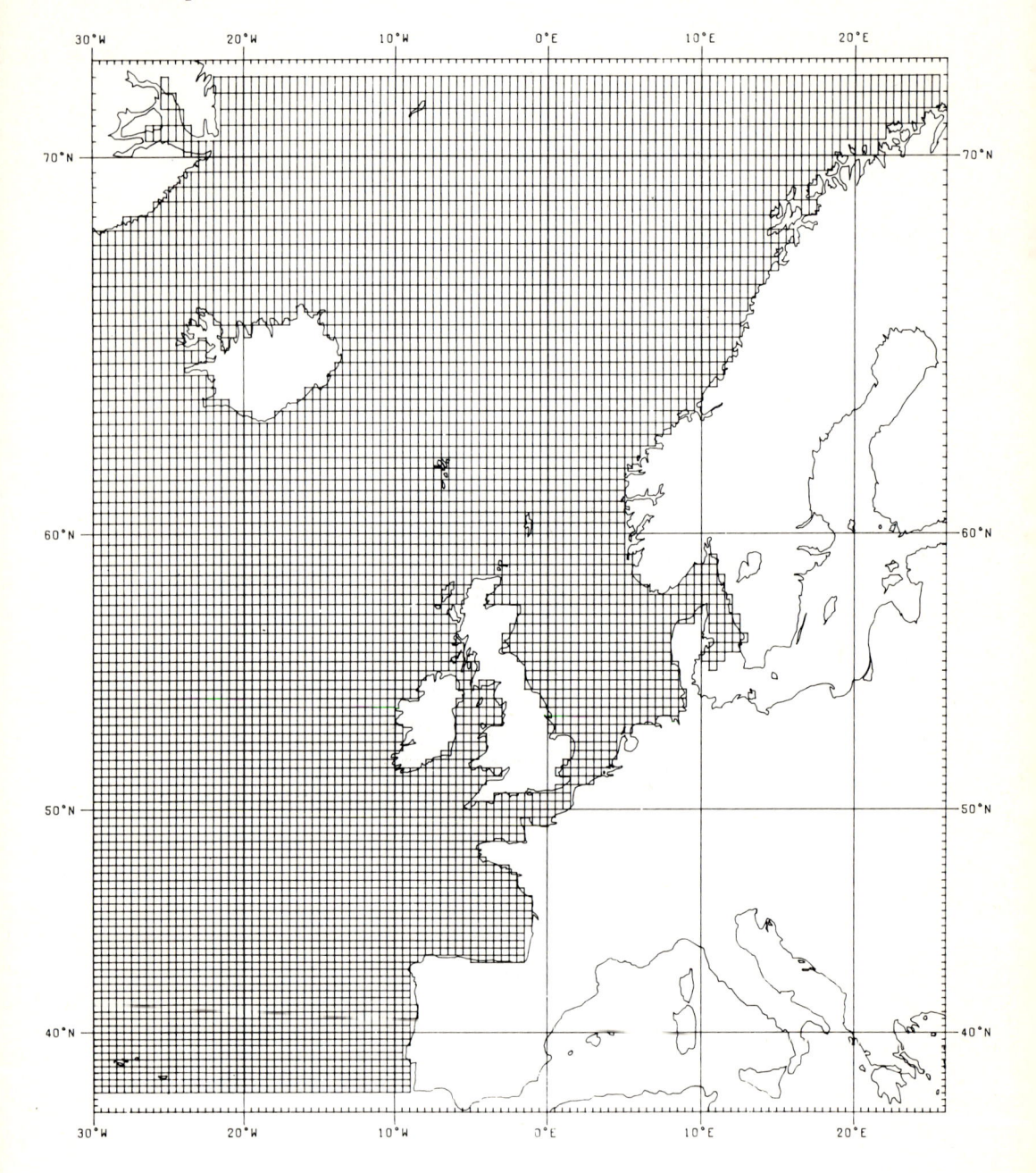

Figure 2.5 Finite difference mesh of the North East Atlantic Model. From Flather (1981).

Tides, Storm Surges and Coastal Circulations

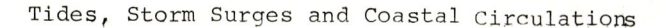

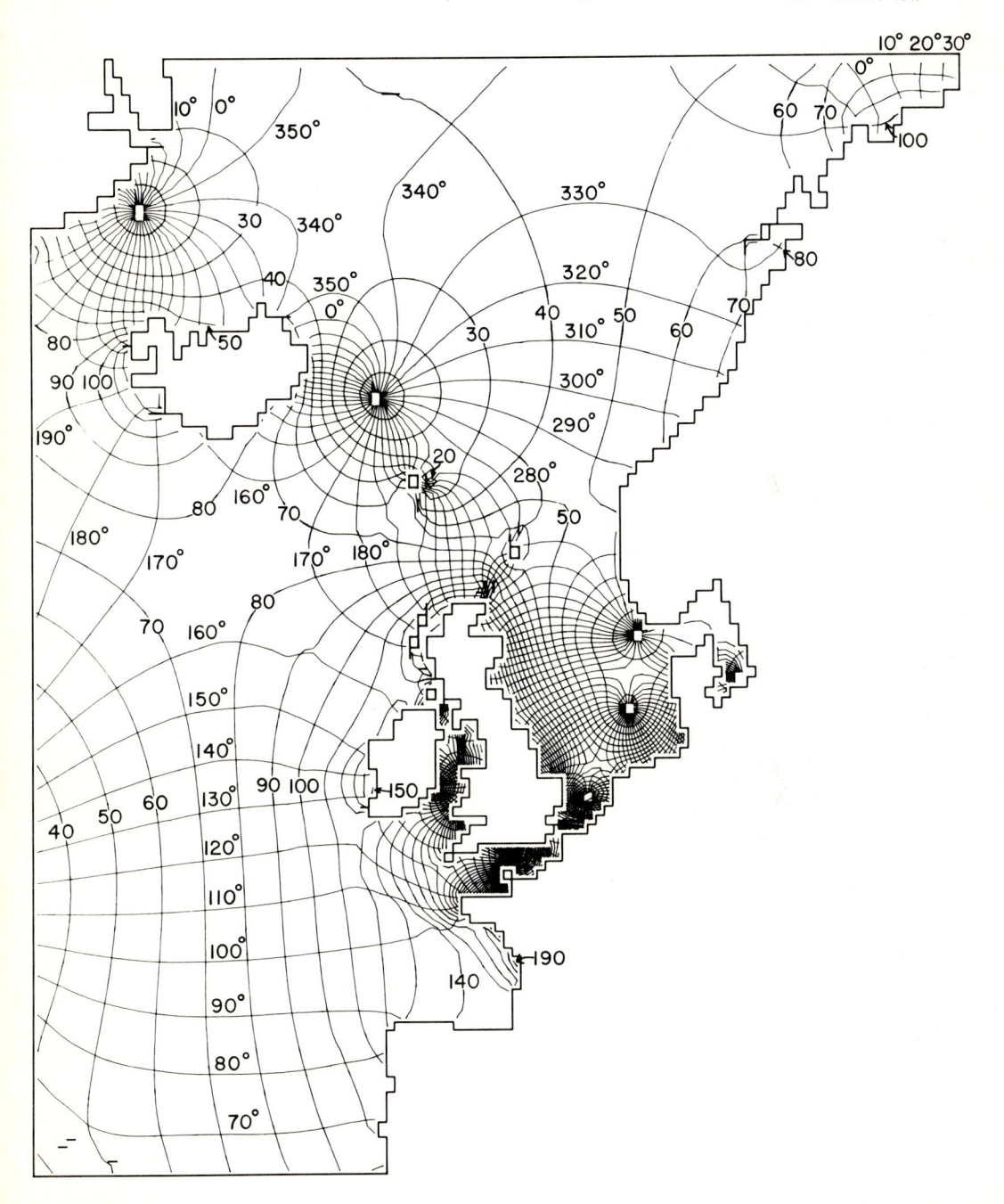

Figure 2.6 Chart of the M_2 co-oscillating tide showing the variation of phase in degrees and amplitude in centimetres. From Flather (1981).

N.S. Heaps

the open boundary to allow disturbances generated in the interior of the model to pass outwards across the boundary without reflection. A horizontal eddy viscosity equal to 10^3 m^2s^{-1} is employed. Further work with this model has included the effects of Earth tides, Earth-tide loading, and the tide-generating forces. Also, several more tidal constituents have been evaluated including S_2, N_2, K_2, K_1, O_1, P_1 and Q_1.

A contour map of near-surface tidal current amplitudes around the British Isles, at mean springs (M_2+S_2), is reproduced in figure 2.7. The map is based on observations of current. Small scale variations in current amplitude in the vicinity of headlands and islands, which can lead to large currents, are omitted. Note the relatively high currents in the western channels and in the sourthern part of the North Sea : currents between 125 and 150cm^{-1} are attained in those areas. A three-dimensional numerical model of the Northwest European Continental Shelf has been used by Davies and Furnes (1980) to compute M_2 tidal currents. The numerical solution of the hydrodynamic equations involved a finite-difference grid in the horizontal, as in figure 5 but covering the shelf only. The components of horizontal current were expanded through the vertical in terms of a set of cosine functions and the coefficients in those expansions were determined by means of the Galerkin Method. Vertical eddy viscosity was assumed to be uniform through the depth and proportional to the square of the depth-mean current. A quadratic law of bottom friction was employed. The computed M_2 current ellipse parameters, namely the semi-major and semi-minor axes, the orientation of the ellipse and the sense of rotation, were found to be in good agreement with observed values in the North Sea. Sea-surface and sea-bed distributions of the computed ellipse parameters are shown in figures 2.8 and 2.9.

2.3 North Sea Storm Surges

Few areas are more susceptible to large dangerous storm surges than the southern part of the North Sea. The east coast of England, including the Thames Estuary, is particularly threatened by sea floods. The Thames Barrage has been constructed to protect London against this danger.

A permanent network of tide guages around the shores of the North Sea monitors the propagation of surges within that basin (figure 2.10).

Tides, Storm Surges and Coastal Circulations

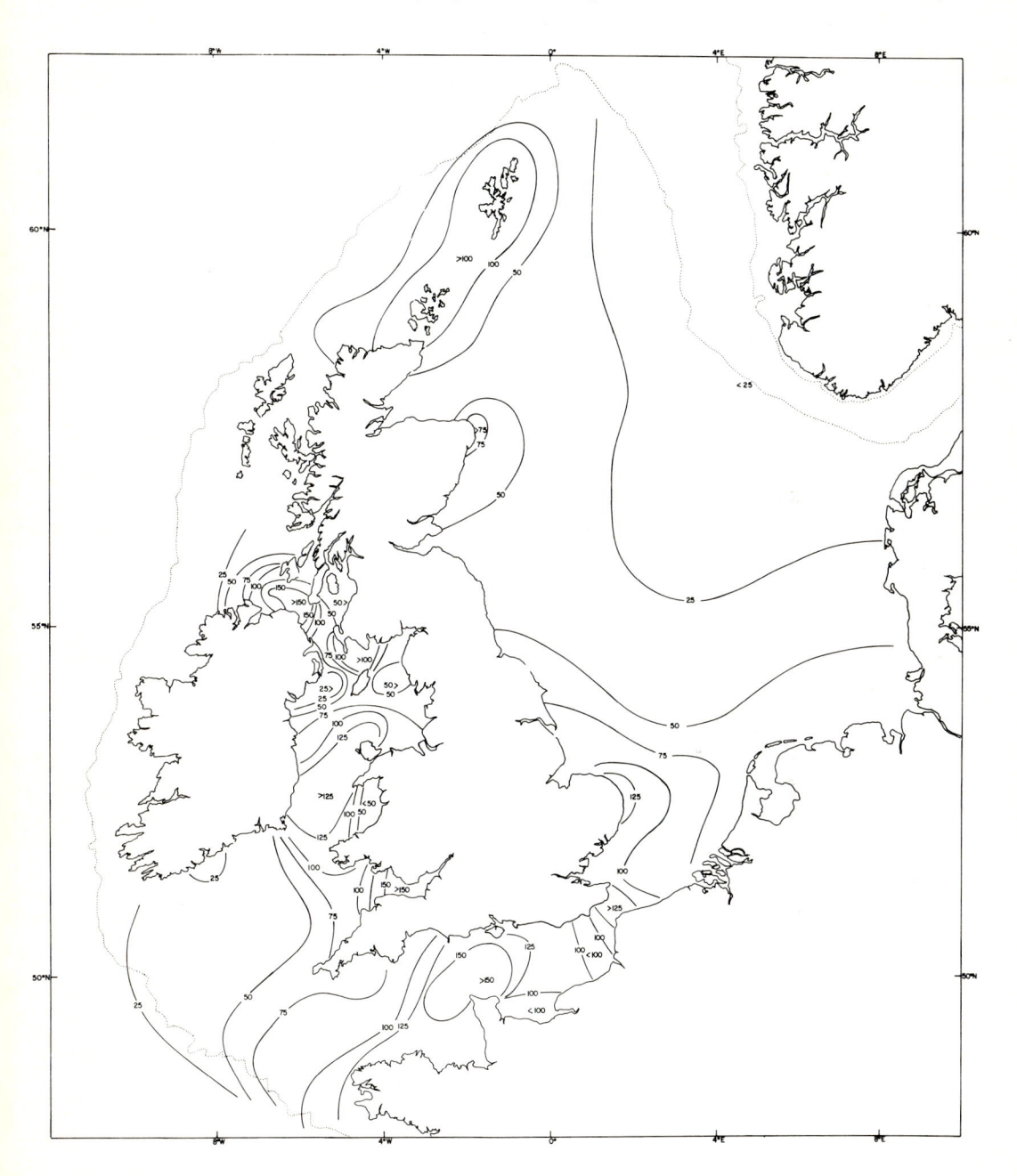

Figure 2.7 Mean spring near-surface tidal current amplitudes in cms^{-1}. From Howarth and Pugh (1983).

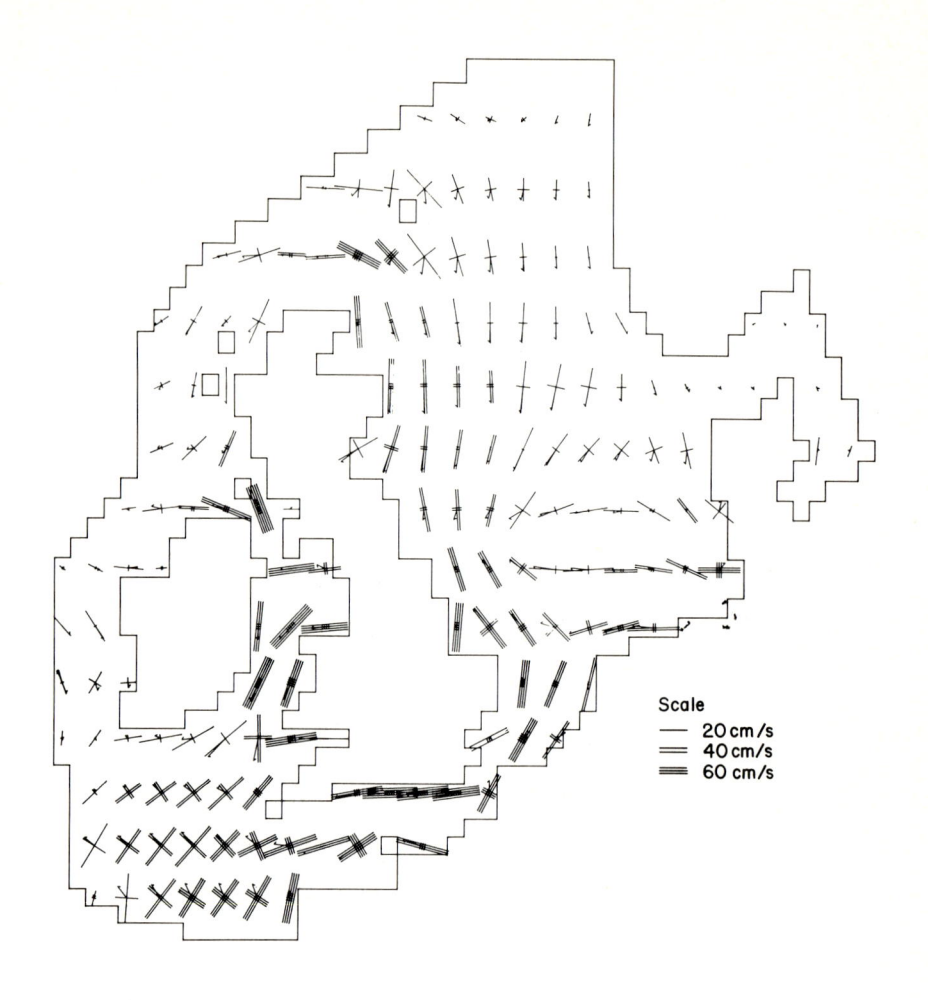

Figure 2.8 Distribution over the continental shelf of M_2 current ellipses at sea surface. From Davies and Furnes (1980).

Tides, Storm Surges and Coastal Circulations

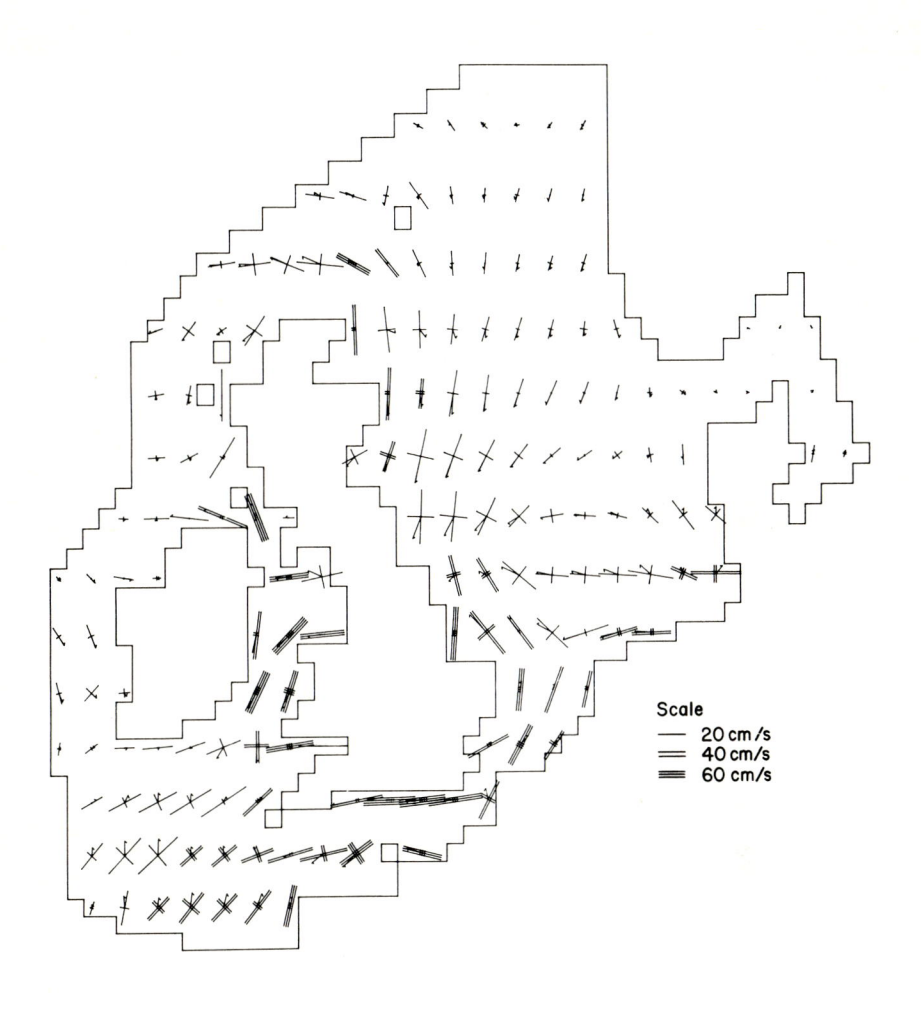

Figure 2.9 Distribution over the continental shelf of M_2 current ellipses at sea bed. From Davies and Furnes (1980).

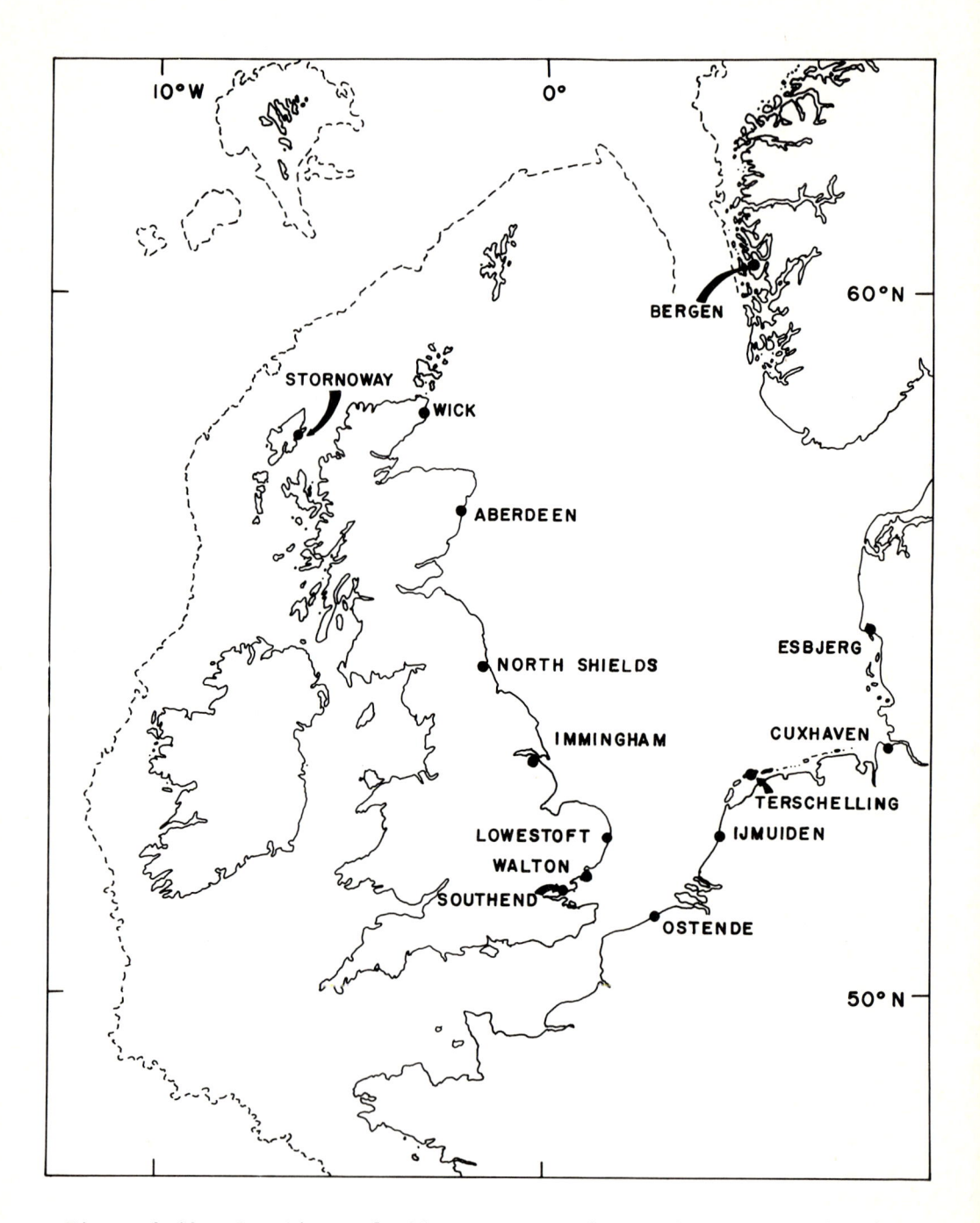

Figure 2.10 Locations of tide gauges in the North Sea. Sea levels recorded at these positions are used in the derivation of surge heights. ------ 100 fathom depth contour.

Tides, Storm Surges and Coastal Circulations

For example, as described by Corkan (1950), surge levels at ports on the east coast of Great Britain, from 6 to 10 January 1949, showed a depression in level on the 7th followed by a rise in level on the 8th travelling southwards down the coast (figure 2.11). These disturbances were generated mainly by the winds of a depression which moved eastwards across the northern entrance of the North Sea (figure 2.12). Southerly winds forced water northwards producing the observed lowerings of level. Then strong northwesterly winds in the right rear quadrant of the depression forced water southwards (a motion reinforced by a flowing back under gravity of water previously expelled) producing the observed rises in level. Southwards wave progression down the east coast, of surge disturbances primarily generated in the northern regions, is an important feature of North Sea dynamics. Thus, surges generated by storms outside and in the northern areas of the North Sea tend to propogate southwards against the Scottish and English coasts. The U.K. Storm Tide Warning Service at the Meteorological Office, Bracknell, employs tide gauge repeaters to indicate the surge progression between Stornoway in the north west of Scotland and Southend at the mouth of the Thames Estuary. Then surge height at tidal high water at a southern port is predicted empirically in terms of earlier surge heights at more northerly ports and forecast winds over coastal sub-areas of the North Sea. A major surge which caused disastrous flooding along the south east coast of England and in Holland occurred during 31 January to 1 February 1953 (Rossiter 1954). Surge levels and the meteorological situation associated with this event are shown in figures 2.13 and 2.14. Prediction of the surge at Southend in this case, using a statistical regression formula involving the surge at Dunbar nine hours earlier and atmospheric pressure gradients over the sea surface at points A, C, D and B in the North Sea (figure 2.15), was only partially successful. The discrepancies in the prediction were due to a large flow of water through the Dover Strait, not accounted for by the statistical formula.

2.4   Two-Dimensional Numerical Storm-surge Models

A series of numerical storm-surge models has been developed at the Institute of Oceanographic Sciences, Bidston, during recent years. The first of these (Heaps 1969) was based on the linearised vertically-integrated hydrodynamic equations in spherical polar coordinates, namely latitude and longitude, solved on a finite difference grid covering the North Sea and shelf regions surrounding the British Isles,

N.S. Heaps

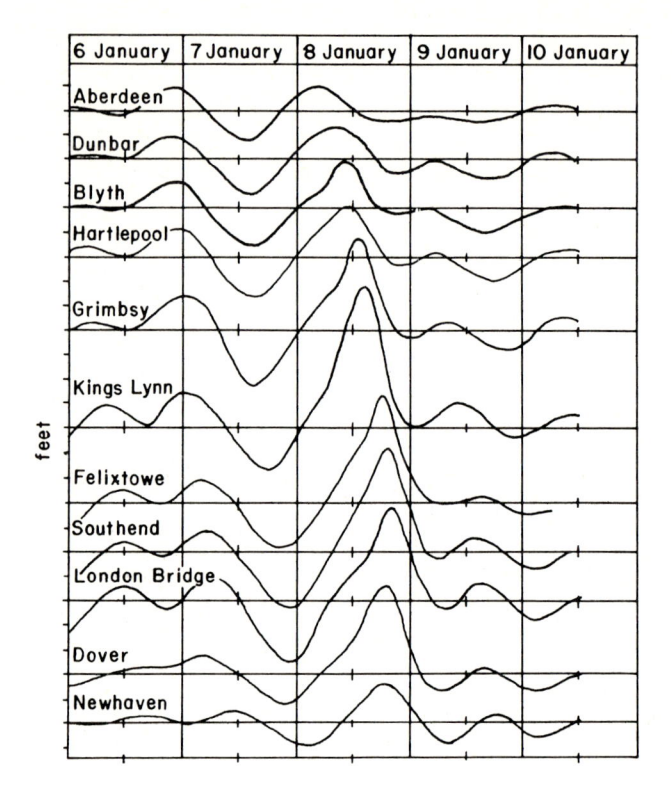

Figure 2.11   Surge levels at ports on the east coast of Great Britain. 6 to 10 January 1949.

Tides, Storm Surges and Coastal Circulations

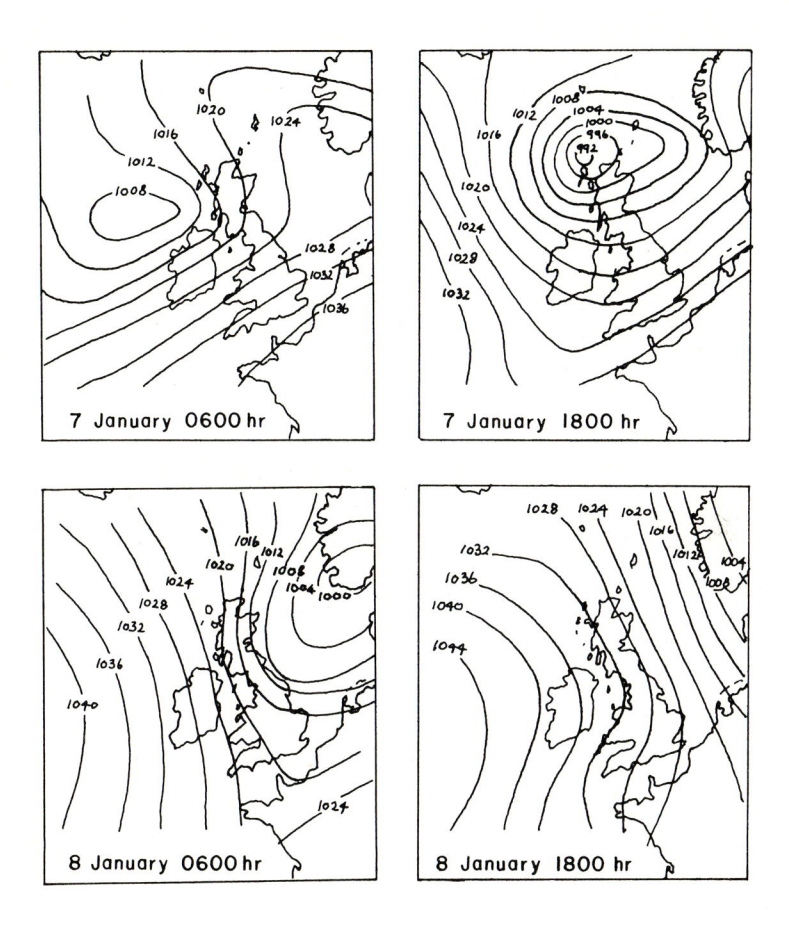

Figure 2.12 Weather charts for the storm surge of 7 to 8 January 1949.

N.S. Heaps

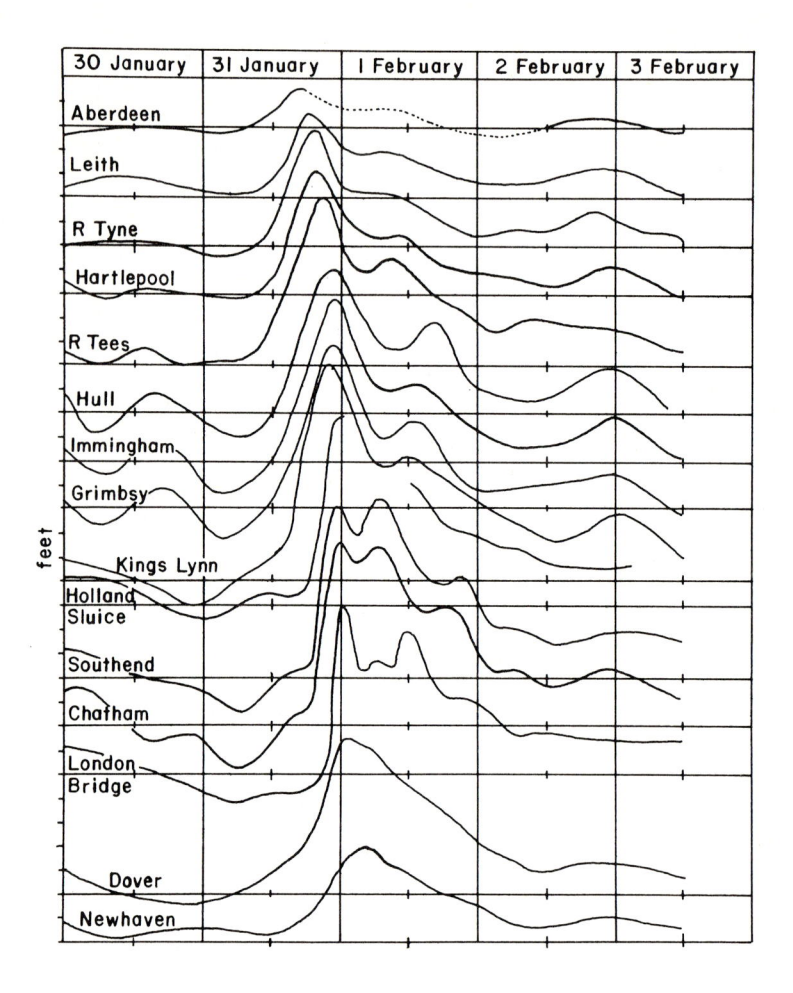

Figure 2.13 Surge levels at ports from north to south on the east coast of Great Britain, 30 January to 3 February 1953.

Tides, Storm Surges and Coastal Circulations

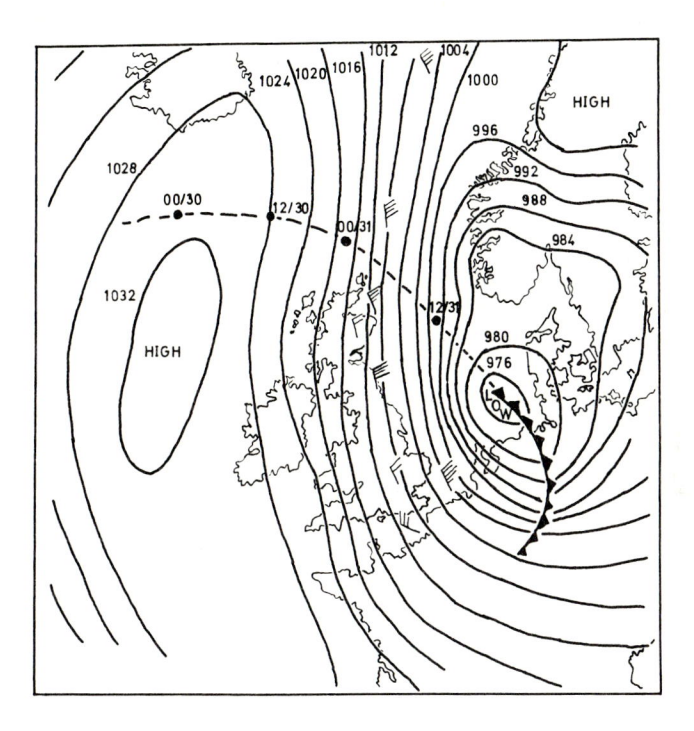

Figure 2.14 Weather chart for the major North Sea storm surge of 31 January to 1 February 1953.

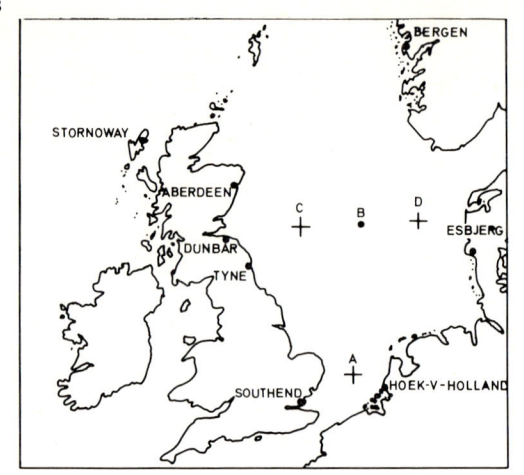

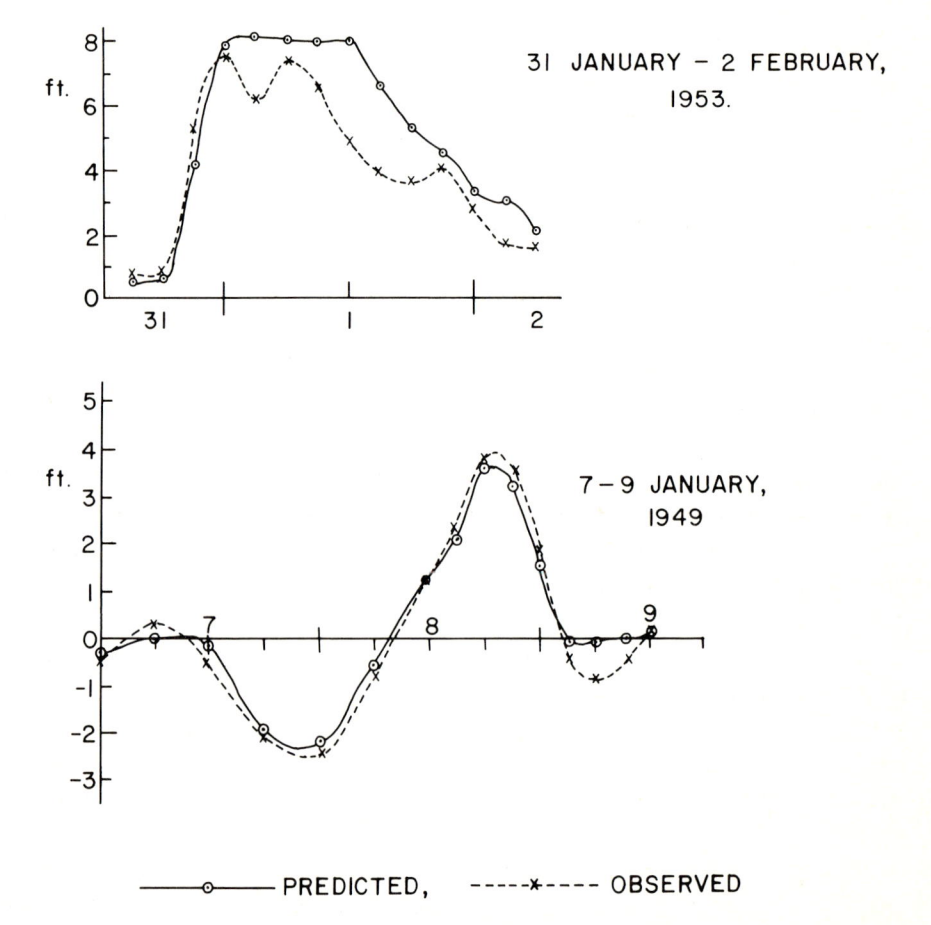

Figure 2.15 Storm surges at Southend in the southern North Sea, showing results obtained from a statistical regression formula.

Tides, Storm Surges and Coastal Circulations

as shown in figure 2.16. The model calculated surges in response to winds specified variously at two-hourly intervals over a set of constituent sub-areas. The role played by the sea-model calculations in determining the variations in sea level at a coastal site is explained in figure 2.17. Thus, the total elevation of the sea surface was assumed to consist of the predicted tide ζ_T, the wind surge ζ_W derived from the model computations, and the barometric surge ζ_B estimated from the statical law of pressure. Non-linear interactions between tide and surge were therefore ignored, as were the dynamic effects produced by changes in barometric pressure over the sea. The open boundary of the model was chosen to follow the edge of the continental shelf and surge elevation along it was assumed to be permanently zero. Computations with the model were carried out for three periods, each of two to three days, covering different types of North Sea surge. For each period, comparisons were made between computed and observed surge profiles at a number of representative North Sea ports. The results showed that the sea-model calculations reproduced to a good first approximation the surge heights derived from sea-level records.

For a detailed examination of storm-surge effects in the southern North Sea and River Thames, Banks (1974) formulated a two-dimensional model of that sea area linked dynamically to a one-dimensional model of the River. A similar more recent model system, due to Prandle (1975), has a grid network extending through the Strait of Dover into the English Channel as indicated in figure 2.18 and a schematic representation of the River Thames as shown in figure 2.19. These models are of finer mesh than the shelf model and include nonlinear terms with tide as well as surge prescribed along the open boundaries. Thus, Prandle specified tidal elevations along the northern and southern boundaries in terms of seven of the more important harmonic constituents, including M_2 and S_2. The disastrous surge of 31 January to 1 February, 1953, was reproduced satisfactorily by Prandle's model, specifying surge values from earlier estimates along the open boundaries. Both wind stress and atmospheric pressure gradient were included as forcing terms in the equations of motion. The role of flow through the Dover Strait in modifying surge levels in the Southern Bight was examined and the effects of a Thames barrier closure on water levels along the River were determined.

Banks (1974) used her nonlinear model of the southern North Sea and Thames Estuary to simulate the major 'Hamburg' surge of 15-17 February,

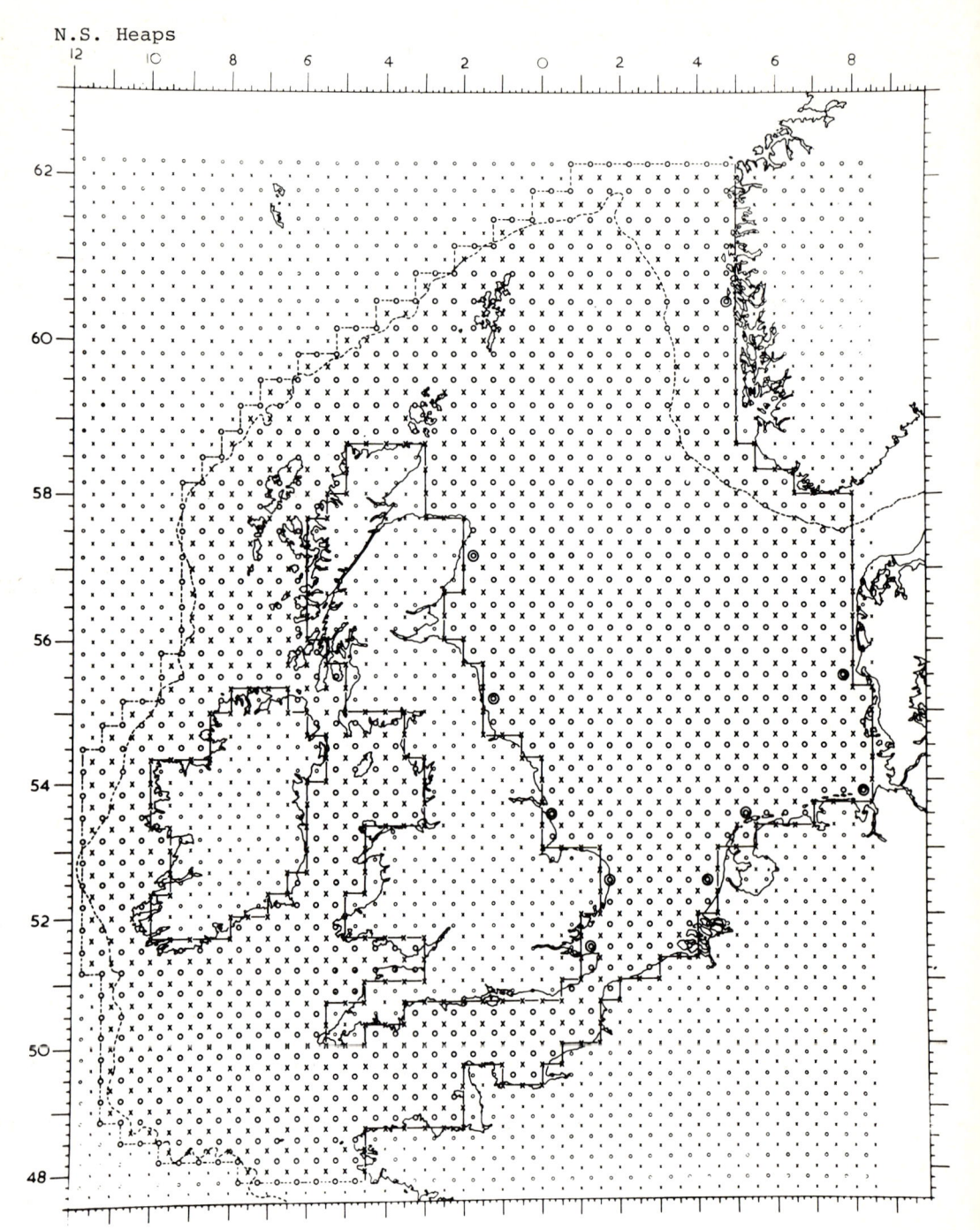

Figure 2.16 Linear shelf model covering the North Sea and shelf regions
surrounding the British Isles: --o--o--o--o-- open boundary;
——x——x—— closed boundary; ----- 100 fathom depth contour;
⊚ point at which calculated surge elevations are compared
with surge elevations derived from observations at a nearby
port; o elevation point, x current point in the finite
difference scheme. From Heaps (1969).

Tides, Storm Surges and Coastal Circulations

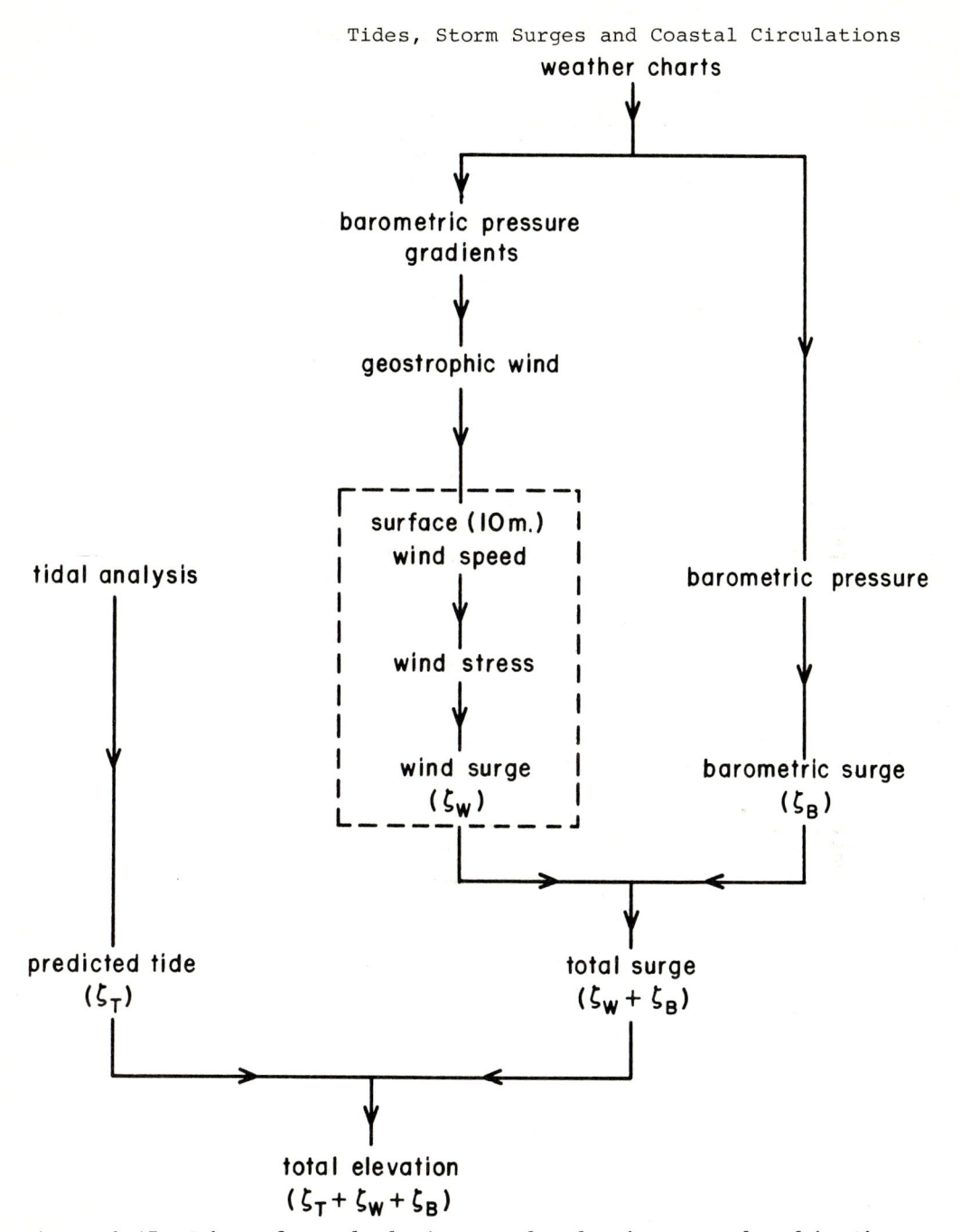

```
                    weather charts
                          │
          ┌───────────────┼───────────────┐
          ▼                               │
    barometric pressure                   │
       gradients                          │
          │                               │
          ▼                               │
    geostrophic wind                      │
          │                               ▼
   ┌──────┼──────────┐           barometric pressure
   │      ▼          │                    │
   │  surface (10m.) │                    │
   │  wind speed     │                    │
   │      │          │                    │
   │      ▼          │                    │
   │  wind stress    │                    │
   │      │          │                    │
   │      ▼          │                    ▼
   │  wind surge     │           barometric surge
   │    ($\zeta_W$)  │               ($\zeta_B$)
   └──────┼──────────┘                    │
          └──────────────►◄───────────────┘
                         │
                         ▼
                    total surge
                  ($\zeta_W + \zeta_B$)
```

tidal analysis

predicted tide
(ζ_T)

total elevation
($\zeta_T + \zeta_W + \zeta_B$)

Figure 2.17 Scheme for calculating sea level; the part played by the linear shelf model calculations is enclosed by the dashed lines. From Heaps (1969).

N.S. Heaps

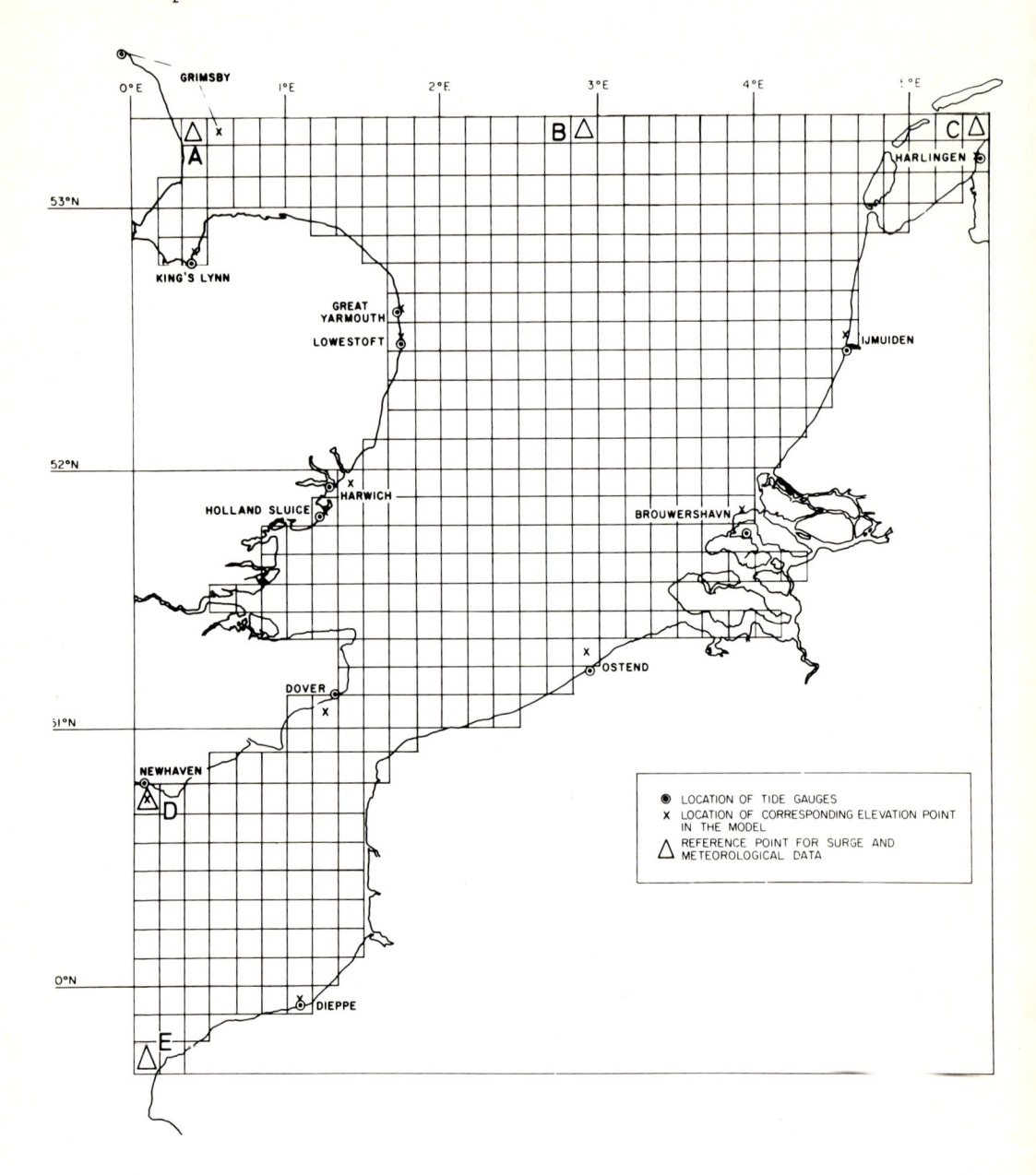

Figure 2.18 Schematic representation of the southern North Sea. From Prandle (1975).

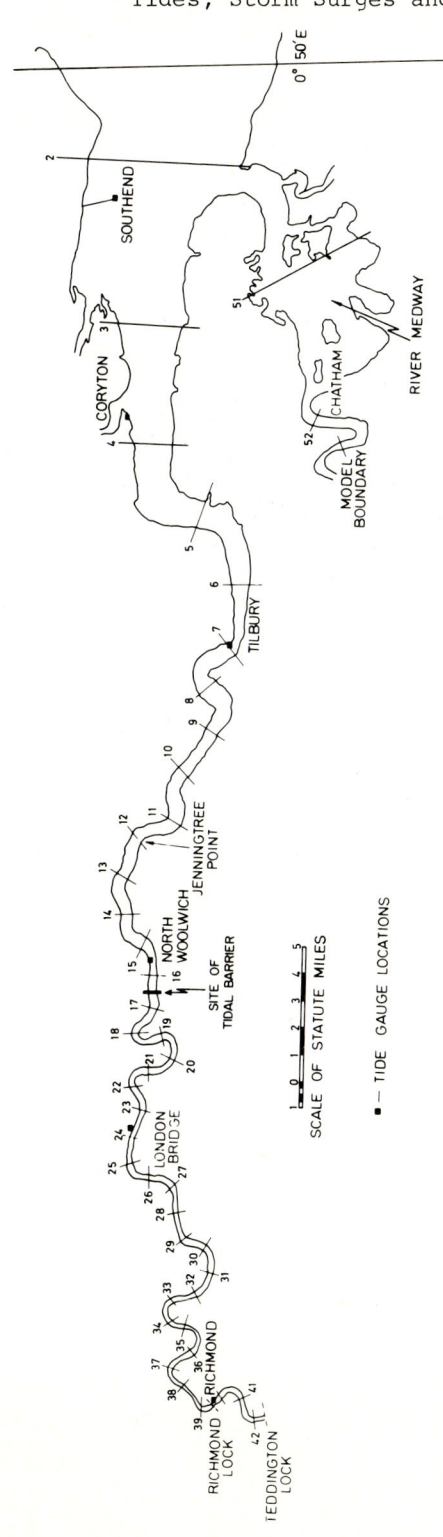

Figure 2.19 River Thames - schematic presentation from Prandle (1975)

N.S. Heaps

1962. Elevation of the sea surface ζ_{T+S} due to tide and surge combined was generated throughout the model. A separate computer run evaluated ζ_T due to tide alone (no wind or boundary surge input). Satisfactory agreement was obtained between the computed wind surge $\zeta_{S+I} = \zeta_{T+S} - \zeta_T$ and the observed residual elevation after removal of the barometric surge. The comparison for Southend is given in figure 2.20, where the corresponding surge derived from the linear shelf model is also shown. Clearly, non-linear interaction between tide and surge produced three significant surge peaks out of one. The peaks occurred on the rising tide and the intermediate troughs near to tidal high water, forming a semidiurnal oscillation. These interaction effects increased with distance travelled up the Thames and were present, in diminished form, across the entire width of the southern North Sea. The tendency for positive surge peaks at Southend to occur on the rising tide and avoid tidal high water is confirmed by a long series of observations (figure 2.21).

2.5 Surge Forecasting

In the U.K. a new system of surge forecasting, mainly for the North Sea, became operational in the autumn of 1978. The system is based on a numerical model of the North-West European Continental Shelf which has been run twice daily at the British Meteorological Office in conjunction with weather forecasts provided by a 10-level model of the atmosphere. Thereby, 30-hour surge forecasts have been obtained every twelve hours, corresponding to successive 36 hour forecasts of atmospheric pressure and wind (figure 2.22).

The continental shelf model used in the work is a development of the linear shelf model already described. The finite difference grid of the model, with sea surface elevation evaluated at the centre of each mesh element, is shown in figure 2.23. Also shown in that figure are the grid points of the 10-level model of the atmosphere. The hydrodynamic equations solved are in vertically-integrated form, include the advective terms, and assume quadratic friction at the sea bed. Tide and surge are computed together so as to include nonlinear interaction between them; tide computed alone is then subtracted to yield the surge forecast. A radiation condition is employed along the open boundaries in which tidal elevation ($M_2 + S_2$) is prescribed from observational data while surge elevation is assumed to be given by the hydrostatic law.

Tides, Storm Surges and Coastal Circulations

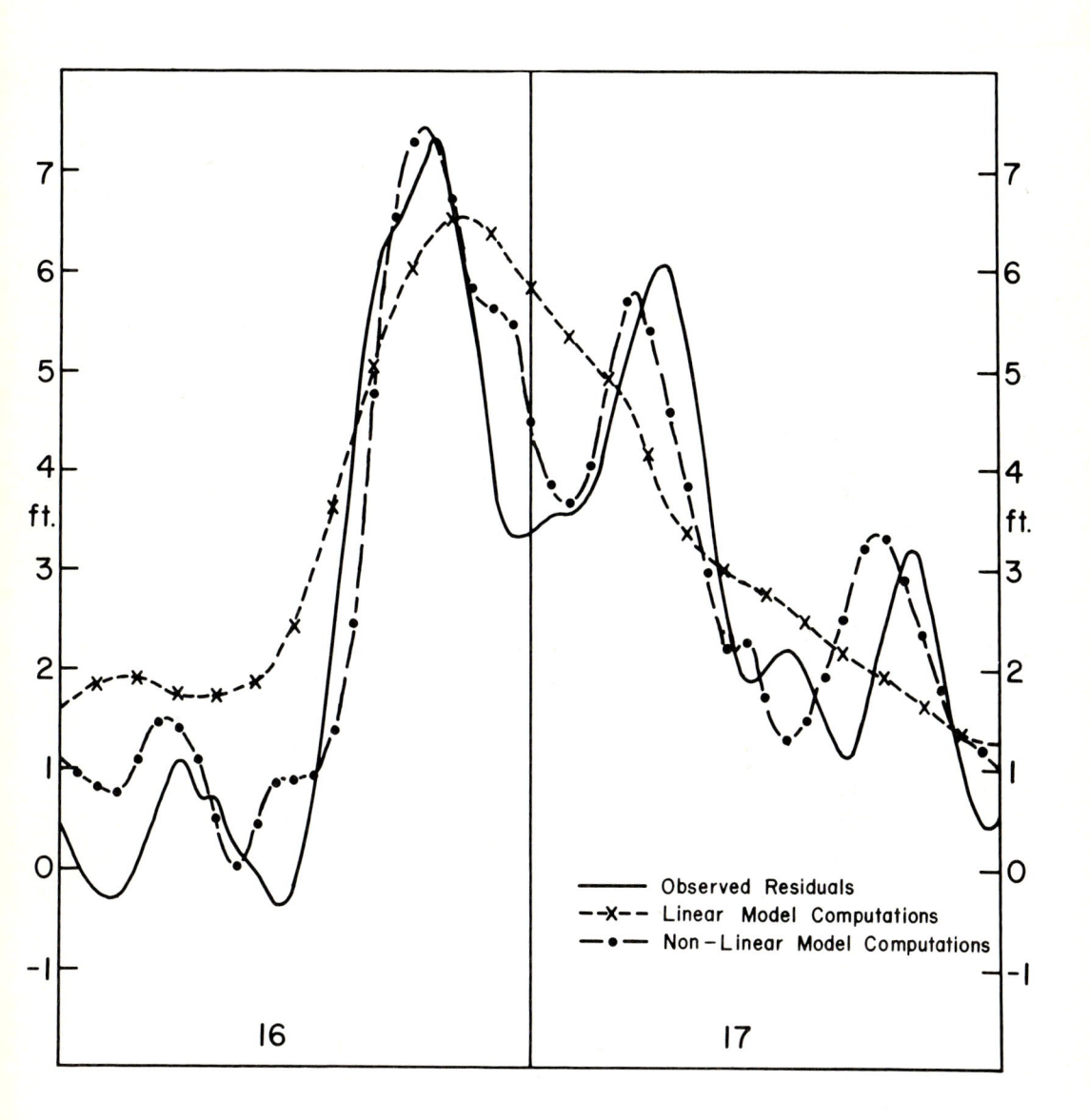

Figure 2.20   Wind-induced surge levels at Southend, 16-17 February, 1962. From Banks (1974).

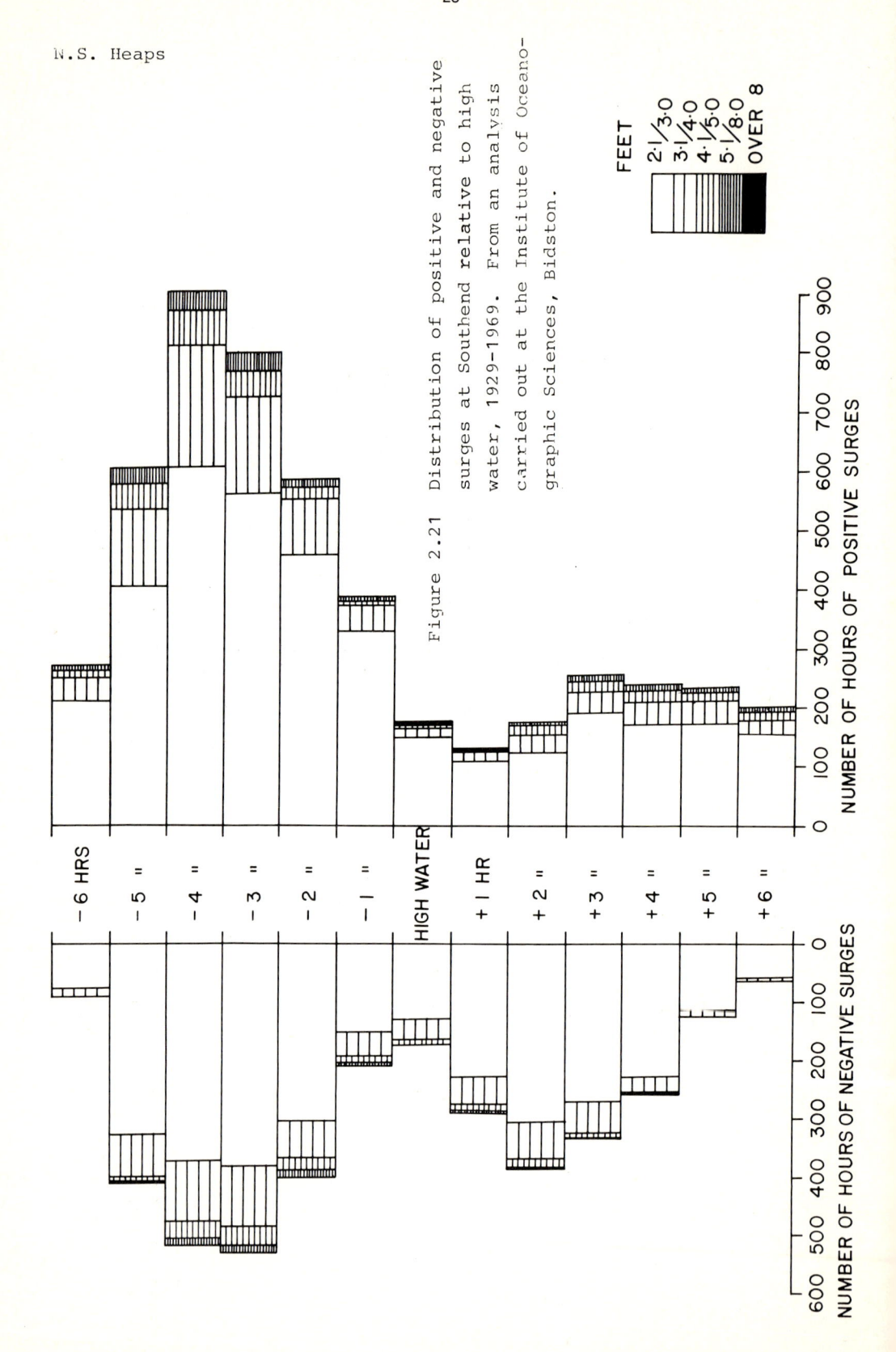

Figure 2.21 Distribution of positive and negative surges at Southend relative to high water, 1929-1969. From an analysis carried out at the Institute of Oceanographic Sciences, Bidston.

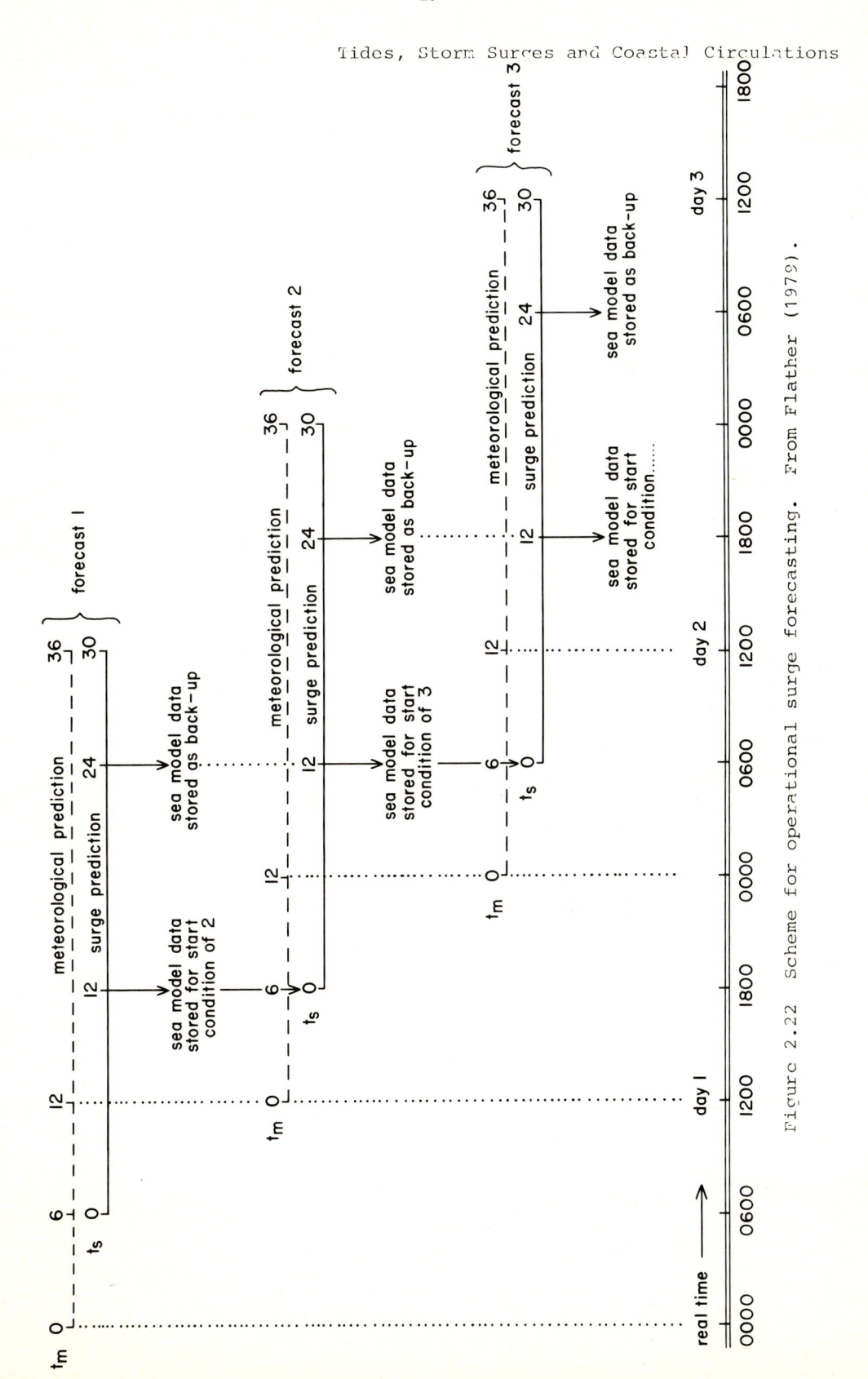

Figure 2.22 Scheme for operational surge forecasting. From Flather (1979).

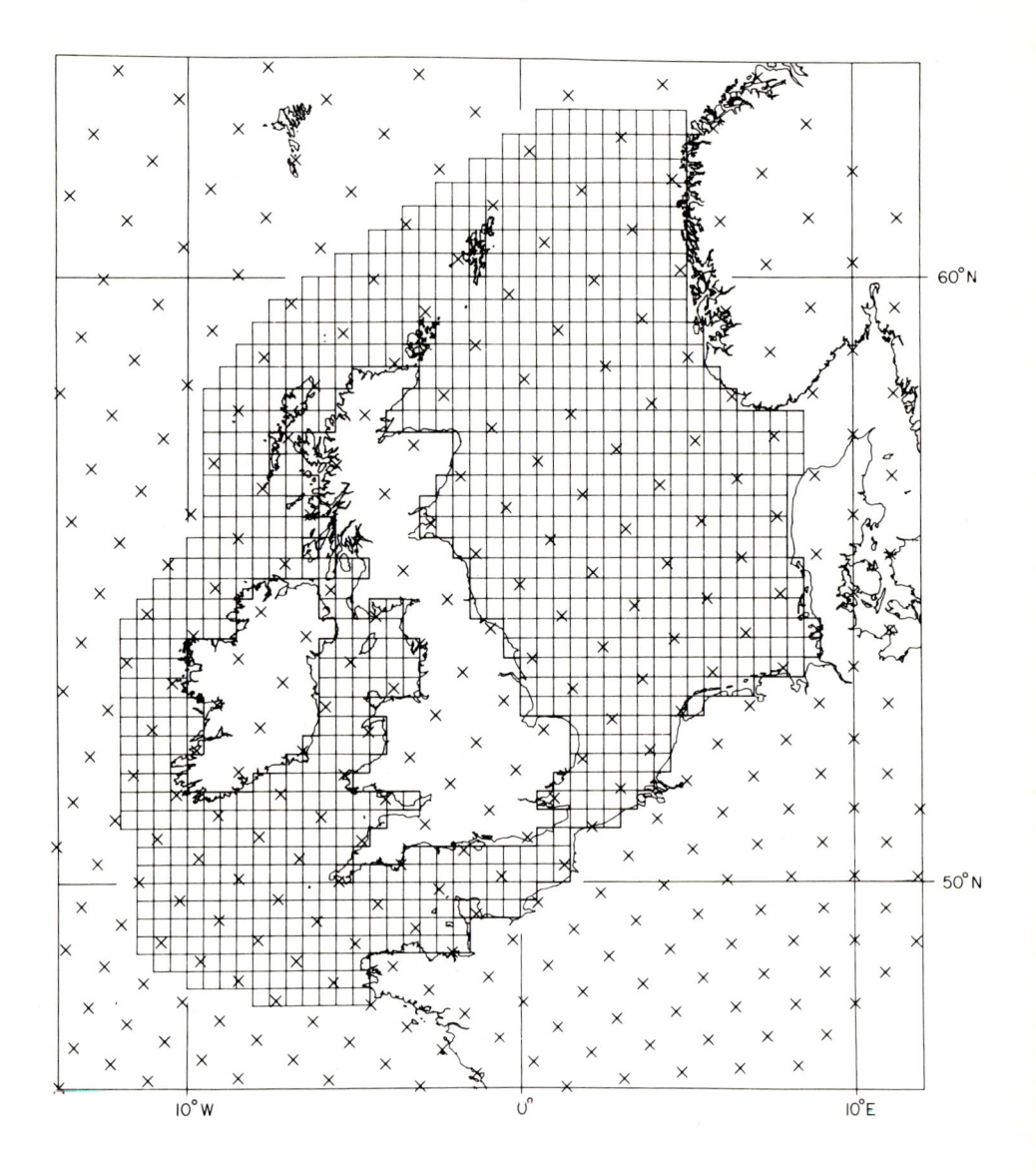

Figure 2.23 Finite difference mesh of the continental shelf model with grid points (X) of the 10-level model of the atmosphere. From Flather (1979).

Tides, Storm Surges and Coastal Circulations

Wind stress $\underset{\sim}{\tau}^{(s)}$ and atmospheric pressure gradient ∇p_a, acting on the sea surface, are the meteorological forcing terms which have to be evaluated through time in the sea-model computations. Four methods of evaluation have been investigated, illustrated in figure 2.24. Each calculates $\underset{\sim}{\tau}^{(s)}$ from estimates of surface wind speed using a quadratic law with a prescribed drag coefficient C_D which varies with the wind speed (figure 2.25). The C_D introduced by Heaps (1965), denoted by H, has been used mostly. Methods 1,2,3 differ in the way they estimate the surface wind. Thus, in Method 1, after atmospheric pressure p_a at the sea surface has been interpolated from the grid of the 10-level model to the grid of the sea model, ∇p_a is evaluated on the latter grid by differencing p_a, then the geostrophic wind $\underset{\sim}{w}_g$ is deduced from ∇p_a, and then the surface wind $\underset{\sim}{w}$ is deduced from $\underset{\sim}{w}_g$ using an empirical formula due to Hasse and Wagner (1971) and a cross-isobar angle of 20°. In method 2 the deduction of $\underset{\sim}{w}$ from $\underset{\sim}{w}_g$ involves a dependence on $\Delta T_{a-s} = T_a - T_s$, where T_a denotes air temperature near the sea surface and T_s sea-surface temperature. Both the wind speed and the cross-isobar angle (determining the direction of the wind in relation to the direction of the isobars) depend on ΔT_{a-s} (see figures 2.26, 2.27). The formula due to Hasse (1974) employed in this method may be replaced with advantage by one due to Duun-Christensen (1975). In Method 3, the one actually used in the operational forecasting, components of the surface wind on the grid of the 10-level model are resolved into east and north components and then interpolated on to the sea-model grid.

A major North Sea storm surge occurred on 3 January 1976 as a result of severe north westerly gales associated with a fast moving deep depression which passed between Scotland and Denmark (figure 2.28). The good performance of the continental shelf model in simulating this surge, at ports along the east coast of England, is illustrated in figure 2.29. Specially prepared wind and pressure fields, belonging to the so-called NORSWAM data set, derived from a careful analysis of synoptic weather charts and other supplementary information, were used in the simulation. Figure 2.30 shows contours of surge elevation and surge currents at the peak of the storm, derived from the model. Computed and observed currents covering the storm period, at the JONSIS 2 position (54° 23'N, 01° 06.5'E) off the east coast of England, are compared in figures 2.31 and 2.32. Depth-averaged currents from the model are here being compared with near-surface currents measured 12m below the sea surface, and near-bottom

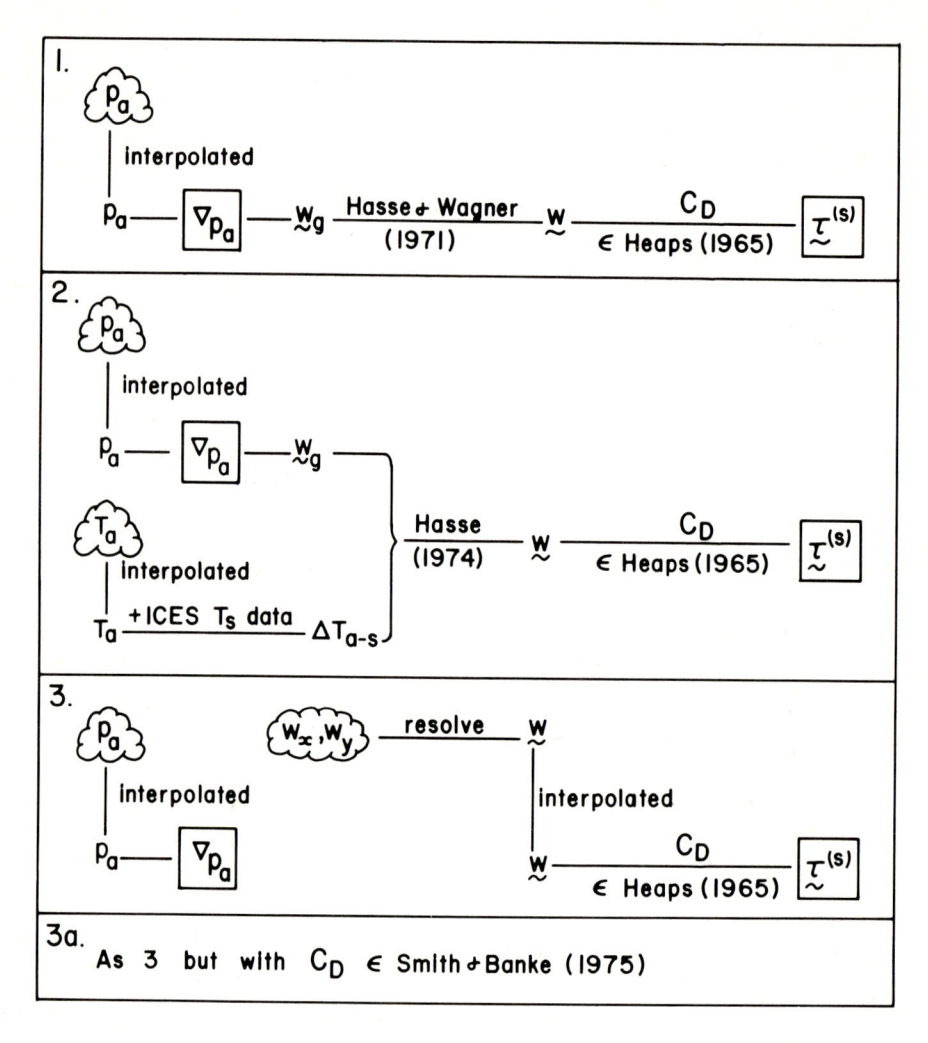

Figure 2.24 Methods used to derive meteorological input data for the numerical shelf model of figure 23. From Flater (1979).

Tides, Storm Surges and Coastal Circulations

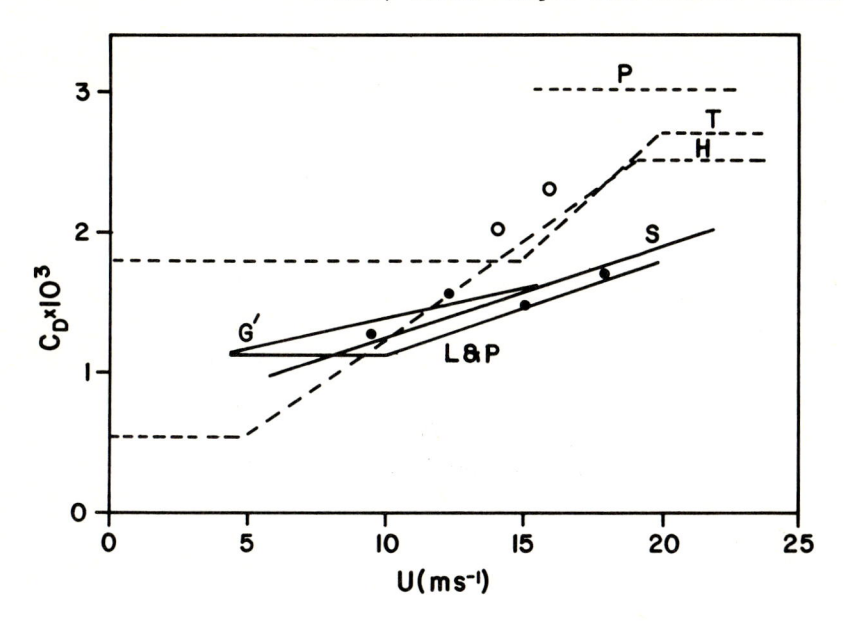

$$H: \quad 10^3 C_D = 0.554 \qquad\qquad U < 4.917$$
$$= -0.12 + 0.137 U \qquad 4.917 < U < 19.221$$
$$= 2.513 \qquad\qquad U > 19.221$$

Figure 2.25   The neutral drag coefficient C_D versus wind speed u from various sources. The solid lines are regression lines from eddy correlation estimates of $\overline{u'w'}$ (S from Smith and Banke 1975); the dashed lines are the formulae adopted by three storm surge modellers; the solid circles are derived from water level fluctuations over several months in Lake Erie; the open circles are derived from the peak storm surge for two storms in Lake Ontario. From Donelan (1982). The variation H, quoted, comes from Heaps (1965). Wind stress= $C_D \rho_A u^2$, ρ_A = density of the air.

Hasse and Wagner (1971):

$$w = 0.56\, w_g + 0.24$$

Hasse (1974):

$$w = a\, w_g + b$$
$$a = 0.54 - 0.012\,(T_a - T_s)$$
$$b = 1.68 - 0.105\,(T_a - T_s)$$

Duun-Christensen (1975):

$$w = a_2 \sqrt{a\, w_g + b} + b_2$$
$$a_2 = 6.82$$
$$b_2 = -11$$

Figure 2.26 Formulae for surface wind in ms^{-1}.

Tides, Storm Surges and Coastal Circulations

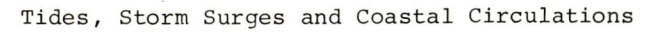

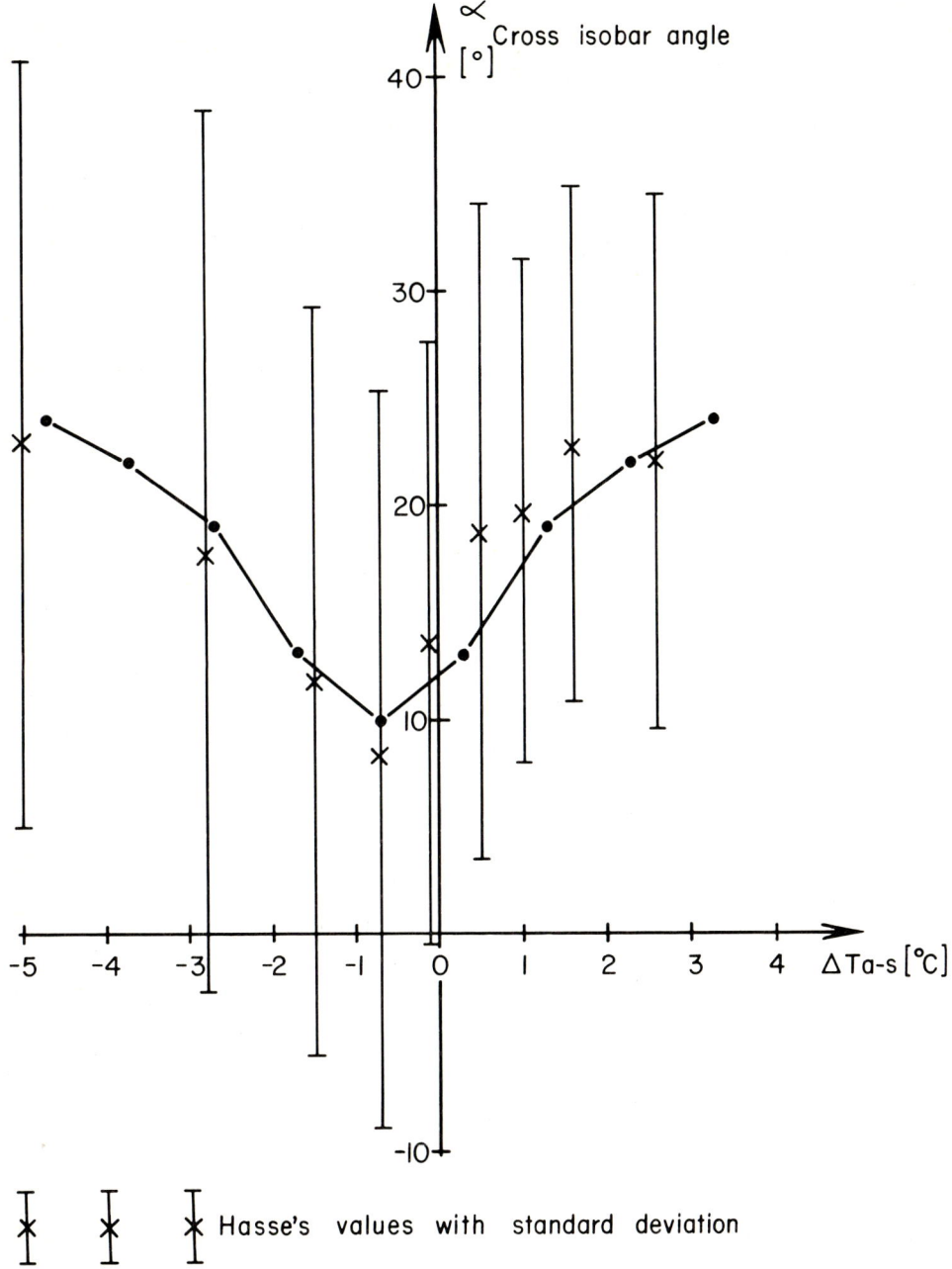

The dependence of the cross isobar angle from the temperature difference air-sea

Figure 2.27 Cross isobar angle.

N.S. Heaps

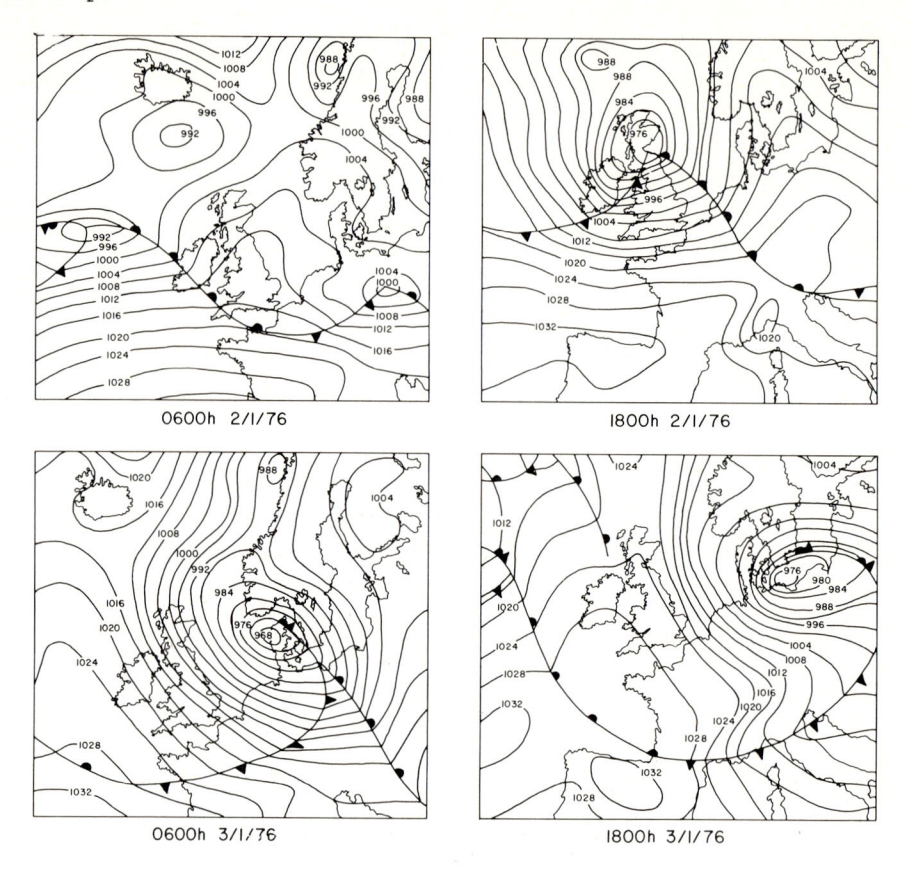

Figure 2.28 Weather charts for the storm surge of 2-3 January 1976.

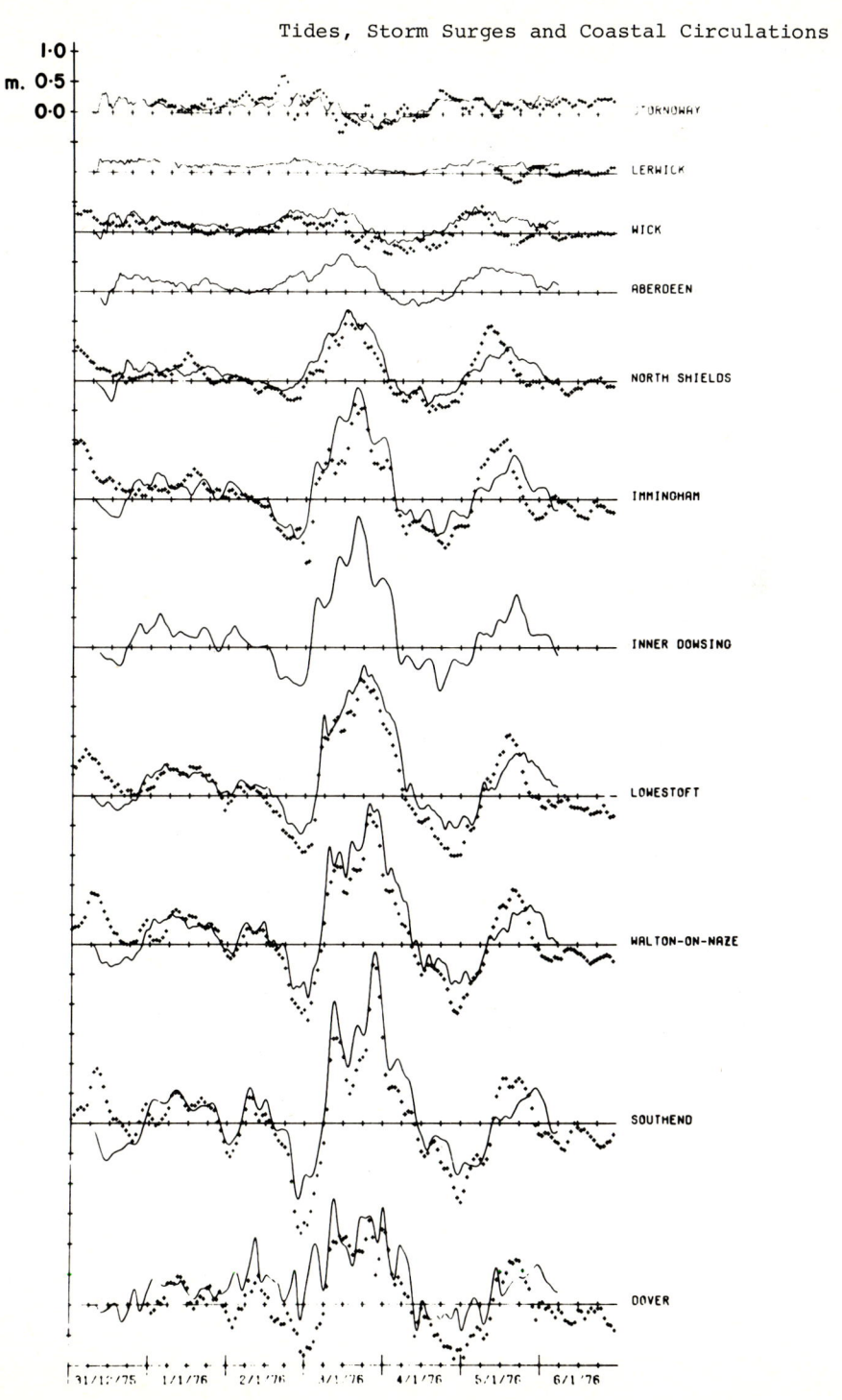

Figure 2.29 Storm surges at U.K. ports for 31 December-6 January, 1976, as computed with the shelf model using the 'NORSWAM' data (———) and as observed (+++++++). From Flather and Davies (1978).

N.S. Heaps

Figure 2.30 Contours of surge elevation (cm) and depth-mean currents at 0900 hours on 3 January, 1976, as computed with the shelf model using the 'NORSWAM' data. From Flather and Davies (1978).

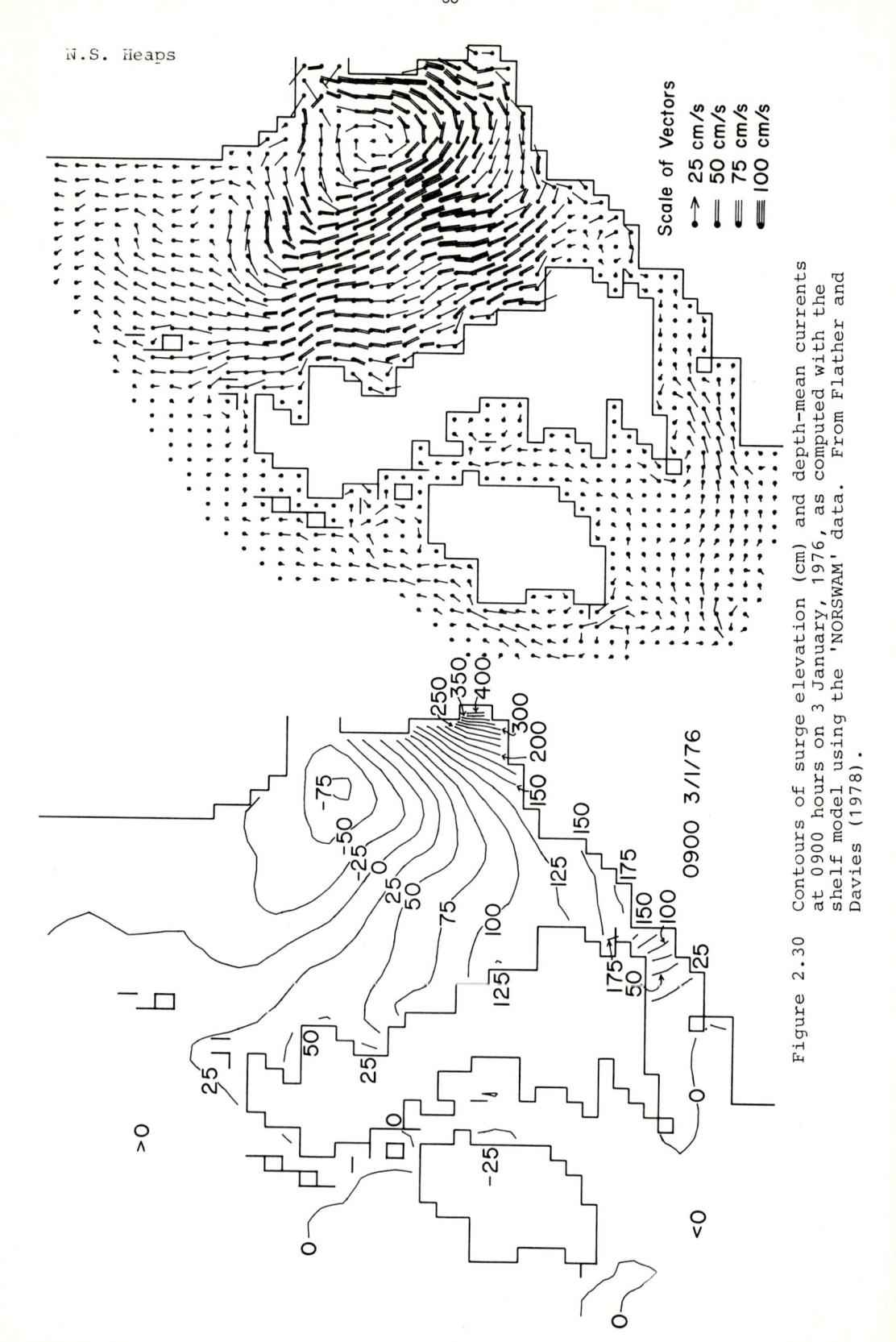

Tides, Storm Surges and Coastal Circulations

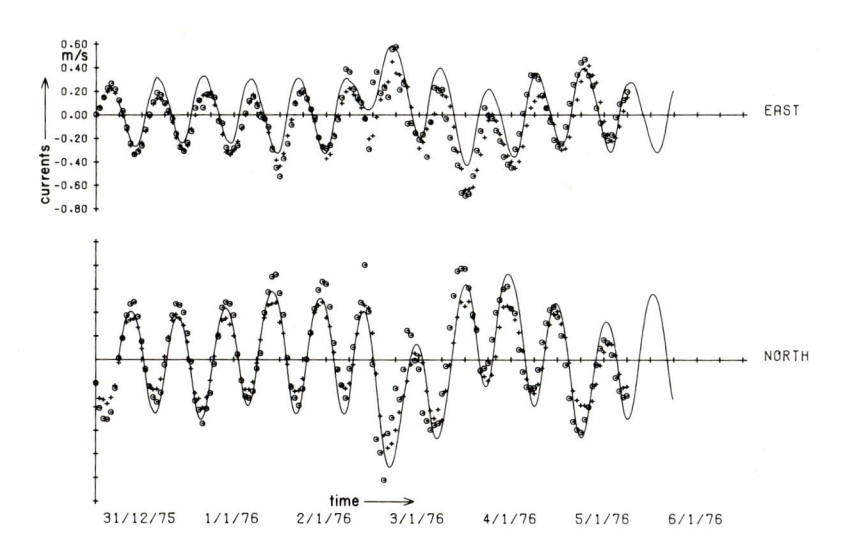

Figure 2.31   Comparison between total current from the continental shelf model (————) and hourly means of observations from near-surface (°) and near-bottom (+) current meters at JONSIS 2 (54° 23'N, 01°, 06.5'E) for the period 31 December 1975 to 6 January 1976. East- and north-directed components are shown. From Flather and Davies (1978).

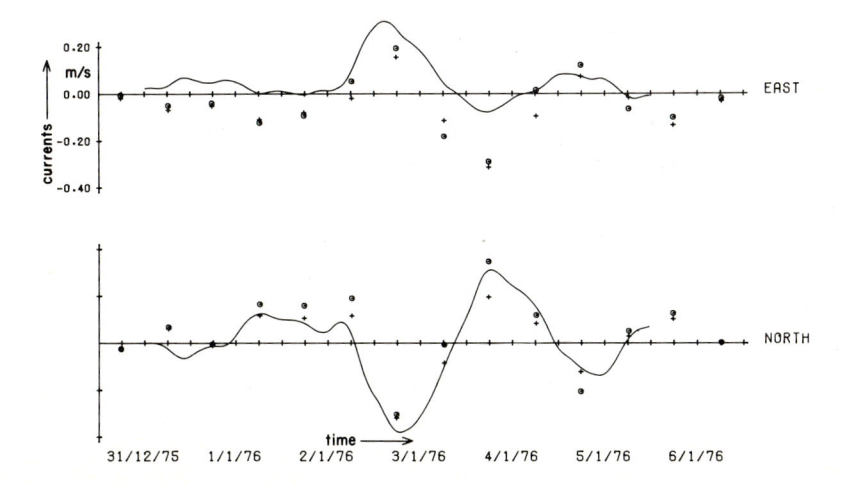

Figure 2.32   Comparison of 12.5 hour means of the computed and observed currents displayed in figure 31. From Flather and Davies (1978).

currents measured 47m below that surface, in 52m of water at low water springs. While vertical variations in the observed surge currents are apparent in figure 2.32, the overall close agreement between the near-top and near-bottom measurements suggests that, during a large storm surge, depth-averaged surge currents may be reasonably representative of surge currents through the vertical water column in the shallower regions of the North Sea.

R. A. Flather at Bidston has simulated the surges due to sixteen North Sea storms using the continental shelf model as described above. The distribution of the maximum computed surge elevation over the shelf was adjusted to give an estimated distribution of the 1 in 50 year surge elevation, employing statistical evaluations of extreme surge height at coastal stations (Pugh and Vassie, 1980) as standards of reference. Flather also extracted maximum surge currents over the shelf from the results of the 16 storm surge simulations with the shelf model. These currents may be adjusted to 1 in 50 year values, again by reference to appropriate elevation statistics at the coastal stations. The model tides (M_2 and S_2) may be used as a basis for the estimation of maximum tidal currents and, in combination with the extreme surge currents, can yield estimates of extreme total current.

2.6 Three-Dimensional Models

Some results from a three-dimensional numerical model, with a finite difference grid in the horizontal and current expanded through the vertical in terms of eigenfunctions, are now presented. The model is linear and vertical eddy viscosity, μ, is time-independent, uniform through the depth, and varies positionally in proportion to the depth. The basic design was introduced by Heaps (1972) and a first numerical experiment considered motion in a closed rectangular sea basin with dimensions and rotation representative of the North Sea (depth = 65m, length = 800km, breadth = 400km), subjected to a uniform longitudinal wind stress from the north of 15 dyn cm^{-2}. Figure 2.33 shows the calculation grid and figure 2.34 the vertical distribution of current at the centre position of the basin in the essentially steady state attained 300 hours after the onset of the wind. Results corresponding to μ = 130, 325, 650, 1300 and 2600 $cm^2 s^{-1}$ are given. The v-profiles indicate currents in the wind direction near the sea surface and return currents at greater depths. The u-profiles show a corresponding two-

Tides, Storm Surges and Coastal Circulations

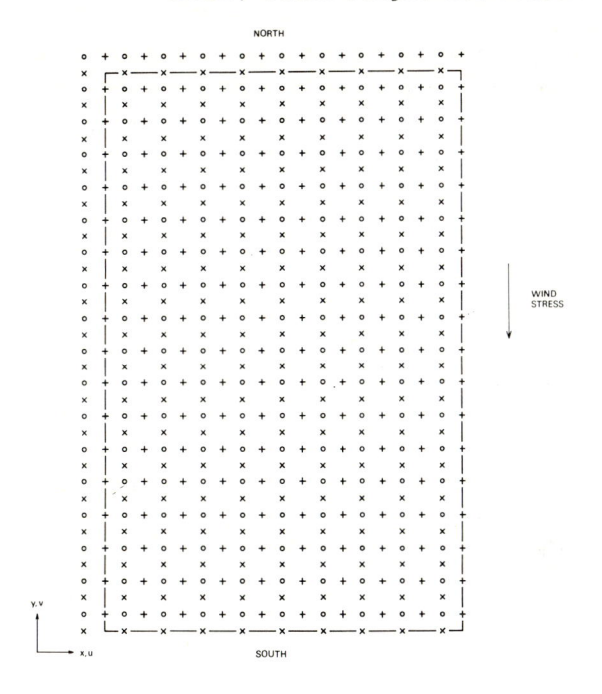

Figure 2.33 Closed rectangular sea basin with dimensions and rotation representative of the North Sea, subjected to a uniform longitudinal wind stress from the north. The resulting elevations of the sea surface (ζ) and current components (u,v) at any depth are evaluated at the grid points shown. (———) land boundary, o = n point, + = u point, x = v point. From Heaps (1972).

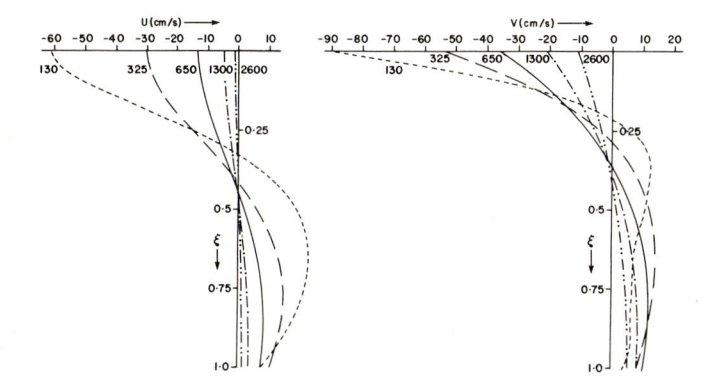

Figure 2.34 Vertical distributions of u and v current at the central position of the rectangular basic after the establishment of a steady wind-driven circulation; ζ denotes fractional depth. From Heaps (1972).

N.S. Heaps

layer flow across the breadth, induced by the Earth's rotation. The changes in μ clearly have a critical influence on the magnitude of the currents. Results from an Irish Sea Model, shown in figure 2.35, are presented in figures 2.36 and 2.37 (Heaps 1979, Heaps and Jones 1977). Clearly, the circulation patterns due to annual mean wind stress are radically different to those due to horizontal density gradients. This is particularly true in Liverpool Bay where observational data has confirmed the existence of the different circulations.

The JONSDAP '76 Oceanographic Experiment took place in the North Sea during the spring of 1976 and provided field data to test and verify North Sea models. Using the three-dimensional model of the North West European Continental Shelf already mentioned, Davies (1983) computed motion due to tide, wind and atmospheric pressure through the period 1-9 April of the Experiment. The changing distribution of meteorologically induced current during that period was thereby determined. In particular, the computed depth-mean flow pattern in the North Sea at 1800h on 6 April (figure 2.38) compares well with a horizontal distribution of currents inferred from the current meter measurements (figure 2.39).

2.7 Vertical Structure of Current

Results from Ekman's theory, which assumes constant μ, are shown in figure 2.40: a uniform steady wind stress F_s acts over the surface of an infinite ocean producing a current which decreases exponentially with depth and rotates in a spiral from being at 45° to the wind direction at the surface. When the depth of water h becomes small in comparison with the depth of frictional influence, D, the deflection of the current vector from the wind direction is reduced (figure 2.41). Generally, however, on the basis of results obtained from surface drifter experiments, the angle of deflection at the surface is about 10° which conflicts with Ekman's 45°. A logarithmic boundary layer at the surface, with the eddy viscosity reducing linearly to a small value as the surface is approached, yields the required smaller angle and a surface drift approximately 3% of the surface wind speed as observed (Madsen, 1977). Eddy viscosity distributions with such a surface wall layer are illustrated in figures 2.42 and 2.43.

Tides, Storm Surges and Coastal Circulations

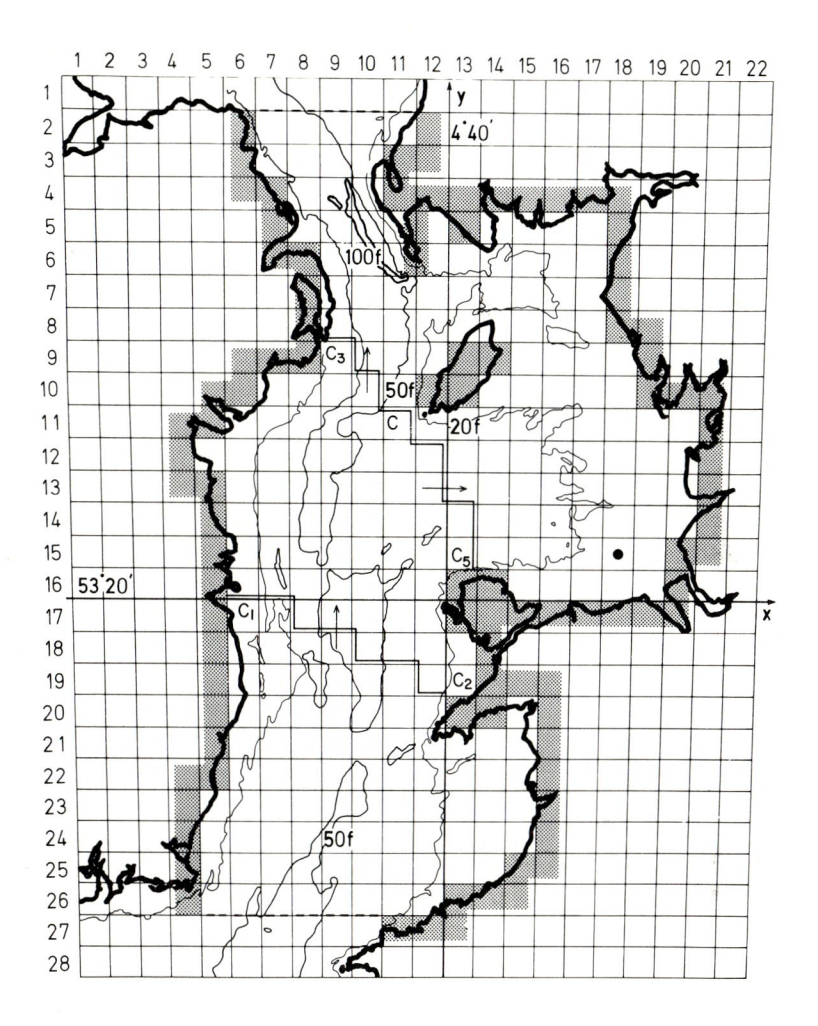

Figure 2.35 Irish Sea model with a grid of side 7.5 nautical miles.

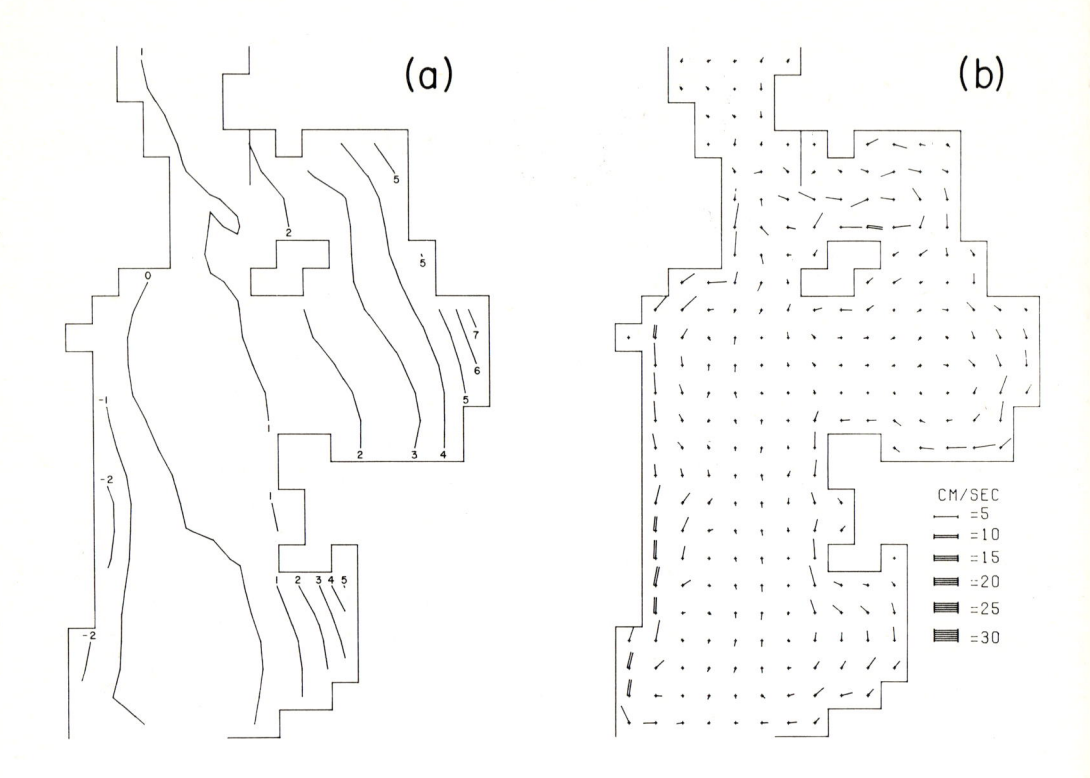

Figure 2.36   Residuals due to annual mean wind stress (1.0 dyn cm^{-2} to the east, 0.6 dyn cm^{-2} to the north):

(a)   elevation contours in centimetres,
(b)   depth-mean currents,
(c)   surface currents,
(d)   bottom-currents.

From Heaps (1979)

(c) and (d) overleaf

Tides, Storm Surges and Coastal Circulations

Figure 2.36 (continued)

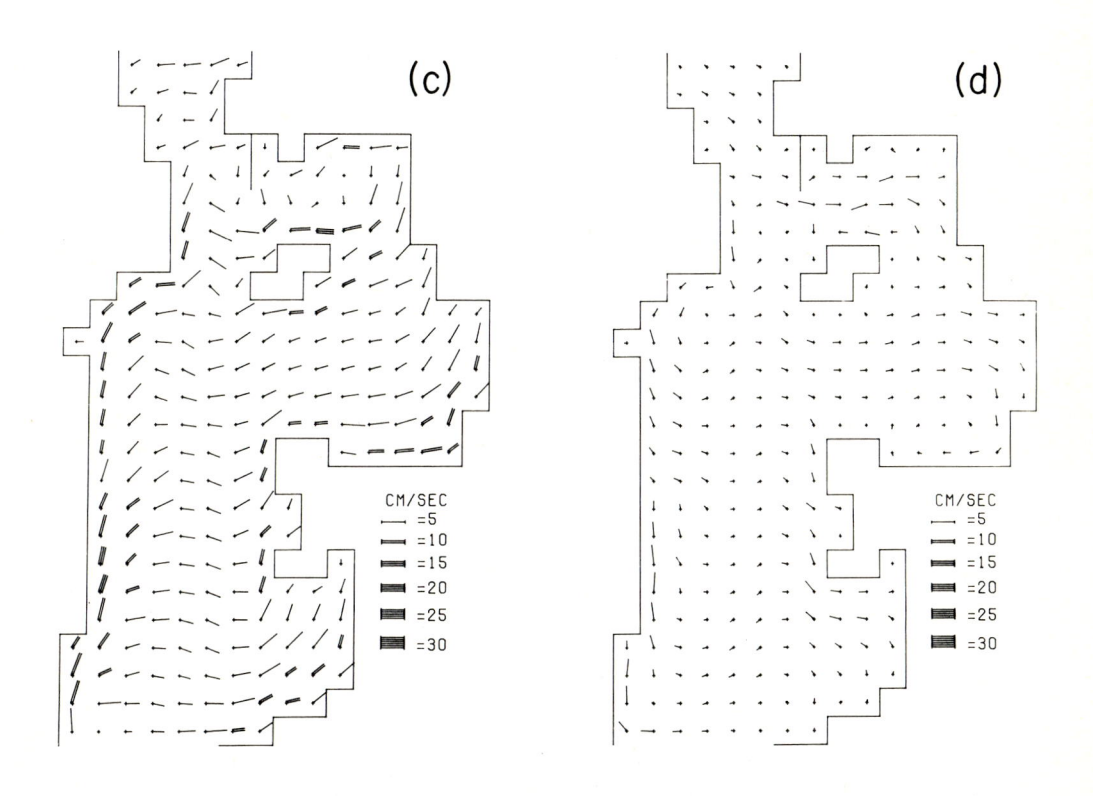

2.36(c)   surface currents   (d)   bottom-currents.

N.S. Heaps

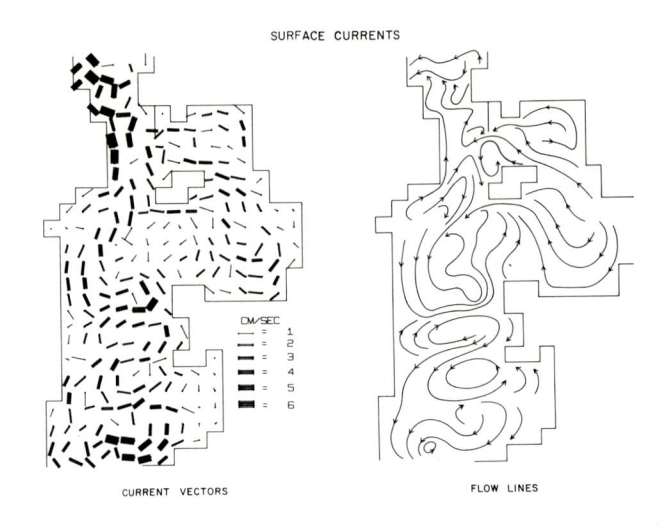

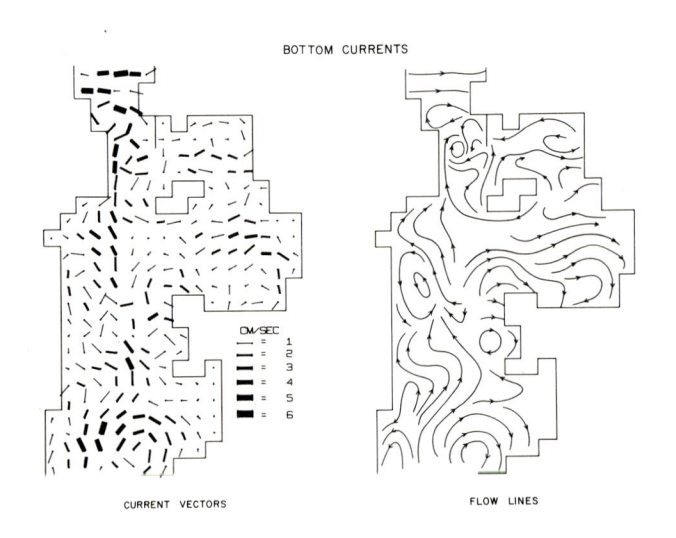

Figure 2.37　Density currents corresponding to the period 6-20 September 1971. From Heaps and Jones (1977).　(continued overleaf)

Tides, Storm Surges and Coastal Circulations

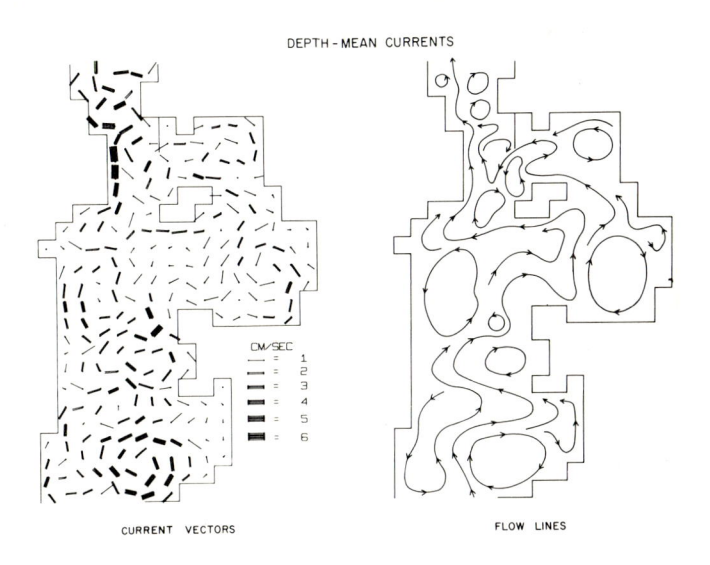

Figure 2.37 (continued)

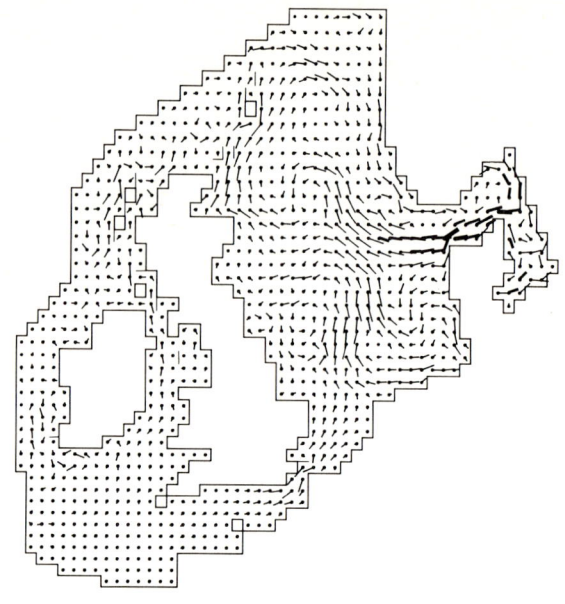

Figure 2.38 Distribution of depth-averaged meteorologically-induced currents at 1800h on 6 April, 1976; from a three-dimensional shelf model. From Davies (1983).

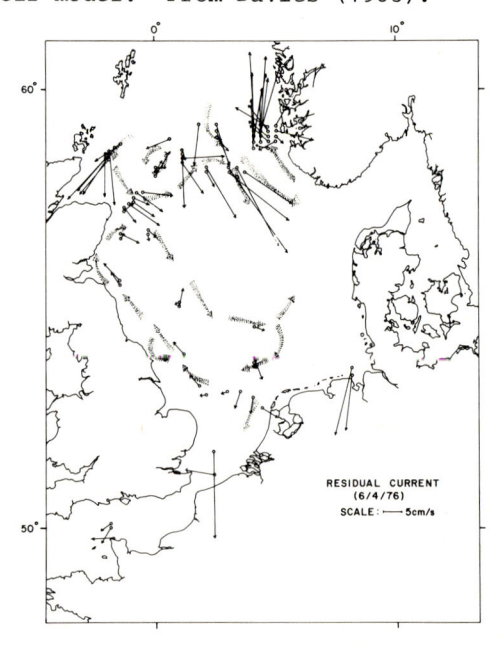

Figure 2.39 Residual current vectors and postulated residual currents (dotted arrows) for 6 April, 1976; from current meter measurements. From Davies (1983).

Tides, Storm Surges and Coastal Circulations

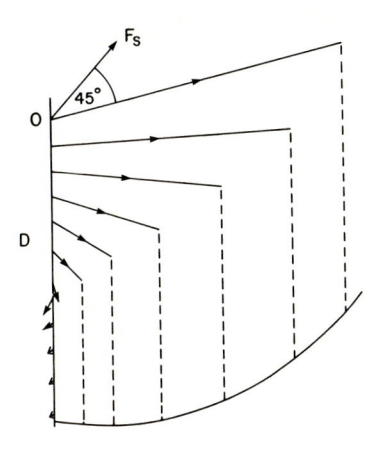

Figure 2.40   The Ekman Spiral.

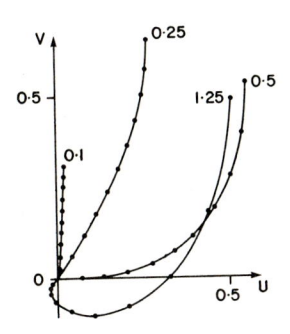

Figure 2.41   Vertical structure of wind-induced current for h/D = 0.1, 0.25, 0.5 : marked in steps of 0.1 h from the surface down to the bottom. Horizontal components of current u, v are oriented such that v lies in the wind direction.

N.S. Heaps

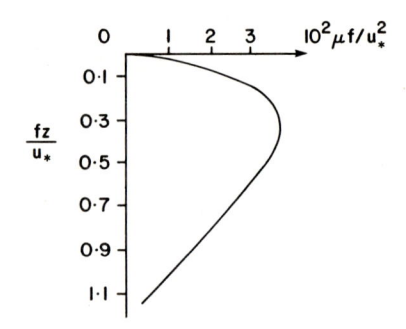

Figure 2.42 Eddy viscosity distribution from a turbulence model. From Svensson (1979).

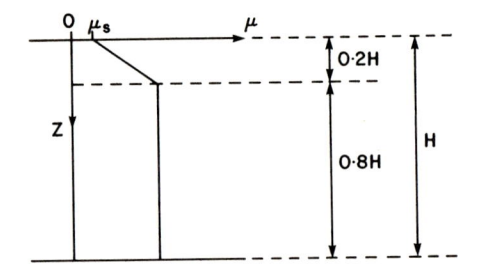

Figure 2.43 Eddy viscosity distribution with a surface wall layer. Pearce and Cooper (1981).

Tides, Storm Surges and Coastal Circulations

An alternative viewpoint to the one presented above is that, for wind-driven currents in homogeneous shallow water for which $h \sim D$, eddy viscosity may be expected to increase continuously from the sea bottom up to the sea surface since (a) vertical eddying motions are inhibited near the rigid bottom, and (b) the driving source of the turbulence, the wind stress, acts at the surface (Thomas, 1975). This denies a decrease in μ as the surface is approached from below. While such a decrease seems most likely on the weight of evidence so far available, there appears to be little or no observational evidence to confirm it for a full range of wind speeds. The general uncertainty is compounded by another viewpoint (Gordon 1982) favouring the use of 'slab' models (which assume that currents are uniform through the depth in the surface mixed layer) for the response of coastal seas to severe storm winds. The role of Stokes' drift in assessing near-surface currents from surface drifter experiments also needs to be borne in mind. Much needs to be done in testing observationally the various hypotheses concerning the variation of μ through the vertical for wind-induced motion.

N.S. Heaps

References

Banks, J. E., 1974.
Phil.Trans.R.Soc., A, 275, 567-609.

Cartwright, D. E., Edden, A. C., Spencer, R. and Vassie, J. M., 1980.
Phil.Trans.R.Soc., A, 298, 87-139.

Corkan, R. J., 1950.
Phil.Trans.R.Soc., A, 242, 493-525.

Davies, A. M., 1983.
In Physical Oceanography of Coastal and Shelf Seas, ed. B. Johns, 357-386, Elsevier, Amsterdam.

Davies, A. M. and Furnes, G. K., 1980.
J.Phys.Oceanogr., 10, 237-257.

Donelan, M. A., 1982.
In First Int. Conf. on Meteorology and Air-Sea Interaction of the Coastal Zone, The Hague, 381-387, American Meteorological Society, Massachusetts (unpublished).

Duun-Christensen, J. T., 1975.
Dt. hydrogr. Z., 28, 97-116.

Flather, R. A., 1976.
Mém. Soc. r. Sci. Liège, 6 Ser., 10, 141-164.

Flather, R. A., 1979.
In Marine Forecasting, ed. J.C.J. Nihoul, 385-409, Elsevier, Amsterdam.

Flather, R. A., 1981.
In Vol. 2, The Norwegian Coastal Current, ed. R. Saetre and M. Mork, 427-458, University of Bergen.

Flather, R. A. and Davies, A. M., 1978.
Dt. hydrogr. Z. Ergänzungsheft, A., No. 15, 51 pp.

Gordon, R. L., 1982.
J. Geophys. Res., 87, 1939,-1951.

Hasse, L., 1974.
Beit. Phys. Atmos., 47, 45-55.

Hasse, L. and Wagner, V., 1971.
Mon. Weath, Rev., 99, 255-260.

Heaps, N. S., 1965.
Phil. Trans. R. Soc., A, 257, 351-383.

Heaps, N. S., 1969.
Phil. Trans, R. Soc., A, 265, 93-137.

Heaps, N. S., 1972.
Mém. Soc. r. sci. Liège, ser. 6, 2, 143-180.

Heaps, N. S., 1979.
In Proc. Sixteenth Coastal Engineering Conf., 1978, Hamburg, Germany, 3, 2671-2686, American Society of Civil Engineers, New York.

Heaps, N. S. and Jones, J. E., 1977.
Geophys. J. R. astr. Soc., 51, 393-429.

Howarth, M. J. and Pugh, D. T., 1983.
In Physical Oceanography of Coastal and Shelf Seas, ed. B. Johns, 135-188, Elsevier, Amsterdam.

Madsen, O. S., 1977.
J. Phys. Oceanogr., 7, 248-255.

Pearce, B. R. and Cooper, C. K., 1981.
J. Hydraul. Div. Proc. Am. Soc. civ. Engrs, 107 (HY3), 285-302.

Prandle, D., 1975.
Proc. R. Soc., A, 344, 509-539.

Proudman, J. and Doodson, A. T., 1924.
Phil. Trans. R. Soc., A, 224, 185-219.

Pugh, D. T. and Vassie, J. M., 1980.
Proc. Inst. civ. Engr., 69, 959-975.

Robinson, I. S., 1979.
Geophys. J. R. astr. Soc., 56, 159-197.

Rossiter, J. R. 1954.
Phil. Trans. R. Soc., 246, 371-400.

Smith, S. D. and Banke, E. G., 1975.
Q. Jl. R. met. Soc., 101, 665-673.

Svensson, U., 1979.
Tellus, 31, 340-350.

Thomas, J. H., 1975.
J. Phys. Oceanogr., 5, 136-142.

3. MODELLING STORM SURGE CURRENT STRUCTURE

A.M. Davies,
Institute of Oceanographic Sciences,
Bidston Observatory,
Birkenhead, Merseyside, L43 7RA, England.

3.1 Introduction

With the development of offshore oil exploration, the need for an accurate determination of environmental conditions during extreme storms, for design purposes has increased. The design of offshore structures has to take account of both the forces due to waves and those exerted by storm driven currents. In the case of storm currents, the current's profile is a significant factor in calculating design forces. Storm induced currents are generally a maximum at the sea surface, and, consequently, except for wave effects, have the largest contribution to design forces.

The need to determine accurately wind induced currents, over the North-West European shelf during extreme wind conditions, has given additional momentum to research on three dimensional storm surge models. In the majority of these models, a coefficient of vertical eddy viscosity is used to parameterize the transfer of the wind's momentum to depth. An excellent review of the needs of the offshore oil industry, and of the limitations of earlier models is given by Gordon (1982).

The importance of the magnitude and vertical distribution of eddy viscosity in determining wind induced current profiles is considered in this paper, using a modal model developed in the next section. The relationship between vertical eddy viscosity and wind/wave conditions together with tidal and storm surge currents is examined. Also the spatial variability of currents over the shelf is considered using a three dimensional shelf model.

In essence this paper gives a brief description of some of the research which is currently in progress on the development of three dimensional models of storm surge currents in the North Sea.

3.2 Spectral Model Formulation

In this section we briefly consider the development of a linear spectral model; although the non-linear terms can also be included (Davies 1980).

A. M. Davies

For simplicity Cartesian co-ordinates are used, although the formulation in spherical co-ordinates can be found in Davies and Furness (1980).

The linear hydrodynamic equations in Cartesian co-ordinates are,

$$\frac{\partial \zeta}{\partial t} + \frac{\partial}{\partial x} \int_0^h u\, dz + \frac{\partial}{\partial y} \int_0^h v\, dz = 0 \qquad (3.2.1)$$

$$\frac{\partial u}{\partial t} - \gamma v = -g \frac{\partial \zeta}{\partial x} + \frac{\partial}{\partial z}\left(\mu \frac{\partial u}{\partial z}\right) \qquad (3.2.2)$$

$$\frac{\partial v}{\partial t} + \gamma u = -g \frac{\partial \zeta}{\partial y} + \frac{\partial}{\partial z}\left(\mu \frac{\partial v}{\partial z}\right) \qquad (3.2.3)$$

The notation used in these equations is, t denotes time, x, y, z Cartesian co-ordinates, with z the depth below the undisturbed surface, h denotes the depth of water, and ζ the elevation of the free surface. Also u, v are the x, y components of current at depth z, with g the acceleration due to gravity, and γ the geostrophic coefficient. The coefficient of vertical eddy viscosity μ in general depends upon x, y z and t.

In this paper we shall be concerned with tidal and wind induced motion, in which case the sea surface boundary condition is given by,

$$-\rho \left(\mu \frac{\partial u}{\partial z}\right)_0 = \tau_x^s \quad , \quad -\rho \left(\mu \frac{\partial v}{\partial z}\right)_0 = \tau_y^s \qquad (3.2.4)$$

Where τ_x^s, τ_y^s denote components of surface stress, and density ρ is constant.

Applying a sea bed stress condition, gives

$$-\rho \left(\mu \frac{\partial u}{\partial z}\right)_h = \tau_x^h \quad , \quad -\rho \left(\mu \frac{\partial v}{\partial z}\right)_h = \tau_y^h \qquad (3.2.5)$$

with τ_x^h, τ_y^h components of bottom stress.

We now consider the solution of equations (3.2.1) to (3.2.3) using an expansion of the two components of velocity u, v in terms of m depth dependent functions f_r (the basis functions) and coefficients $A_r(x,y,t)$ and $B_r(x,y,t)$

Modelling Storm Surge Current Structure

Thus

$$u = \sum_{r=1}^{m} A_r f_r(\sigma) \qquad (3.2.6a)$$

$$v = \sum_{r=1}^{m} B_r f_r(\sigma) \qquad (3.2.6b)$$

with,

$$\sigma = z/h \qquad (3.2.7)$$

a dimensionless depth.

In theory the choice of functions f_r is arbitrary. However in this paper, we confine our attention to the case in which the eddy viscosity μ can be written as

$$\mu(x,y,\sigma,t) = \alpha(x,y,t)\psi(\sigma) \qquad (3.2.8)$$

In equation (3.2.8) α represents the time and horizontal spatial variation of μ with ψ giving its vertical variation. In this particular case it is advantageous to choose the functions f_r as eigenfunctions of

$$\frac{d}{d\sigma}\left[\psi \frac{df}{d\sigma}\right] = -\varepsilon f \qquad (3.2.9)$$

where ε are the associated eigenvalues. With this particular choice of f_r, a set of essentially uncoupled modal equations can be obtained.

The derivation of these equations involves a number of steps which will not be given here; the interested reader is referred to Davies (1983a) for detail. In essence, these steps involve an initial transformation of equations (3.2.1) to (3.2.3) to sigma co-ordinates using (3.2.7). The Galerkin method is then applied and boundary conditions (3.2.4) and (3.2.5) are incorporated. Expansions (3.2.6a) and (3.2.6b) are then used to derive modal equations for the coefficients A_r and B_r. For the case in which μ is given by equation (3.2.8), the resulting modal equations are,

A. M. Davies

$$\frac{\partial \zeta}{\partial t} + \sum_{r=1}^{m} \left(\frac{\partial}{\partial x}(A_r h) + \frac{\partial}{\partial y}(B_r h) \right) \int_0^1 f_r \, d\sigma = 0 \qquad (3.2.10)$$

$$\frac{\partial A_r}{\partial t} = \gamma B_r - g \frac{\partial \zeta}{\partial x} a_r \Phi_r - \frac{f_r(1)}{\rho h} \tau_x^h \Phi_r + \frac{\tau_x^s}{\rho h} \Phi_r - \frac{\alpha}{h^2} A_r \varepsilon_r \qquad (3.2.11)$$

$$\frac{\partial B_r}{\partial t} = -\gamma A_r - g \frac{\partial \zeta}{\partial y} a_r \Phi_r - \frac{f_r(1)}{\rho h} \tau_y^h \Phi_r + \frac{\tau_y^s}{\rho h} \Phi_r - \frac{\alpha}{h^2} B_r \varepsilon_r \qquad (3.2.12)$$

where $r = 1, 2, \ldots, m$

with

$$a_r = \int_0^1 f_r \, d\sigma \qquad \text{and} \qquad \Phi_r = 1 \bigg/ \int_0^1 f_r^2 \, d\sigma \qquad (3.2.13)$$

Also the eigenfunctions have been normalized such that

$$f_r(0) = 1 \qquad (3.2.14)$$

For certain idealized vertical variations of eddy viscosity the eigenfunctions and eigenvalues of (3.2.9) can be determined analytically (Heaps, 1972 & 1981, Baker and Jordan, 1981). However for the more physically realistic vertical variations found in nature (Wolf, 1980), equation (3.2.8) can be accurately solved using either an expansion of spline functions (Davies, 1983a) or by an iterative method involving the Runge-Kutta technique (Davies and Furnes, 1984).

Equations (3.2.10) to (3.2.12) are the working equations which have to be integrated through time in order to determine the variation of the coefficients A_r, B_r within a sea area. Spatial discretization of these equations is performed using a finite difference grid in the horizontal (see for example figure 3.1). Details of the differencing methods used to solve these equations can be found in Heaps (1972).

3.3 Form of Vertical Eddy Viscosity

In this section we consider briefly what form vertical eddy viscosity should take for wind driven currents in a tidal sea region, such as the North Sea.

Modelling Storm Surge Current Structure

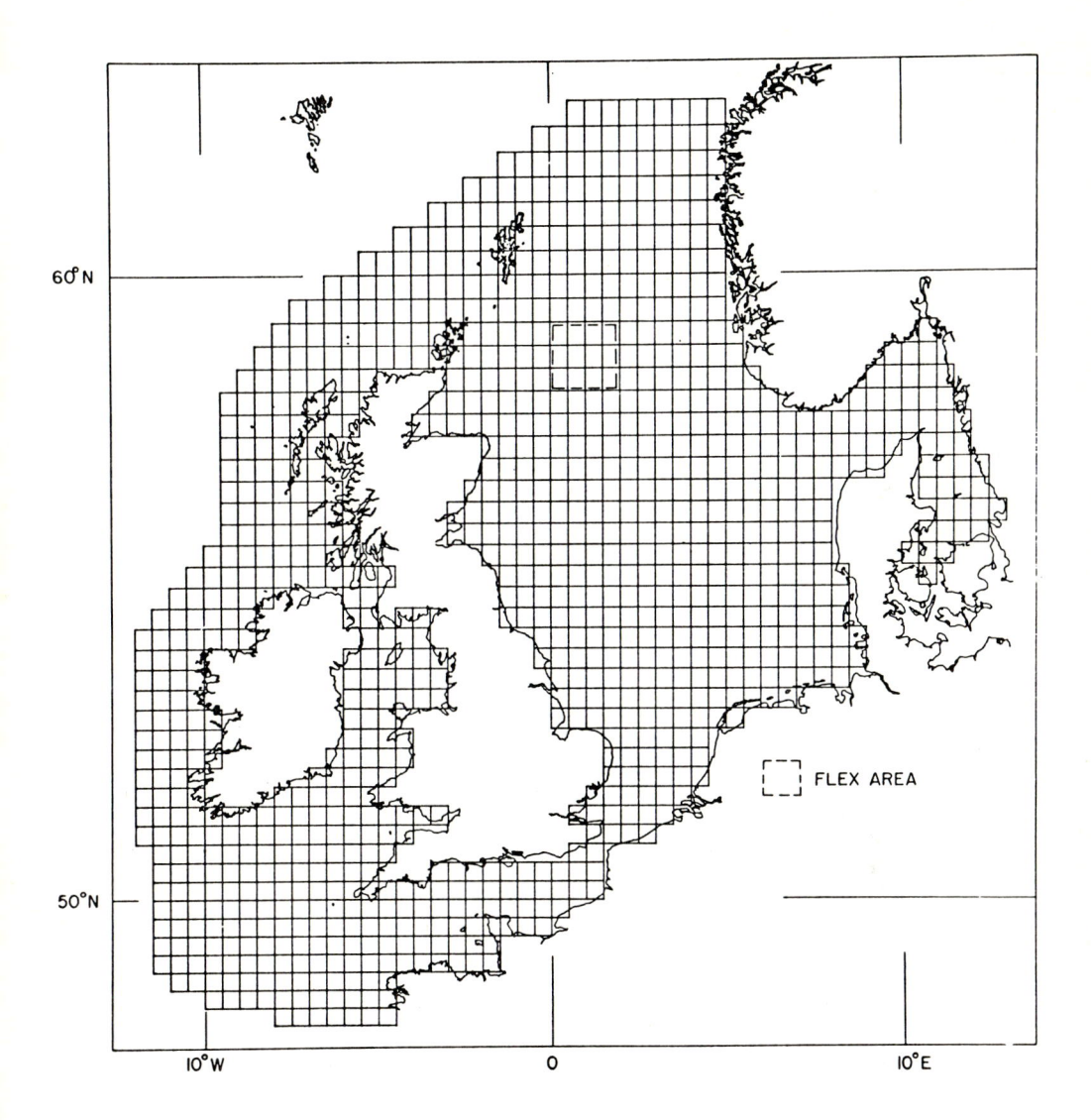

Figure 3.1   Three dimensional shelf model: finite difference grid.

A. M. Davies

Eddy viscosity values inferred from current measurements in the sea (Bowden et al, 1959, Wolf, 1980) and recent work on turbulence theory (Smith, 1982, Csanady and Shaw, 1980) has shown that eddy viscosity is a physical property of the flow, depending upon current velocity and wind strength. A constant value of eddy viscosity does not therefore appear physically realistic for wind driven currents in a tidal sea.

Davies and Furnes (1980) suggested a formulation of eddy viscosity μ_1, away from sea surface and sea bed regions, of the form,

$$\mu_1 = \frac{K\bar{u}^2}{\sigma} \tag{3.3.1}$$

with $\bar{u}$ depth mean current velocity and $K = 2.0 \times 10^{-5}$, a dimensionless coefficient, with $\sigma = 10^{-4}$ s^{-1}, a frequency comparable with the Coriolis coefficient. Equation (3.3.1) implies that eddy viscosity is constant through the vertical, though its magnitude varies with horizontal position and time through the depth mean current $\bar{u}$. Such a formulation does not directly take into account variations in eddy viscosity with wind strength, except in so far as changing wind strength affects the value of $\bar{u}$. In the central northern North Sea, tidal and strong wind driven currents are typically of the order of 75cm/s, giving from (3.3.1) a μ_1 value of order 1000cm^2/s during major wind events.

At the sea surface, wind's energy together with wind induced waves are a source of turbulence. Neumann and Pierson (1964) give an empirical formula for the effective surface eddy viscosity due to wind and wave action, μ_0, namely

$$\mu_0 \text{ (cm}^{-1} \text{ gr s}^{-1}\text{)} = 0.1825 \times 10^{-4} W^{5/2} \text{ (W in cm/s)} \tag{3.3.2}$$

An alternative formulation due to Ichiye (1967), derived by calculating the Reynolds stresses due to surface waves, is

$$\mu_0 = 0.028 \, H^2/T \tag{3.2.3}$$

with T and H average period and significant height of surface waves.

Modelling Storm Surge Current Structure

In a fully developed sea state, T and H can be related to the wind speed (Carter, 1982), giving the following relationships, namely

$$H = 0.0248 \, W^2 \qquad (3.3.4)$$

$$T = 0.5660 \, W \qquad (3.3.5)$$

Substituting (3.3.4) and (3.3.5) into (3.3.3), gives,

$$\mu_0 \, (m^2 s^{-1}) = 0.3043 \times 10^{-4} W^3 \quad (W \text{ in m/s}) \qquad (3.3.6)$$

The importance of wind fetch upon significant wave height and period has been examined by Carter (1982). In fetch limited condition (see Carter (1982) for equations determining these conditions), T and H are related to fetch L(km), by,

$$H = 0.0163 \, L^{0.5} \, W \qquad (3.3.7)$$

$$T = 0.439 \, L^{0.3} \, W^{0.4} \qquad (3.3.8)$$

Substituting (3.3.7) and (3.3.8) into (3.3.3), gives

$$\mu_0 \, (m^2 s^{-1}) = 0.1695 \times 10^{-4} \, L^{0.7} \, W^{1.6} \qquad (3.3.9)$$

Values of near surface eddy viscosity μ_0 computed with equations (3.3.2), (3.3.6) and (3.3.9), for a range of wind speeds and fetch, are shown in Table 3.1. It is evident from this table that values of μ_0 computed using equations (3.3.2) and (3.3.6) are comparable, when the fetch is large. However, once the effect of fetch is considered (equation 3.3.9), values of μ_0 are significantly lower than those computed using (3.3.6) which is only appropriate for fully developed wave conditions.

At the sea bed, eddy viscosity depends upon the roughness length Z_0 and current velocity (Bowden et al, 1959). From observations, Channon and Hamilton (1971) found a range of Z_0 values, typically from 7.0cm to 0.44cm. Using this range of Z_0 values with strong tidal and wind driven North Sea currents, Davies (1983a) suggested that eddy viscosity at the sea bed could vary from 100cm^2/s to below 1cm^2/s.

A. M. Davies

Eddy viscosity value μ_0 (cm^2/s)

Wind W (m/s)	A	B	C L=600km	D L=200km	E L=50km	F L=10km	G L=5km
5	100	40	40	40	34	11	7
10	563	316	316	275	104	34	21
15	1551	1069	1069	527	200	65	40
20	3184	2535	1800	834	316	103	63
25	5563	4951	2573	1192	452	146	90

Table 3.1 Values of surface eddy viscosity for a range of wind speed computed using (A) an empirical relationship of Neumann and Pierson, (B) an equation taking into account the surface wave field in open ocean conditions, (C)-(G) an equation taking into account the surface wave field but in fetch limited conditions with fetch L given in time.

	Homogeneous		Stratified
	μ_0=50cm^2/s μ_1=1000cm^2/s μ_2=50cm^2/s	μ_0=5000cm^2/s μ_1=1000cm^2/s μ_2=50cm^2/s	μ_0=5000cm^2/s μ_1=1000cm^2/s μ_1'=50cm^2/s μ_2=50cm^2/s
	(A)	(B)	(C)
$\alpha\varepsilon_1$	0.169	0.169	0.126
$\alpha\varepsilon_2$	1.629	1.643	0.456
$\alpha\varepsilon_3$	4.819	4.946	2.564
$\alpha\varepsilon_4$	9.767	10.295	4.652
$\alpha\varepsilon_5$	16.274	17.719	8.703
$\alpha\varepsilon_6$	24.157	27.125	14.824
ϕ_1	1.697	1.663	3.494
ϕ_2	2.050	1.696	1.947
ϕ_3	2.698	1.580	0.325
ϕ_4	3.753	1.368	0.764
ϕ_5	5.151	1.152	0.110
ϕ_6	6.273	0.972	0.384

Table 3.2 Values of $\alpha\varepsilon_r$ and ϕ_r for the first six modes computed using the viscosity profiles shown in Figures 3.2(a), (b), (c) for a range of μ_0, μ_1, μ_1' and μ_2 values.

Modelling Storm Surge Current Structure

What form the vertical variation of eddy viscosity should take in a wind driven tidal sea is difficult to determine. Figure 3.2(a) shows a typical variation which may be appropriate in a strong tidal or wind driven current regime ($\bar{u}$ = 75cm/s giving μ_1 of order 1000cm^2/s, equation (3.3.1)); at a low wind speed (W $\approx$ 5m/s) or in limited fetch conditions (μ_0 of order 50cm^2/s, Table 3.1). A value of eddy viscosity at the sea bed, μ_2 = 50cm^2/s, lying within the range of 100cm^2/s to 1cm^2/s suggested by Davies (1983a) was assumed in Figure 3.2. Bowden et al (1959) suggested that the distance h_2 over which μ increases linearly from its sea bed value would be of order 0.1h. A linear variation, near the sea surface over an equal distance i.e. h_1 = h_2 = 0.1h is assumed in Figure 3.2(a). In a water depth h = 100m, this gives h_1 = h_2 = 10m.

At high wind speed (W $\approx$ 22m/s) in open sea conditions, equation (3.3.6) gives $\mu_0 \approx$ 5000cm^2/s, and a profile reflecting this variation is shown in Figure 3.2(b).

The profiles shown in Figure 3.2(a), (b), are typical of vertical eddy viscosity variations in homogeneous regions. In a stratified sea, a reduction in eddy viscosity within the thermocline due to enhanced vertical stability (Mortimer,1952) will occur and this has to be taken into account in the assumed vertical variation of viscosity. A vertical profile of viscosity reflecting a wind induced turbulent surface layer of thickness h_1, with a linear reduction of μ from μ_0 to μ_1 is shown in Figure 3.2(c). Below this surface layer eddy viscosity decreases over a region h_3^* to a low value μ_1^* applying within the thermocline. A linear increase to the value μ_1, over h_3 is assumed below the thermocline. At high wind speeds (W $\approx$ 22m/s) μ_0 will be of order 5000cm^2/s, with μ_1 = 1000cm^2/s and μ_2 = 50cm^2/s. Typical values (Davies,1981) of h_3^* and h_3 are 2m and 6m respectively, giving a total thermocline thickness of 10m. Under strongly stratified conditions, μ_1^* will be of order 50cm^2/s.

In order to examine the influence of bottom topography, eddy viscosity and wind strength upon wind induced currents, a number of calculations have been performed using the modal model developed here. These calculations have involved two types of model. Initially a simulation model, which contains realistic bottom topography and wind fields, and secondly a mechanistic model which uses an idealized bottom topography and wind field.

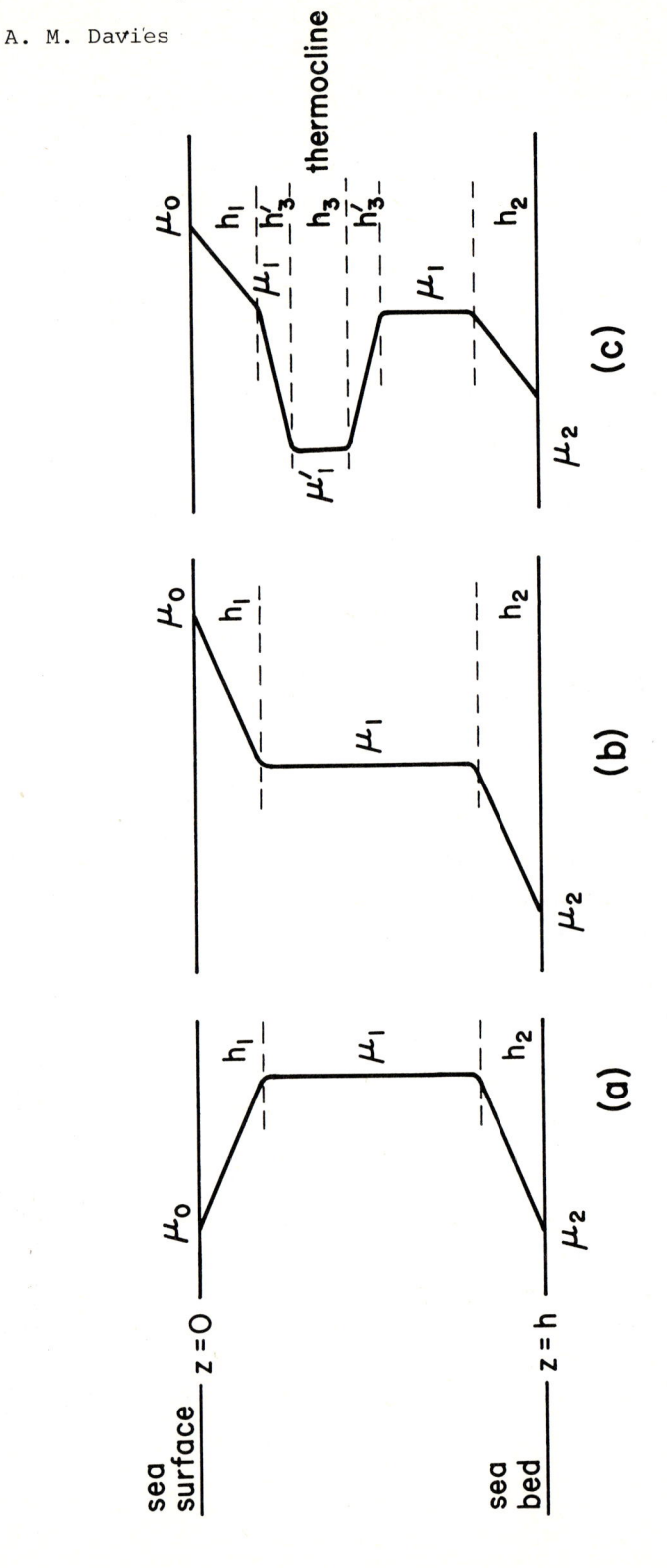

A. M. Davies

Figure 3.2 Schematic representation of eddy viscosity profiles used in the wind driven model. Profiles (a) and (b) are for a homogeneous sea, with profile (c) incorporating a reduction in turbulence within the thermocline due to enhanced vertical stability.

Modelling Storm Surge Current Structure

In the next section some results are presented from the simulation model of the shelf. In the subsequent section the influence of the profile of eddy viscosity upon wind induced current structure is examined using a mechanistic model in which the wind field is uniform and depth constant.

3.4 A Three Dimensional Simulation Model of Surge Currents on the North-West European Shelf

During the spring of 1976, a large scale observational programme (JONSDAP '76) took place in the North Sea. One part of this experiment was to examine the spatial and temporal variation of wind induced currents in the area. The observational experiment was complemented by a programme of numerical modelling a part of which involved the development, using the method described in section 3.2, of a three dimensional model covering the area shown in Figure 3.1. The grid resolution used in this model was $1/3^\circ$ latitude by $1/2^\circ$ longitude. Bottom stress was parameterized using a quadratic law of bottom friction.

Eddy viscosity was determined in the model using equation (3.3.1). Since the tidal currents are particularly important on the shelf in determining the level of mixing, it is important to include them within the model in order to determine the correct level of turbulence from (3.3.1) (see Davies 1983b for details). Consequently eddy viscosity varied both spatially and temporally within the model, with a magnitude determined by both tidal and storm surge current strength. Recent observational evidence (Provis and Lennon, 1983) has confirmed the importance of including tidally induced turbulence in a model of wind driven currents. This point will be considered in detail later in this paper.

Using this model, with hourly wind stress data derived from an atmospheric model (see Davies(1983b)for details), the meteorologically induced circulation for 30 days during March/April was simulated (Davies, 1983b). This was a period of maximum deployment of instruments and also a time of strong winds.

A. M. Davies

One wind event of particular interest, occurred on the 6th April 1976; a time of strong notherly winds over the North Sea (see Figure 3.3), associated with a depression which moved from Iceland over to Scandinavia (Figure 3.3). At 0600 GMT on April 6th, strong north-westerly winds of up to 35 knots persisted over the northern North Sea and 20 knot onshore winds were present in the German Bight.

These winds produced strong southerly surface currents of between 40cm s^{-1} and 100cm s^{-1} over the North Sea at this time (see Figure 3.4(a)). However in the English Channel, where the wind field was light, surface currents were only of the order of 5cm s^{-1}. The on-shore winds in the German Bight induced an easterly sea surface current of between 40cm s^{-1} and 80cm s^{-1}. An offshore flow at mid-depth (Figure 3.4(b)) and at the sea bed (Figure 3.4(c)) occurred in this region. It is evident from Figures 3.4(b), (c) that a northerly flow occurred at depth along the west coast of Denmark, which opposed the strong south easterly surface current.

It is apparent from Figures 3.4(a), (b), (c) that wind induced currents computed with the model, exhibit a maximum value at the sea surface, and decrease rapidly with distance below the surface (Davies, 1983b). At the sea surface, away from coasts, the current is essentially in the direction of the wind, whereas at depth, the current is significantly affected by bottom topography. In many cases the direction of current at depth is significantly different from that at the sea surface. The complex spatial variations of current at depth found in the model were in good agreement with observations (Riepma, 1978). The interested reader is referred to Davies (1983b), for a detailed discussion of these comparisons.

The ratio between surface current velocity and surface wind (the wind factor) was also computed with the model (Davies, 1983b). A value of the order of 15% was found in an area to the west of Denmark, associated with surface current of the order of 150cm s^{-1}. This high ratio was limited to an area of high wind velocity. In sheltered areas, or regions where the winds were light, the wind factor was below 1%. Davies (1983b) suggested that these large variations of wind factor may have been due to the absence of a wind dependent near surface eddy viscosity within the model. Observational evidence (Ambjorn, 1981, Smith, 1979) does in

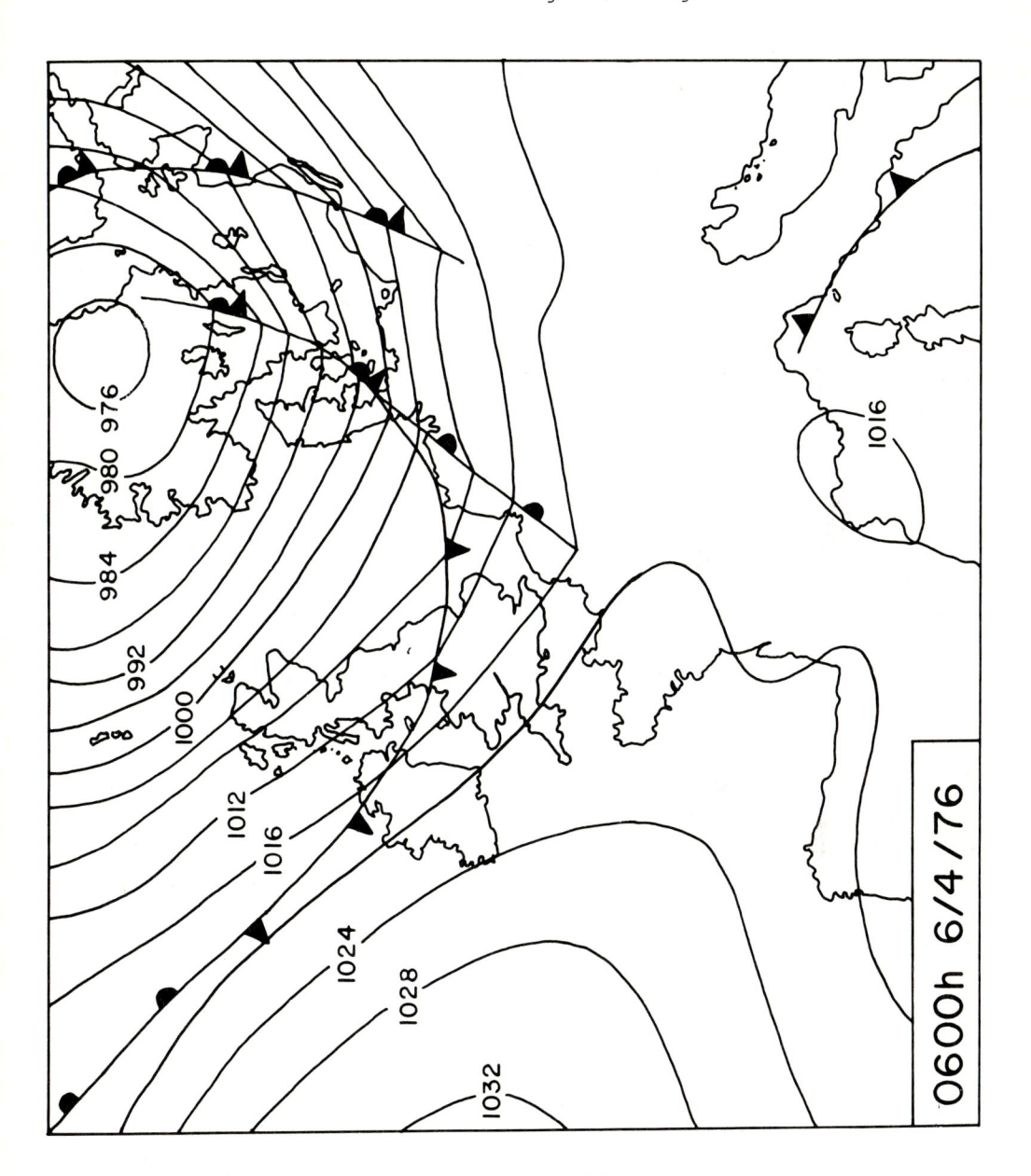

Figure 3.3   Weather Chart for April 6th 1976.

A. M. Davies

SURFACE (0600h 6/4/76)

Figure 3.4 (a)   Computed current vectors:
Sea Surface

The convention used is that the dot represents
the grid point in the model, with current flowing
away from this point.
(Scale of vectors - 20cm s^{-1}, = 40cm s^{-1}, ≡ 60cm s^{-1})

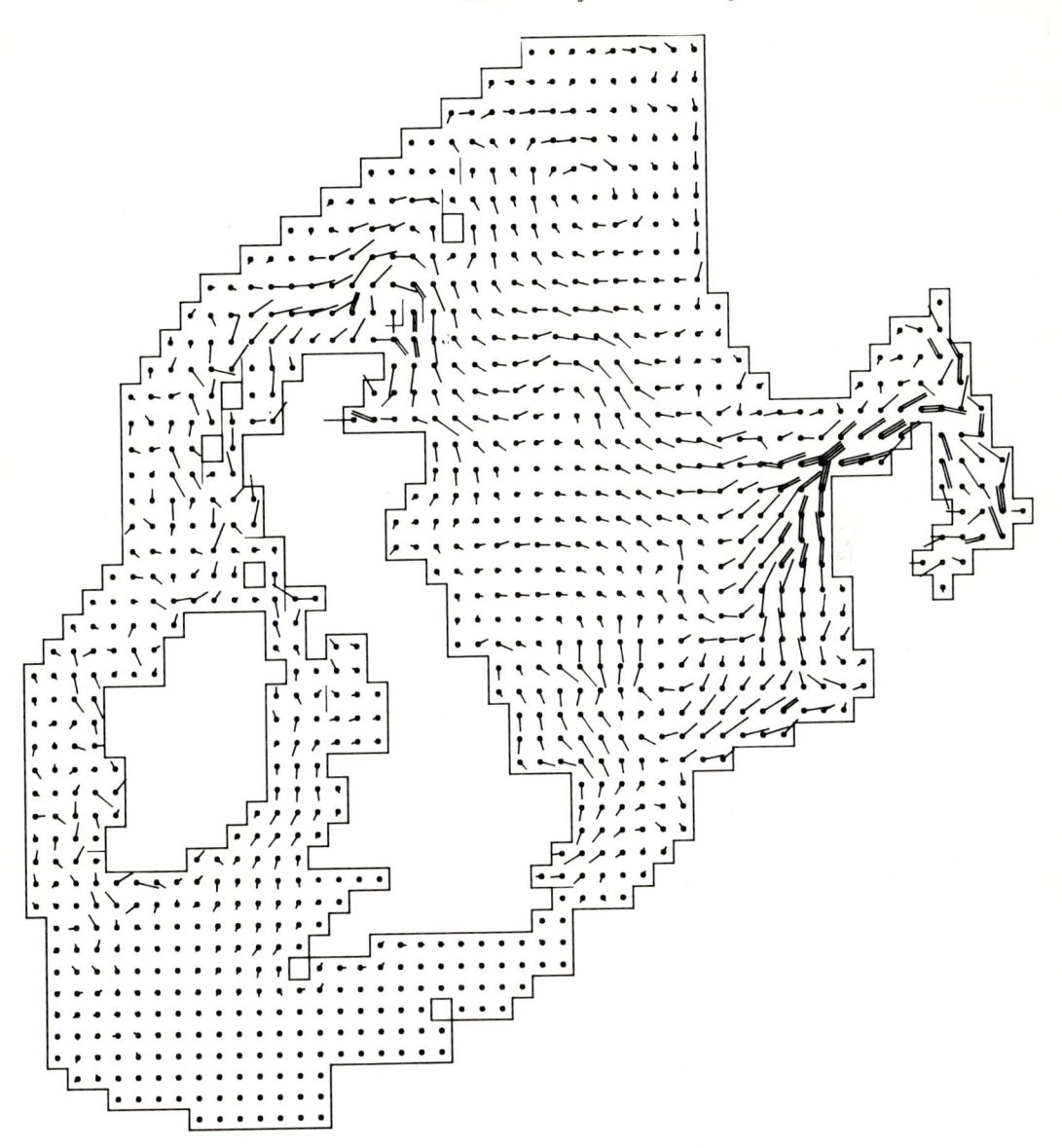

Modelling Storm Surge Current Structure

MID-DEPTH (0600h 6/4/76)

Figure 3.4 (b) Computer current vectors:
Mid-depth

The convention used is that the dot represents
the grid point in the model, with current flowing
away from this point.
(Scale of vectors - 20cm s^{-1}, = 40cm s^{-1}, ≡ 60cm s^{-1})

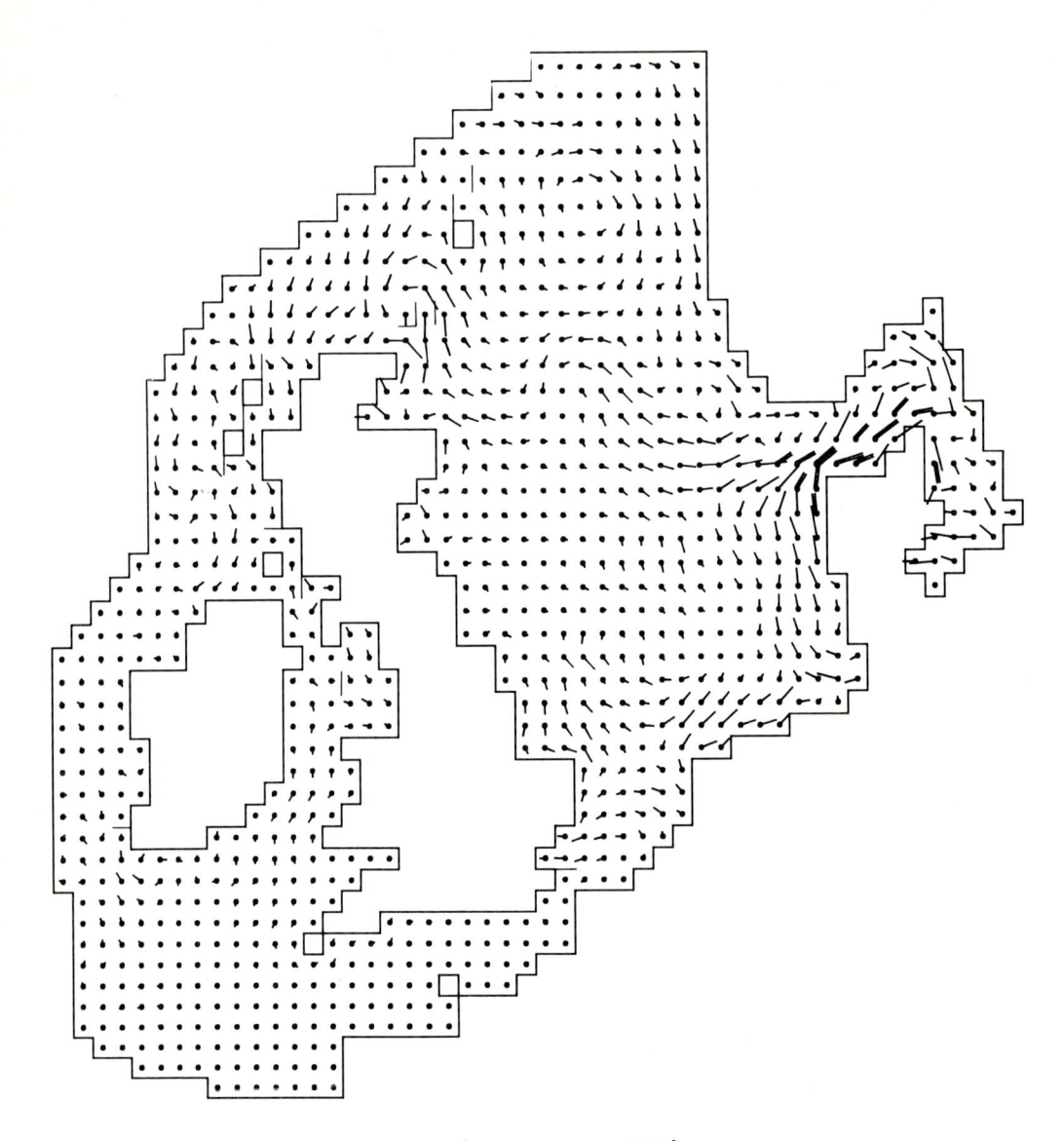

A. M. Davies

BOTTOM (0600h 6/4/76)

Figure 3.4 (c) Computer current vectors:
Sea bed at 0600 on the 6th April 1976

The convention used is that the dot represents
the grid point in the model, with current flowing
away from this point.
(Scale of vectors - 20cm s^{-1}, = 40cm s^{-1}, ≡ 60cm s^{-1})

Modelling Storm Surge Current Structure

fact suggested a decrease of wind factor with wind speed, the opposite of that found in the model. This suggests that it is particularly important for the determination of surface currents to include a wind/wave dependent near surface eddy viscosity within the model. This point will be discussed further in the next section.

Over the majority of the North Sea a ratio of the order of 3% was obtained. This value is in reasonable agreement with the experimental results of Hughes (1956), a ratio of 2% in the open ocean; and Tomczak (1964), a ratio of 4.2% in the North Sea.

3.5 A Mechanistic Model of Wind Induced Current Profiles

The circulation patterns from the continental shelf model, clearly show a complex spatial variation of wind induced current, due to the influence of coastline and bottom topography. In this model eddy viscosity was constant in the vertical and its value was not directly affected by the wind/wave field, although equations (3.3.6) and (3.3.9), suggest that near surface eddy viscosity should be influenced by these parameters. However viscosity did vary with horizontal position, time and magnitude of tidal and wind induced current.

In order to understand the time evolution of current profiles, in response to an applied wind, and the effect upon them of changes in the vertical profile of viscosity, it is necessary to consider a mechanistic model, which is significantly simpler than the simulation model described previously. To this end we consider the time dependent Ekman problem of an unbounded sea of constant depth h. Motion is generated by a suddenly applied and maintained uniform wind stress, τ_y = 10 dyn/cm^2, corresponding to a northerly wind (the direction of the strongest winds during the JONSDAP period, in particular on the 6th April 1976).

In the absence of land boundaries, and with constant depth and uniform wind stress, no gradients of sea surface elevation occur. In the calculations described here, a no-slip bottom boundary condition was used. Under these conditions, the working equations are (3.2.11) and (3.2.12), with $\partial \zeta/\partial x$, $\partial \zeta/\partial y$ and τ_x^h, τ_y^h set to zero. In this case the modes in (3.2.11) and (3.2.12) are uncoupled, except for coupling due to the Coriolis term, and it is therefore possible to use these equations to examine the model response of the sea region to a suddenly applied wind stress.

A. M. Davies

It is evident from (3.2.11) that Φ_k determines the initial distribution of the wind stress τ^s between the modes, with the parameter $\alpha\varepsilon_k$ producing the damping of the modes.

Values of $\alpha\varepsilon_k$ and Φ_k computed using the eddy viscosity profile shown in Figure 3.2(a), with $\mu_0 = \mu_1' = 50 \text{cm}^2/\text{s}$, $\mu_1 = 1000 \text{cm}^2/\text{s}$, and $h = 100\text{m}$, $h_1 = h_2 = 10\text{m}$ are given in Table 3.2, distribution A. Also given (Distribution B) are values of $\alpha\varepsilon_r$ and Φ_r computed with μ_0 increased to $5000 \text{cm}^2/\text{s}$ (the profile shown in Figure 3.2(b)). Values computed using the profile shown in Figure 3.2(c), with $\mu_0 = 5000 \text{cm}^2/\text{s}$, $\mu_1 = 1000 \text{cm}^2/\text{s}$, $\mu_1' = 50 \text{cm}^2/\text{s}$ and $\mu_2 = 50 \text{cm}^2/\text{s}$ are included in the table (Table 3.2, Distribution C). The vertical variation of the first four modes computed with these viscosity profiles are shown in Figure 3.5.

It is evident from Table 3.2, that the $\alpha\varepsilon_r$ values increase with mode number, and hence the damping of the higher modes is significantly greater than the lower ones. However values of Φ_r do not exhibit such a uniform variation. It is apparent from Table 3.2, that for the case in which $\mu_0 = 50 \text{cm}^2/\text{s}$, the higher Φ_r values exceed the lower ones. However when μ_0 is increased to $5000 \text{cm}/\text{s}$, Φ_r values are comparable in magnitude. In the case of the vertical variation of viscosity representing stratified conditions (Figure 3.2(c)), Φ_r values of the lower modes, exceed those of the higher ones.

It is evident from Figure 3.5, that when $\mu_0 = 50 \text{cm}^2/\text{s}$, the higher order modes exhibit increased near surface shear as the mode number increases. However when $\mu_0 = 5000 \text{cm}^2/\text{s}$, near surface shear is reduced, and in the case of the modes of the stratified problem (Figure 3.5(c)), there is little near surface shear, although a high shear region occurs across the thermocline.

It is apparent from equations (3.2.11) and (3.2.12), that the initial response to a suddenly applied wind stress is the excitation of all vertical modes. The distribution of the wind's stress between the modes is determined by the weighting term Φ_r. It is evident from Table 2, that in the case in which $\mu_0 = 50 \text{cm}^2/\text{s}$, because Φ_r is greater for the higher modes then these are excited in preference to the lower ones. Since near surface shear is higher in these modes (see Figure 3.5) then the initial response in this case is the generation of a high shear near surface layer.

Modelling Storm Surge Current Structure

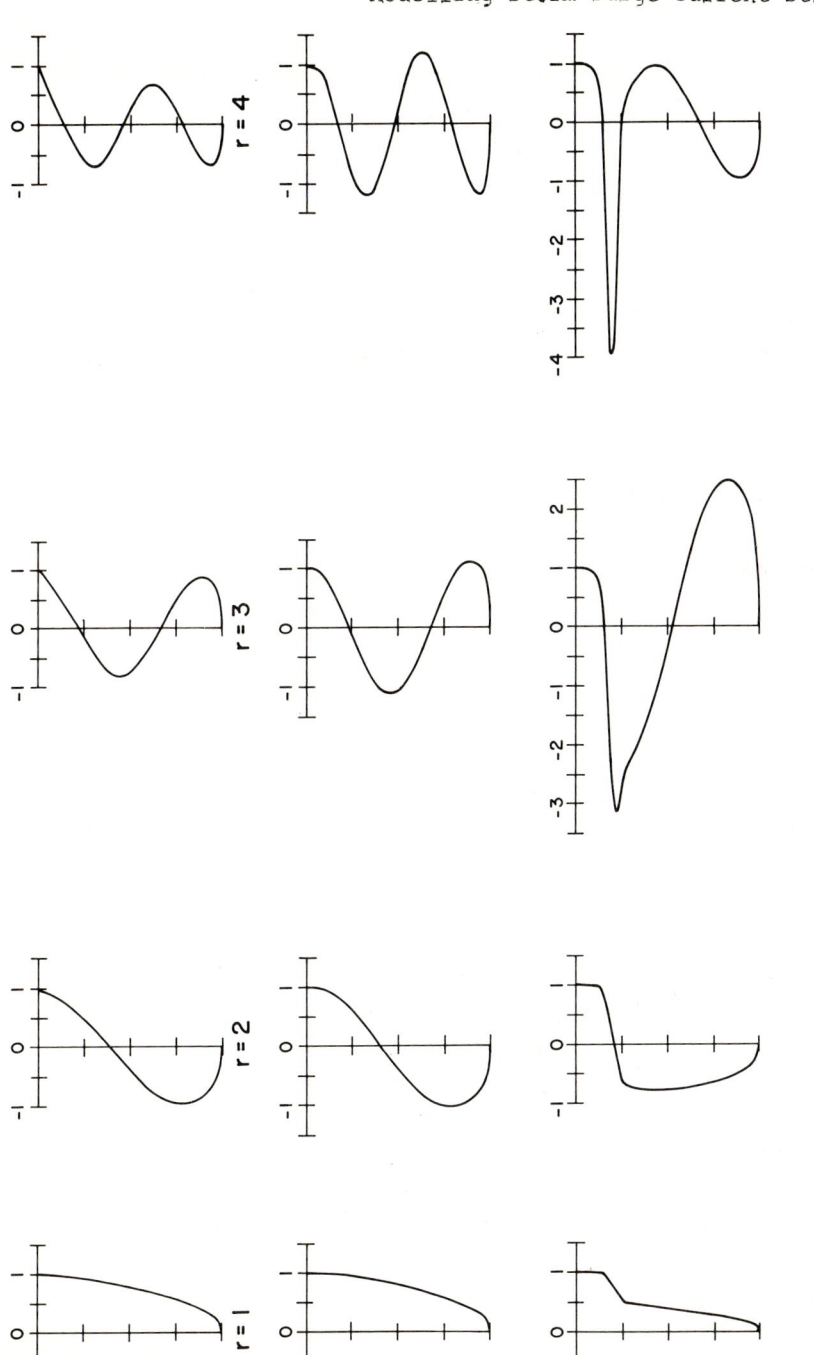

Figure 3.5 Vertical variation of the first four modes in a water depth h=100m, computed with
(A) Viscosity profile given in Figure 1(a) with μ_0=50cm²/s.
(B) Viscosity profile given in Figure 1(b) with μ_0=5000cm²/s.
(C) Viscosity profile given in Figure 1(c), see text for values of viscosity.

A. M. Davies

In the case in which $\mu_0 = 5000\text{cm}^2/\text{s}$, near surface shear is significantly less, and for the viscosity distribution shown in Figure 3.2(c), the modes are essentially constant above the thermocline, with a high shear region across it.

After their initial excitation, the modes are damped, with damping determined by the term $\alpha\varepsilon_r/h^2$. Consequently the higher order modes are damped most rapidly, which produces a reduction in near surface shear, and an increase in current at depth. A thorough discussion of the modal response of a sea region to a suddenly applied wind stress is beyond the scope of this paper, but is discussed in detail in Davies (1984a,b).

Steady state U and V components of current computed with the viscosity profiles shown in Figure 3.2(a), (b), (c) and μ uniform through the vertical at $1000\text{cm}^2/\text{s}$ are given in Figure 3.6(a), (b), (c), (d). (Note the V component is wind driven).

Considering initially velocity profiles computed in the homogeneous case, it is evident that as μ_0 is increased, surface current magnitude is reduced. When μ_0 is low ($\mu_0 = 50\text{cm}^2/\text{s}$) there is significant near surface shear in the V component of current (the component in the wind direction). As μ_0 is increased this shear is significantly reduced reflecting the variation in vertical structure of the modes (Figure 3.5). It is apparent that below the near surface region, velocities are not significantly different. This suggests that a model in which μ is constant through the vertical with its value determined from tidal and wind induced current may be reasonably accurate away from the near surface and near bed layers. Indeed a comparison (Davies 1983b) of circulation patterns from the simulation model with observations taken during JONSDAP '76 tends to confirm this. However in the near surface region currents shown in Figure 3.6(a), (b), (c) are significantly different. Unfortunately during JONSDAP '76, no observations were made close to the sea surface, and hence no comparisons with the simulation model are possible in this region.

Equations (3.3.2), (3.3.6) and (3.3.9) do however suggest that in open sea conditions at high wind speeds (W of order 20m/s), then μ_0 should be of order $5000\text{cm}^2/\text{s}$. It is evident from Figure (a), (b), (c) that this

Modelling Storm Surge Current Structure

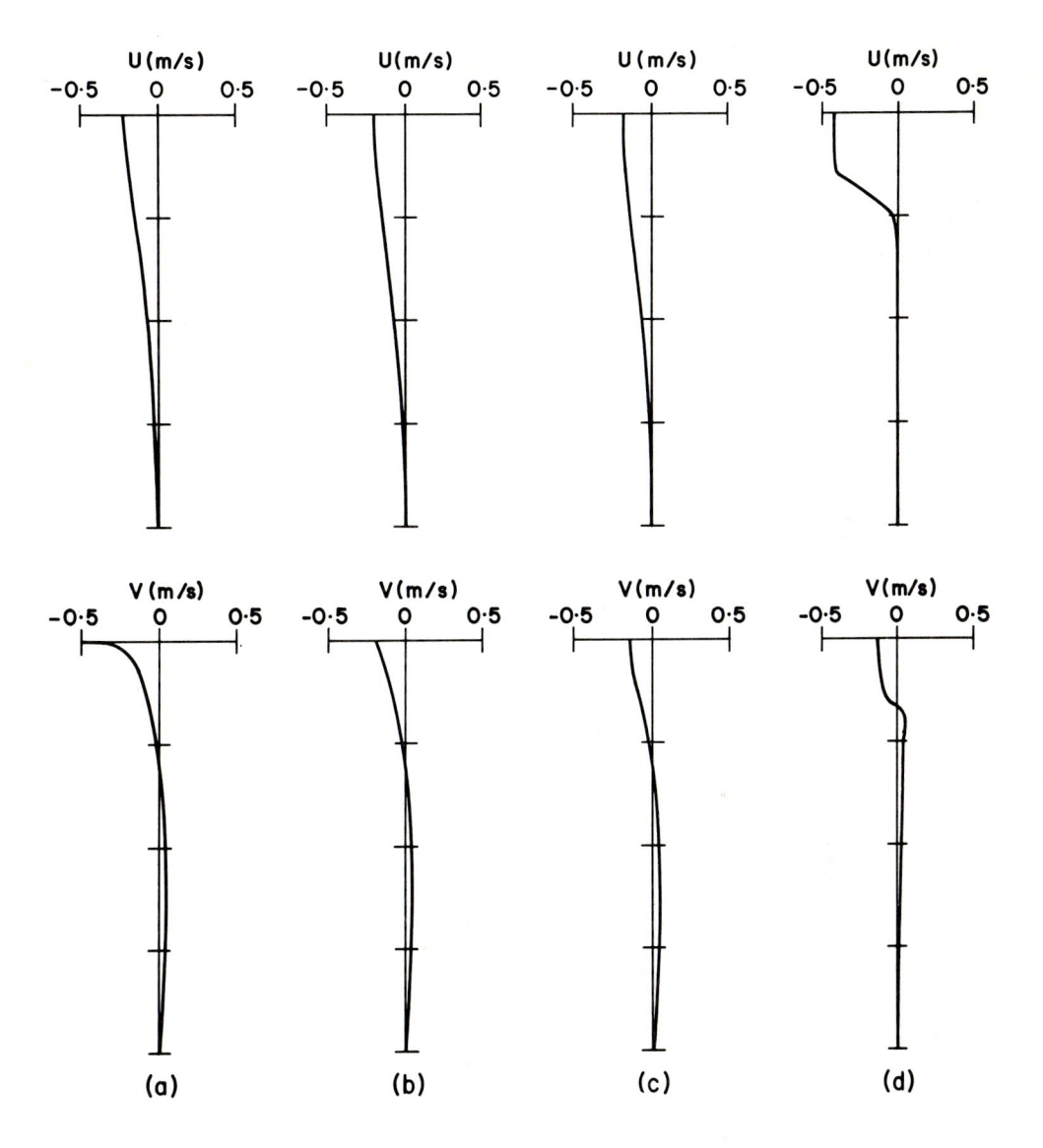

Figure 3.6 Steady state U and V components of drift current computed
with a wind stress $\tau_x=0.0$, $\tau_y=-10$dyn/cm^2, using
(a) the viscosity profile shown in Figure 1(a).
(b) eddy viscosity constant in the vertical.
(c) the viscosity profile shown in Figure 1(b).
(d) the viscosity profile shown in Figure 1(c).

A. M. Davies

would signficantly reduce the near surface currents below those computed with µ constant at 1000cm^2/s. If an increase in near surface eddy viscosity with wind speed was included in the simulation model described previously then the ratio of 15% (found at high wind speed) between surface current and surface wind (the wind factor) would be reduced. In the case of low wind speeds, with reduced near surface viscosity, the wind factor would increase. Observational evidence (Ambjorn, 1981, Smith, 1979), clearly shows a reduction in wind factor with wind speed.

It is evident from Figure 3.6(d), that in stratified conditions, the steady state surface U component of current for the case considered here, exceeds the V component. Above the thermocline the U component is nearly constant, with an internal shear region occurring across the thermocline. The magnitude of the component of current is however signficantly influenced by the thickness of the mixed layer and stability of the thermocline (Davies, 1984c).

It is apparent from Figure 3.6(a), (b), (c), (d) that a range of wind induced current profile is possible depending upon near surface eddy viscosity, intensity of stratification and thickness of the surface mixed layer.

In a lake or semi-enclosed sea region, with limited fetch, and at modest wind speed, Table 3.1 suggests a low value of surface eddy viscosity, giving a profile having a high near surface shear (Figure 3.6(a)). Observational evidence from lakes (Bye, 1965), confirms such high shear near the sea surface. At the other extreme in open sea conditions at high wind speed, with associated high surface viscosity (Table 3.1), a near linear variation of current occurs (Figure 3.6(c)). When surface eddy viscosity is high, with stable stratification at depth, a "slab like" behaviour above the thermocline, with little or no current shear (Figure 3.6(d)), appears appropriate. Observations in these conditions (Gordon, 1982) confirm this type of profile.

The profiles shown in Figure 3.6(a), (b), (c) clearly show that the magnitude of near surface eddy viscosity is important in determining the vertical variation and intensity of near surface currents. In these profiles the same values of viscosity at depth, μ_1 and μ_2 were used. However Davies (1984a,b) has shown that viscosity at depth and in the near bed region can significantly influence surface currents in areas such as the North Sea. This is because the rate at which the wind's momentum, which has been imparted to the sea's surface layer, can diffuse to depth and finally dissipate in the bottom boundary layer is critically dependent upon eddy viscosity at depth.

It is evident from equation (3.3.1), that viscosity at depth is determined by total current, and in a region of high tidal currents will depend upon tidal current strength. Consequently spatial and temporal variations of tidal currents will significantly influence the profile and magnitude of wind induced currents (Davies, 1984a,b). Recent observational evidence supporting this conclusion and emphasising the need to relate turbulence at depth to tidal currents has been reported by Provis and Lennon (1983). They found that as tidal currents increase, wind induced currents were reduced.

Also of particular importance is the direction of wind induced surface current with respect to the tidal current. With opposing currents, surface waves will steepen and break. A process which is a very efficient mixing mechanism and could significantly influence the value of μ_0 and hence near surface currents.

3.6 Concluding Remarks

In this paper we have briefly indicated the major steps in the formulation of a three dimensional modal model. By using modes in the vertical rather than a finite difference grid, a continuous current profile from sea surface to sea bed can be computed. The modal model also gives significant insight into the time evolution of wind induced current profiles in response to wind forcing. Computed wind induced currents on the shelf, at depth, show a complex spatial variation in good agreement with observations.

Calculations show that near surface currents are substantially influenced by near surface viscosity, the magnitude of which varies significantly with the wind and wave field. They are also influenced by viscosity at depth (Davies, 1984a,b) with the magnitude of this viscosity depending upon tidal and storm surge currents.

A. M. Davies

By using the Galerkin method, the three dimensional shelf model described here, could be extended to incorporate a wind/wave dependent near surface viscosity with viscosity at depth still determined by the flow field. However without a substantial synoptic set of near surface current measurements, a rigorous validation of computed surface currents is not possible.

The mechanistic model does however indicate that a range of current profiles is possible depending upon environmental conditions, namely, wind and wave field, tidal and storm surge current magnitude and degree of stratification.

By a combination of models and current profile observations, the role of wind, wave and tidally induced turbulence in determining the physical nature of current profiles can be assertained. Such a programme, will lead to verified models which can accurately predict the environmental parameters required in the design of offshore structures.

References

Ambjorn, C., 1981
An operational oil drift model for the Baltic, in Proceedings of the Symposium on the Mechanics of Oil Slicks, Paris, France.

Baker, J.R. and Jordan, T.F., 1981
Vertical structure of time-dependent flow for viscosity that depends on both depth and time. J.Phys.Oceanogr. 11, 1673-1674.

Bowden, K.F., Fairbairn, L.A. and Hughes, P., 1959
The distribution of shearing stresses in a tidal current. Geophysical Journal of the Royal Astronomical Society 2, 288-305.

Bye, J., 1965
Wind-driven circulation in unstratified lakes. Limnol.Oceanogr. 10, 451-458.

Carter, D.J.T., 1982
Prediction of wave height and period for a constant wind velocity using the JONSWAP results. Ocean Engng., 9, 17-33.

Channon, R.D. and Hamilton, D., 1971
Sea Bottom velocity profiles on the Continental Shelf south-west of England, Nature 231, 383-385.

Csanady, G.T. and Shaw, P.T., 1980
The evolution of a turbulent Ekman layer. Journal of Geophysical Research, 85, 1537-1547.

Davies, A.M., 1980
Application of the Galerkin method to the formulation of a three-dimensional non-linear hydrodynamic sea model. Applied Mathematical Modelling, 4, 245-256.

Davies, A.M., 1981
Three dimensional hydrodynamic models. Part 1. A homogeneous ocean-shelf model. Part 2. A Stratified model of the northern North Sea pp 370-426 in Vol 2. The Norwegian Coastal Current (ed R. Saetre and M. Mork) Bergen University 795pp.

Davies, A.M., 1983a
Formulation of a linear three-dimensional hydrodynamic sea model using a Galerkin-Eigenfunction method. International Journal for Numerical Methods in Fluids 3, 33-60.

A. M. Davies

Davies, A.M., 1983b
Comparison of computed and observed residual currents during JONSDAP'76. In: Coastal and Shelf Dynamical Oceanography ed B. Johns, Elsevier Scientific Publishing Company, Amsterdam.

Davies, A.M., 1984a
Spectral models in Continental Shelf Sea Oceanography to appear in SCOR book on Coastal Oceanography ed N.S. Heaps.

Davies, A.M., 1984b
A three dimensional modal model of wind induced flow in a sea region, to appear in Progress in Oceanography.

Davies, A.M., 1984c
Application of a spectral model to the calculation of wind drift currents in a stratified sea. (in preparation).

Davies, A.M. and Furnes, G.K., 1980
Observed and computed M_2 tidal currents in the North Sea. Journal of Physical Oceanography 10, 237-257.

Davies, A.M. and Furnes, G.K., 1984
On the determination of vertical structure functions for time dependent flow problems. (in preparation).

Gordon, R.L., 1982
Coastal Ocean Current Response to storm winds. J.Geophys.Res. 87, 1939-1951.

Heaps, N.S., 1972
On the numerical solution of the three-dimensional hydrodynamical equations for tides and storm surges. Mem.Soc.r.Sci.Liege.Ser 6, 2, 143-180.

Heaps, N.S., 1981
Three-dimensional model for tides and surges with vertical eddy viscosity prescribed in two layers - I. Mathematical Formulation. Geophysical Journal of the Royal Astronomical Society 64, 291-302.

Hughes, P., 1956
A determination of the relation between wind and sea-surface drift. Q.J.R.Met.Soc. 82, 494-502.

Ichiye, T., 1967
Upper ocean boundary-layer flow determined by dye diffusion. Phys. Fluids.Suppl. 10, 270-277.

Mortimer, C.H., 1952
Water movements in lakes during summer stratification; evidence from the distribution of temperature in Windermere, Phil.Trans.B236, 355-404.

Newmann, G. and Pierson, W.J., 1964
Principles of Physical Oceanography, published Prentice-Hall.

Riepma, H.W., 1978
Residual currents in the North Sea during IN/OUT phase of JONSDAP'76 (First results extended). ICES Pap. CM 1978/C:42 Hydrography Committee.

Provis, D.G. and Lennon, G.W., 1983
Eddy viscosity and Tidal Cycles in a Shallow Sea. Estuarine, Coastal and Shelf Science, 16, 351-361.

Smith, I.R., 1979
Hydraulic conditions in isothermal lakes, Freshwater Biology 9, 119-145.

Smith, T.J., 1982
On the representation of Reynolds stress in estuaries and shallow coastal sea, Journal of Physical Oceanography 12, 914-921.

Tomczak, G., 1964
Investigations with drift cards to determine the influence of the wind on surface currents. In: Studies in Oceanography. Tokyo University, Tokyo pp129-139.

Wolf, J., 1980
Estimation of shearing stresses in a tidal current with application to the Irish Sea, in, Marine Turbulence, ed, J.C.J. Nihoul, Elsevier Scientific Publishing Company, Amsterdam.

4. OPTIMALLY CONTROLLED HYDRODYNAMICS FOR TIDAL POWER FROM THE SEVERN ESTUARY

S.C. Ryrie,
Department of Computer Studies and Mathematics,
Bristol Polytechnic.

4.1 Introduction

There has for some time been interest in the building of an energy-extracting barrage across the Severn Estuary, where the tidal range is one of the greatest in the world. In planning such a barrage, it is important to estimate the likely energy output from it, and to design it and its operation so as to maximise this output.

Previous work (Owen and Heaps 1979, Miles 1979, Proctor 1981) has shown that the operation of a barrage is likely to have a significant effect on the tidal dynamics of the estuary. Although this work predicts the likely energy output from the barrage, it does not attempt to find the maximum possible output. Conversely, Bickley (1983) uses optimal control methods to maximise output, but does not model the effect of the barrage on, nor its interaction with, estuarine hydrodynamics. The present work aims to develop a model, of the operation of the barrage and its interaction with tidal dynamics, within which it is possible to find the likely maximum output from the available machinery.

The barrage as at present planned will run from near Weston-Super-Mare to near Cardiff, and will contain 150 sluices, through which the rising tide passes into the basin, and 160 turbines, which generate power on the falling tide. We will present results applicable to such a barrage, with a realistic model of the hydrodynamics of the estuary. A fuller account of this work is given by Ryrie and Bickley (1984).

4.2 Hydrodynamics

We use the linearised one-dimensional long wave equations, with a quadratic friction term:

$$h_t + (hu)_x = 0$$
$$u_t + g\zeta_x + \frac{Cu|u|}{h} = 0$$
(4.1)

S.C. Ryrie

where x is a coordinate measured downstream,

> $h(x)$ is water depth,
> $u(x,t)$ is water velocity,
> $\zeta(x,t)$ is surface elevation above same reference level
> and C is a dimensionless friction coefficient.

In order to use the optimal control methods described later, it is more convenient to deal with a set of ordinary differential equations than with partial differential equations. Accordingly, we write equation (4.1) in the conservation form appropriate to flow in a channel:

$$S_t + Q_x = 0$$
$$Q_t + R_x = 0 \qquad (4.2)$$

where $S(x,t)$ is cross-sectional area of the flow,
> $Q(x,t)$ is volume flux of water,
> $R(x,t) = gS(x)\zeta(x,t)$
> $D(x,t) = gS'_o(x)\zeta(x,t) = S_o(x)C\dfrac{u|u|}{h}$

and $S_o(x)$ is the cross-sectional area of the estuary below the reference level $\zeta = 0$.

The equations (4.2) are to be solved on a finite difference grid, $x = x_1, \ldots, x_m$, with spacing Δx. Taking finite difference approximations to (4.2) we obtain a set of ordinary differential equations

$$\dot{S}_j = -\frac{Q_{j+1} - Q_{j-1}}{2\Delta x} \qquad (2 \leq j \leq m-1)$$

$$\dot{Q}_j = -\frac{R_{j+1} - R_{j-1}}{2\Delta x} + D_j \qquad (2 \leq j \leq m-1)$$

where $S_j(t) = S(x_j,t)$, etc.

Similar one-sided finite difference approximations are used at the boundaries and at the barrage, taken to be at $x = x_k$, where, since there is a discontinuity in water level at the barrage, two variables,

$$S_k^+ = S(x_k^+, t)$$
and
$$S_k^- = S(x_k^-, t), \text{ are used.}$$

Optimally Controlled Hydrodynamics

The boundary conditions used are:

$Q_1 = 0$ (to give no flow through the upstream boundary),

$S_m(t)$ specified, (representing the forcing at the sea boundary),
and periodicity for all the other dependent variables.

The topography used for the results shown here was a simple wedge shape, with depth and breadth given by

$h(x) = sx$ (water depth)

and

$B(x) = bx$ (breadth);

following Robinson (1981), we used $s = 2.36 \times 10^{-4}$ and $b = 0.312$. The origin $x = 0$ is at about Sharpness; the boundaries of the model are at

$x_1 = 20$kms (near the Severn Bridge) and
$x_m = 130$kms (near Ilfracombe).

The grid spacing used was $\Delta x = 5\frac{1}{3}$ kms.

In order to check the validity of this model, it was tested without a barrage present, and results from it were compared to Admiralty predictions for tide levels and times (see figure 4.1). Agreement was acceptably good.

4.3 Optimal Control

In the above model, the barrage discharge, $Q_k(t)$ is a control variable whose variation with time is to be chosen so as to maximise the energy output over one tidal cycle of 12.4 hours. The power output P from the barrage depends on Q_k and on the head H across the barrage, so we are to choose $Q_k(t)$ to maximise

$$\mathcal{E} = \int_0^T P(Y, Q_k)\, dt \qquad (4.3)$$

The dependence during generation of P on H and Q_k is taken from turbine manufacturers' data, and is typically as shown in figure 4.2 for a given head. $Q_f(H)$ and $Q_{max}(H)$ are discharges at which all turbines are operating respectively at maximum efficiency and at maximum discharge. For $Q_f \leq Q_k \leq Q_{max}$, all turbines are operating with equal discharge and

S.C. Ryrie

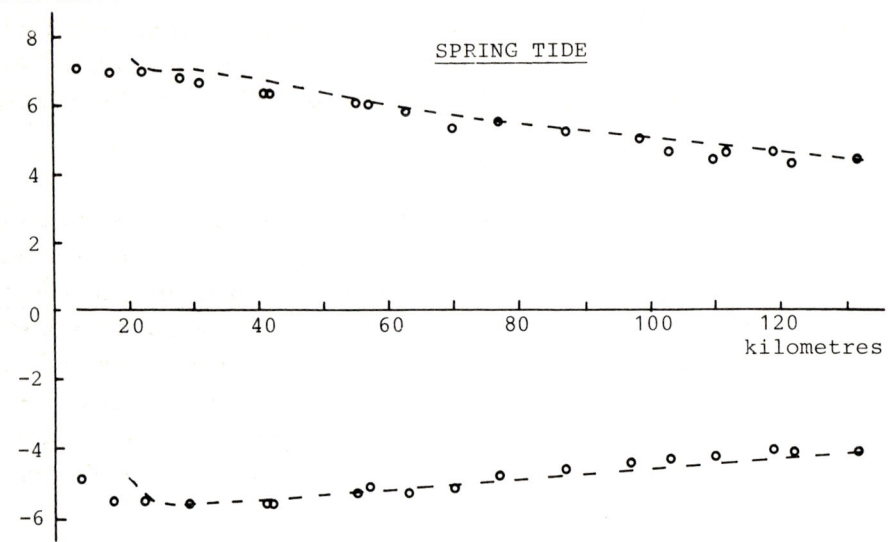

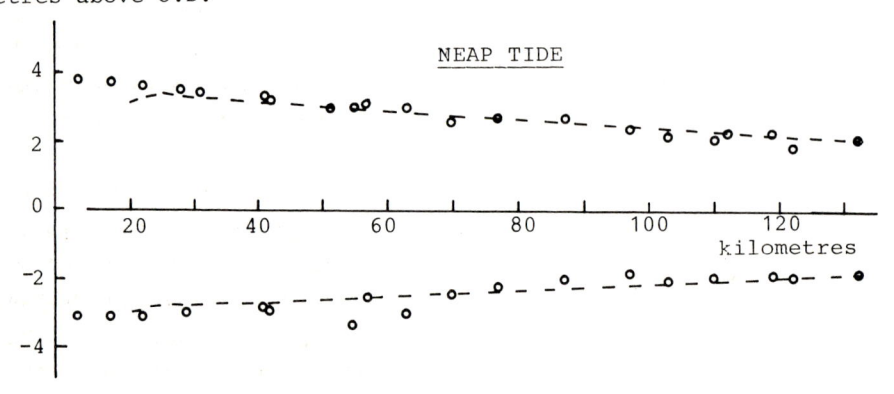

Figure 4.1 (a) Envelopes of high and low tide at points along the estuary measured in kilometres downstream from Sharpness
------ present results
o   Admiralty predictions.

Optimally Controlled Hydrodynamics

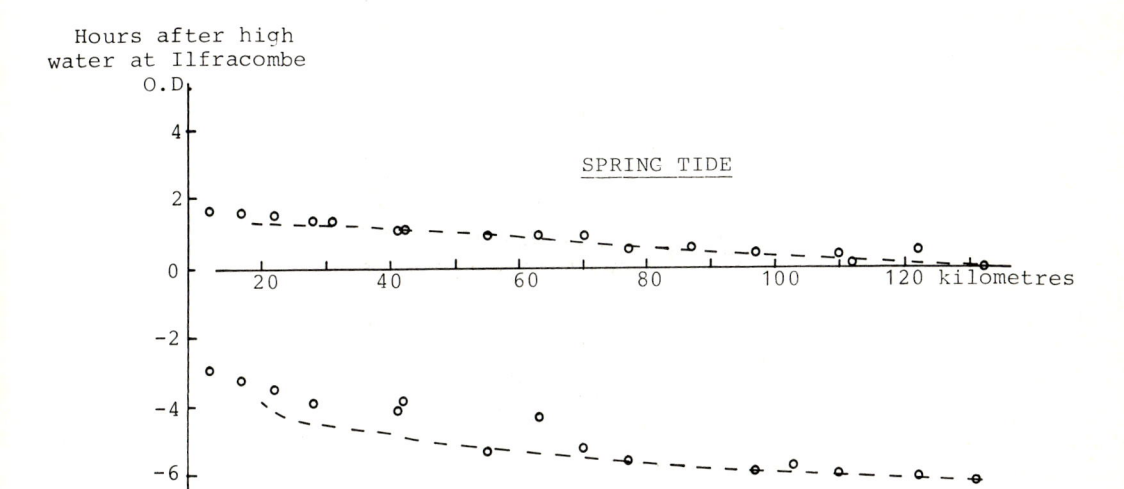

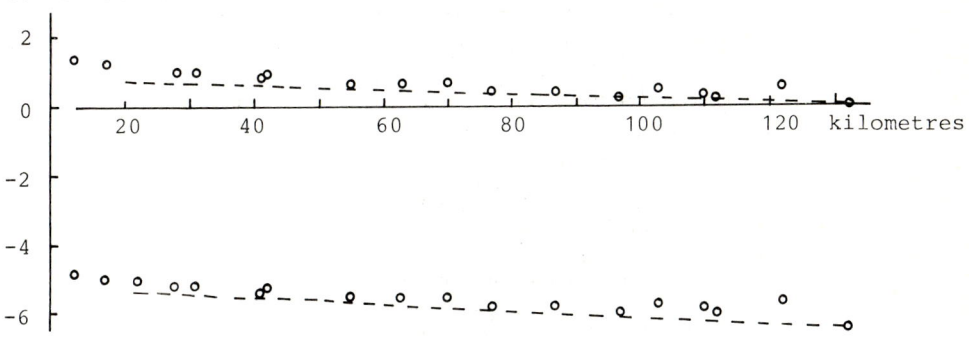

Figure 4.1 (b)   Time, in hours, after high water at Ilfracombe, of high and low tide at points along the estuary measured in kilometres downstream from Sharpness.

-----   present results
o   Admirality predictions.

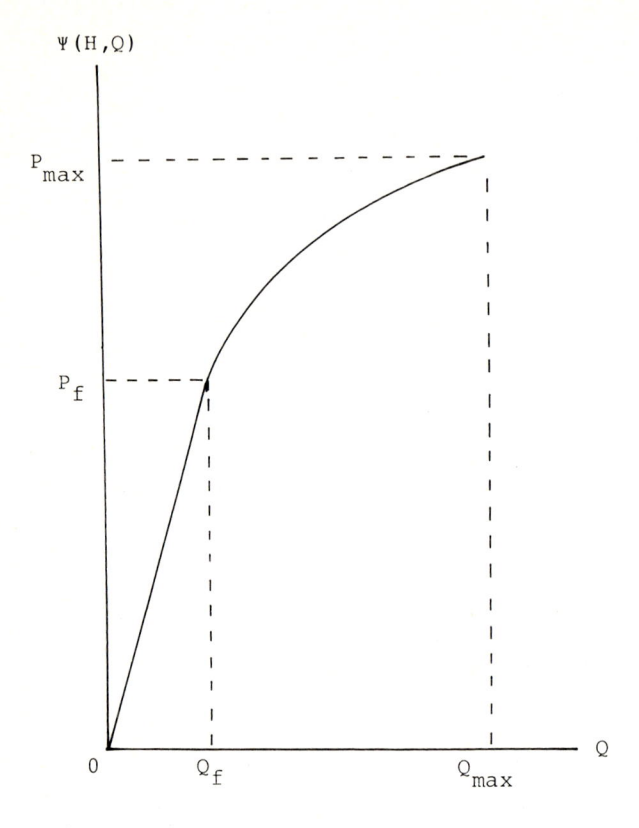

Figure 4.2 The typical form of the variation of power Ψ with discharge Q, for fixed head H. The slope of the graph is constant for $0 < Q < Q_f$, and is decreasing, but is always positive, for $Q_f < Q < Q_{max}$. For sufficiently large head, Q_f may be greater than Q_{max}.

Optimally Controlled Hydrodynamics

power. However, for $0 < Q_k < Q_f$ it is more efficient only to operate an appropriate fraction of the turbines, but at maximum efficiency: this gives

$$P = \frac{P_f}{Q_f} Q_k \qquad \text{as the linear portion of the graph.}$$

During sluicing, Q_k is subject to the constraint

$$-Q_s(H) \leq Q_k \leq 0$$

where $Q_s(H)$ is the maximum possible flow through the sluices.

We use dynamic programming to maximise (4.3): let E be the maximum obtainable energy output in the <u>remaining</u> time, if the optimal control $Q_k = \bar{Q}(t)$, say, is used. Thus,

$$E(t, S_i, Q_j) = \int_t^T P(H, \bar{Q}) \, dt \qquad \begin{pmatrix} i = 1, \ldots, n-1, \\ j = 2, \ldots, n, \\ j \neq k \end{pmatrix}$$

so

$$\frac{dE}{dt} + P(H, Q_k) \begin{cases} = 0 & \text{if } Q_k = \bar{Q} \\ < 0 & \text{otherwise} \end{cases}$$

We must therefore, choose $Q_k(t)$ so as to maximise this expression, i.e. to maximise

$$\sum_i E_{S_i} \dot{S}_i + \sum_j E_{Q_j} \dot{Q}_j + E_t + P(H, Q_k) \tag{4.4}$$

at any given time. This gives a straightforward problem of optimisation of a convex function of one variable. During generation, we find that the optimum, $\bar{Q}$, always satisfies $Q_f \leq \bar{Q} \leq Q_{max}$, so that the turbines always operate at a discharge greater than that for which their efficiency is greatest.

The co-state variables E_{S_i} and E_{Q_j} are found by simultaneously solving a further set of differential equations, which are not shown here, but which are derived by standard methods.

Standard numerical techniques were used to integrate the differential equations and to maximise (4.4) at each time step. Results are shown here for a barrage used for ebb generation with sluicing. Figure 4.3 shows the estimated energy output over one tidal cycle, for varying values of the tidal range at Ilfracombe.

Comparisons are shown with the "flat surface model" of Bickley (1983), who assumes that the water surface inside the barrage remains always flat, and that the water level immediately outside the barrage is a predetermined function of time. For a given tidal range at Ilfracombe, the corresponding tidal range at the barrage site but with no barrage present (the "existing tide") has been used in the flat surface model, giving the results shown.

However, the operation of the barrage will lead to a reduction in the tidal range at the barrage; this reduction is found, in the present work, to be between 7% and 10%. Using this reduced range (the "predicted tide") in the flat surface model gives a reduced output as shown. The fact that this output is very close to the output predicted by the hydrodynamic model suggests that the only major effect on the estuary-barrage system of the tidal dynamics of the estuary, when the barrage is operated for ebb generation, is the reduction in tidal range mentioned above.

However, it has been suggested (Robinson 1981) that, by more flexible operation of the barrage, using pumps for filling the basin and turbines for generation in both directions, the energy output may considerably be increased, by making use of favourable interactions with the tidal dynamics of the estuary. We intend to extend the present work to investigate this, and to use a more realistic model of the estuary topography.

Optimally Controlled Hydrodynamics

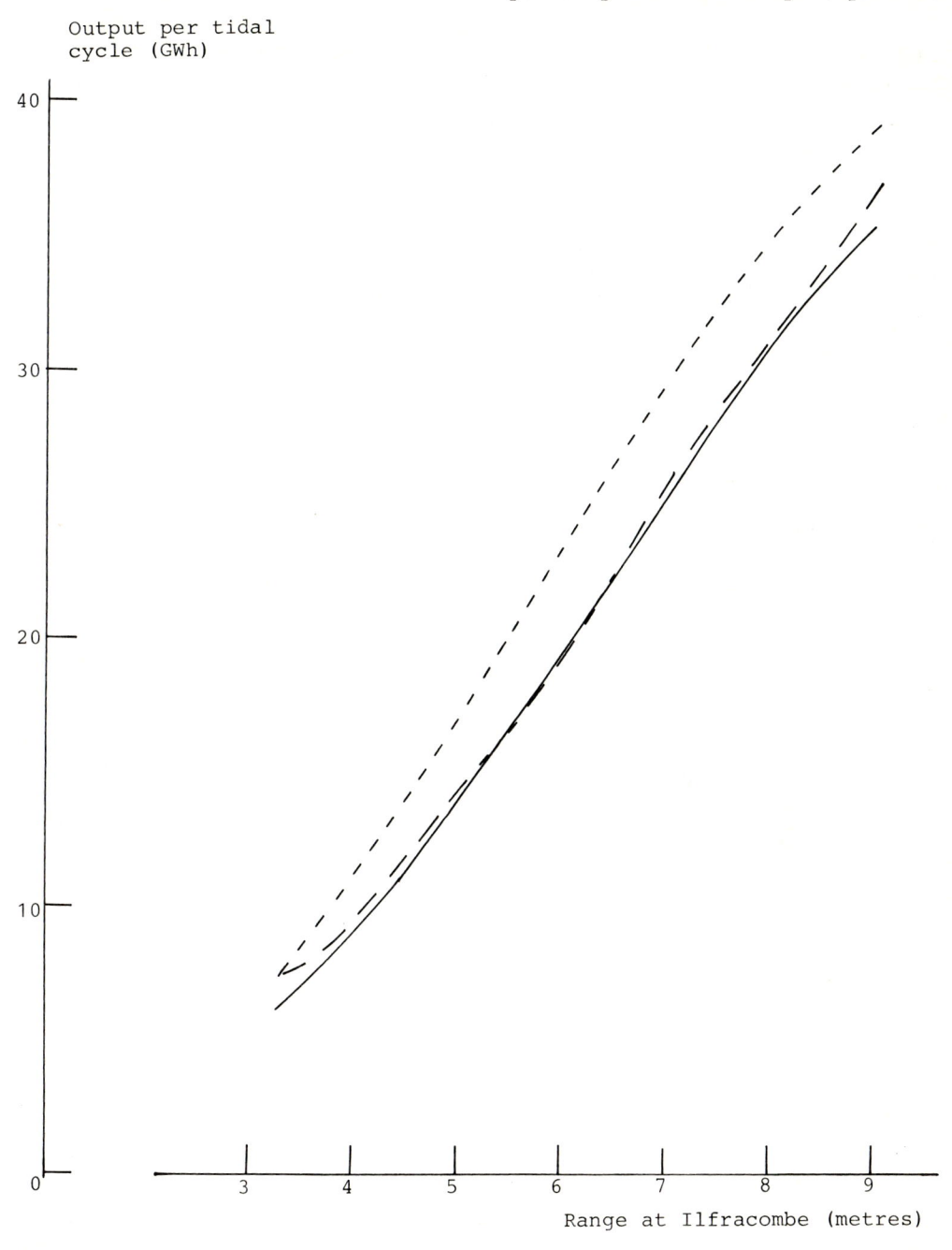

Figure 4.3 Energy output during one tidal cycle
——————— Hydrodynamic model
------ Flat surface model with existing tides
– – – Flat surface model with predicted tides

References

Bickley, D. T., 1983.
Optimal control of a single basin tidal power scheme in various operating modes. University of Bristol, Dept. of Eng. Maths. Report No. EM/TP9.

Miles, G. V., 1979.
Estuarine modelling - Bristol Channel. In "Tidal Power and Estuary Management", Ed R. T. Severn, D. Dineley and L. R. Hawker, Scientechnica.

Owen, A. and Heaps, N. S., 1979.
Some recent model results for tidal barrages in the Bristol Channel. In "Tidal Power and Estuary Management", Ed R. T. Severn, D. Dineley and L. E. Hawker. Scientechnica.

Proctor, R., 1981.
Tidal Barrage Calculations for the Bristol Channel (III). IOS Bidston, Report No. WPTP (81) 482.

Robinson, I. S., 1981.
Tidal power from wedge shaped estuaries - an analytical model with friction, applied to the Bristol Channel. Geophys. J. R. Astro. Soc. 65, 611-626.

Ryrie, S. C. and Bickley, D. T., 1984.
Optimally controlled hydrodynamics for tidal power in the Severn Estuary. Applied Mathematical Modelling (to appear).

5. NUMERICAL MODELLING OF STORM SURGES IN RIVER ESTUARIES

E.H. Twizell,
Department of Mathematics and Statistics,
Brunel University, Uxbridge, Middlesex. UB8 3PH.

5.1 Introduction

Any mathematical model of tidal flow in a given river estuary must satisfy two main objectives if its use is to be justified. The first of these objectives is to provide a reasonably accurate prediction of water level elevations resulting from particular weather conditions, while the second is to provide long-term statistical data on maximum water levels. Such data are essential in the design of efficient flood prevention measures.

Prandle (1975) has noted that one of the chief advantages in compiling a mathematical model of a river estuary is the facility to isolate parameters and thus consider their importance. A mathematical model must be able to reproduce water elevation levels and currents of previous storms. Data relating to these are readily available from sources such as Local Water Authorities, the Coastguard Service, the Admiralty and some Polytechnics and Universities.

Mathematical modelling of phenomena such as storm surges has a number of disadvantages, too. Perhaps the most obvious is the tendency to predict smoother surge-curves than are recorded in reality. The numerical results to be reported in the final section of the present paper provide no exception to this observation. A more fundamental disadvantage lies in the assumption of fixed boundary conditions when changes in water elevation levels and water currents are being computed within the geographical boundaries of the model. There is no doubt that time-dependent boundary conditions would be more realistic. It is well known to numerical analysts, however, that such boundary conditions pose severe problems and they are not considered in the present paper.

5.2 Points to Note in Modelling

As in the mathematical modelling of any system, the modeller of storm surges at sea and in rivers must incorporate those features which lead to the most accurate predictions by the model.

E. H. Twizell

There are many features which have a bearing on the success of a model of storm surges in river estuaries. The most obvious factor is probably the schematic representation of river banks, the river bed, the coastline and the sea bed. The approximating geometry of all of these can be arbitrarily refined to improve accuracy, though, as will be seen later in the paper, geometrical approximations to these geographical features need only be recognizable to lead to acceptable numerical predictions of water elevation levels and currents.

A second important factor to be incorporated in a mathematical model of storm surges is the friction between the water and the river bed or sea bed. Friction is introduced via Chezy coefficients. The inadequacy of using constant Chezy coefficients is outlined in the paper by Brebbia and Partridge (1976) who note that the different materials making up river and sea beds, such as mud, sand or rock, have different frictional resistances as the water depths and currents change. Even to the casual observer, wind is an important factor in the movement of shallow water, such as a river; less obvious, though of equal importance, is the depth of the river bed. For sea models, Brebbia and Partridge (1976) discuss Chezy coefficients in some detail and quote the formula

$$C = 15\log_e(Z + D) \qquad (5.2.1)$$

where C is the Chezy coefficient, Z is the elevation of the water level above the assumed mean water level and D is the depth of the sea bed below this level.

The next point to note by the modeller is the variation in the density of water. In sea models the density may be assumed to be uniform when the temperature is uniform. Heaps (1969) uses the value $\rho = 1025 \text{kg.m}^{-3}$. This value is also applicable to the estuaries and tidal stretches of rivers and may be assumed to fall linearly to the value $\rho = 1000 \text{kg.m}^{-3}$ at tributary inlets and to where the river rises.

In measuring water elevation levels and water velocities, against which numerical values predicted by a mathematical model may be compared, care must be taken in the location of observation points and tide guages. This is particularly important where rapidly changing tidal flows occur such as in Morecambe Bay, the Ribble Estuary and the Great Ouse Estuary.

Numerical Modelling of Storm Surges

Arguably the most important factors governing storm surges are the current meteorological aspects of which wind velocity and atmospheric pressure at least, must be incorporated into every mathematical model of moderate sophistication. Mathematical models of storm surges are usually tested for accuracy against a few notorious storms of which those of 30 January to 02 February 1953 (Prandle, 1975) and 24 to 26 February 1958 (Heaps, 1969) are often quoted.

Weather charts for the latter storm at 0600 hours GMT on each of 25 and 26 February are given in Figures 5.1 and 5.2. The cause of the inclement weather and subsequent storm surge was a deep depression which moved up the English Channel on 25 February before passing into Belgium on 26 February 1958. Strong east-northeasterly winds to the north of the occluded front, which, in Figure 1, is seen to pass from East Anglia across the southern North Sea into Holland, piled water up against the coast of eastern England and Scotland. With the passing of the depression into mainland Europe, the strong winds in south east England veered from westerly to north-northeasterly and the build up of water further north soon affected the Thames estuary. Observed water elevation levels at five points on the east coast relating to this storm are reproduced in Figure 5.3 (Heaps, 1969).

5.3 The Aims of a Mathematical Model of Storm Surges

To a large extent, the aims of every mathematical model of storm surges will have a considerable overlap. The three main aims of the model of the present paper are as follows:-

i)   to formulate the geography of adjacent land in convenient geometrical terms. This will allow rivers, estuaries and sea areas of different shapes, sizes and depths to be modelled by varying certain parameters;

ii) to reproduce successfully storm surges which have been observed and well documented. As Prandle (1975) has noted this gives an appreciation of the mechanics of storm surges and of their propogation. Acceptable computed results relating to previous storm surges also give an indication that any differential equations used in the model and the associated boundary conditions and initial conditions, are the right ones;

E. H. Twizell

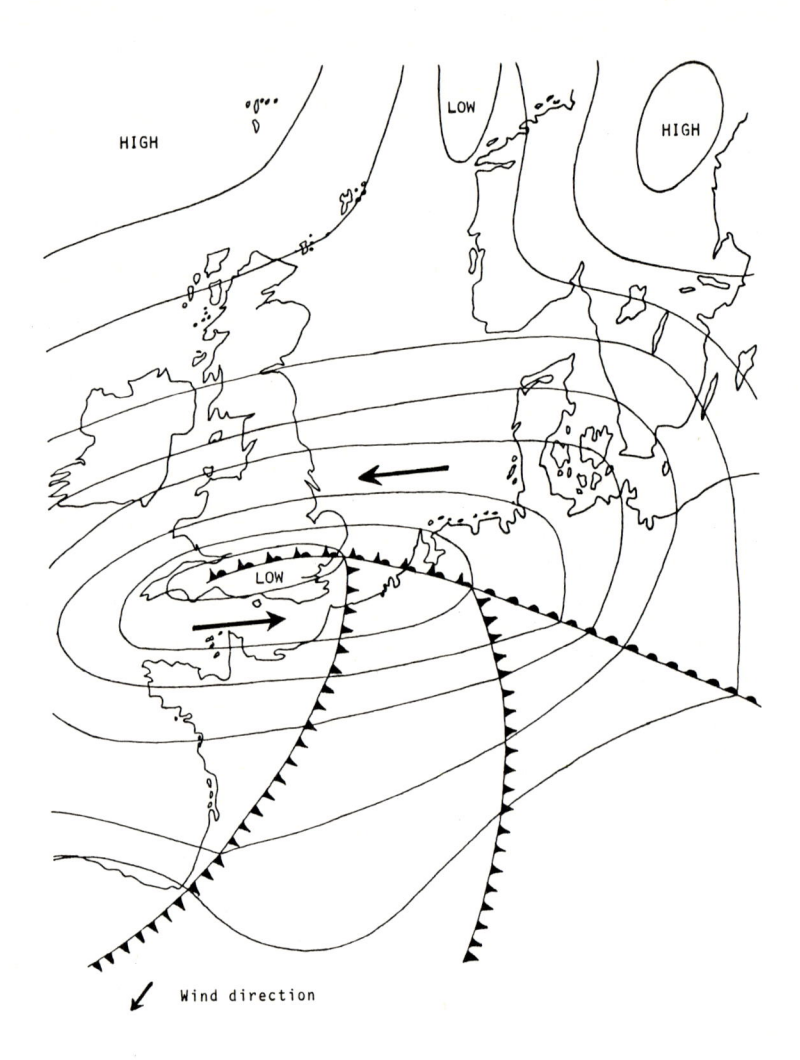

Figure 5.1 Weather chart for 0600 hours GMT on 26 February 1958 (Heaps, 1969).

Numerical Modelling of Storm Surges

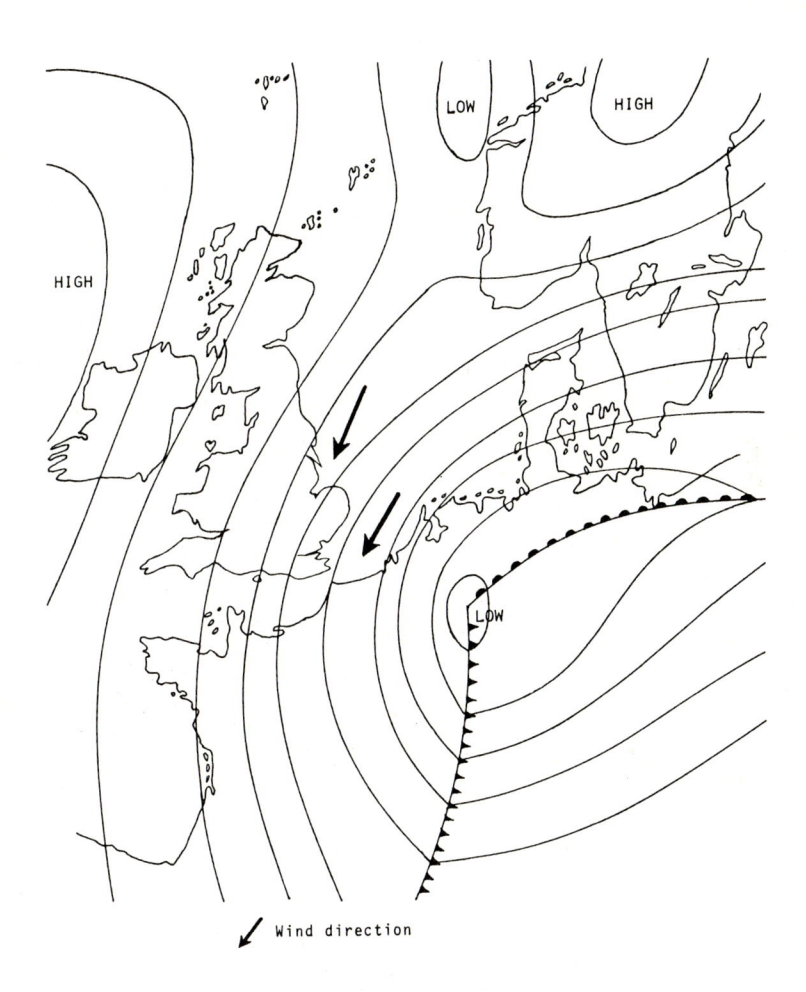

Figure 5.2   Weather chart for 0600 hours GMT on 25 February 1958
(Heaps, 1969).

iii) to provide forecasts of surge elevations, with sufficient warning for flood prevention facilities, if any, to be activated. The water elevation levels following any storm clearly vary with location. Surges of major interest are those which enter the North Sea through its northern opening and then travel southwards affecting sea levels of eastern Scotland and England. The northern North Sea is wide compared to the southern North Sea; consequently a southerly-moving surge is funelled through the narrow Straits of Dover with the inevitable result that surge elevations along the Essex and Kent coastlines are much higher than further north. The funelling effect is shown in Figure 5.4. Prior to the completion of the Thames Barrier, the areas of the county of Greater London adjacent to the riverside were unprotected against major storm surges and fears were held for life, for a number of buildings of national and historic interest (see figure 5.5) and for travel and communication networks. Further north, such dangers are not so great and partial protection can be provided by, for instance, piers and sea walls. A case in point is Sunderland Harbour located well to the north of the Straits of Dover and with a pier at each side of the river mouth. A rough map of the Wear Estuary is given in Figure 5.6 The relationship between location and surge elevation levels can be seen in Figure 5.3 (Heaps,(1969)) which refers to the storm of 24-26 February 1958.

5.4 The Differential Equations of the Model

5.4.1 The River Model

For the river model the relevant partial differential equations are those given by Rossiter and Lennon (1965) and Prandle (1975). The equation of continuity is

$$S \frac{\partial Z}{\partial t} + h \frac{\partial Q}{\partial x} - Q_r = 0 \qquad (5.4.1)$$

and the equation of motion is

$$\frac{\partial U}{\partial t} + g \frac{\partial Z}{\partial x} + g \frac{U|U|}{C^2 M} + \frac{gD_r}{2\rho} \frac{\partial \rho}{\partial x} = 0 \qquad (5.4.2)$$

The nomenclature used in these equations is as follows:-

Numerical Modelling of Storm Surges

S = surface area between adjacent sections of river,

Z = elevation of water surface above assumed mean water level,

t = time,

x = longitudinal axis (positive upstream),

Q = total flow in the positive x-direction,

h = length of river section (increment in x),

Q_r = tributary discharge (Thames : approximately $60 m^3 sec^{-1}$),

U = velocity of water in positive x-direction (mean value in a section),

g = acceleration due to gravity,

C = Chezy coefficient,

M = hydraulic depth,

D_r = depth of water (mean value in a section),

ρ = density of water

5.4.2 The Sea Model

Any mathematical model of the sea must have two space variables, unlike rivers which, in comparison, are thin and can consequently be thought of as being one-dimensional. The dynamical equation are those used by Dronkers, (1969), Heaps (1969) and Prandle (1975) and are given by:-

$$\frac{\partial U}{\partial t} + g\frac{\partial Z}{\partial x} + \frac{1}{\rho}\frac{\partial P}{\partial x} - \frac{F}{\rho(D+Z)} + g\frac{C_0 U(U^2+V^2)^{\frac{1}{2}}}{(D+Z)^{4/3}} - 2\omega\sin\phi . V = 0 \quad (5.4.3)$$

$$\frac{\partial V}{\partial t} + g\frac{\partial Z}{\partial y} + \frac{1}{\rho}\frac{\partial P}{\partial y} - \frac{G}{\rho(D+Z)} + g\frac{C_0 V(U^2+V^2)^{\frac{1}{2}}}{(D+Z)^{4/3}} + 2\omega\sin\phi . U = 0 \quad (5.4.4)$$

$$\frac{\partial Z}{\partial t} + \frac{\partial}{\partial x}\{U(D+Z)\} + \frac{\partial}{\partial y}\{V(D+Z)\} = 0 \quad (5.4.5)$$

The notation used in the above equations is as follows:-

t = time,

x,y = space axes positive to east and north,

U,V = velocities of water in postiive x,y directions,

Z = height of water above assumed mean water level,

P = atmospheric pressure,

E. H. Twizell

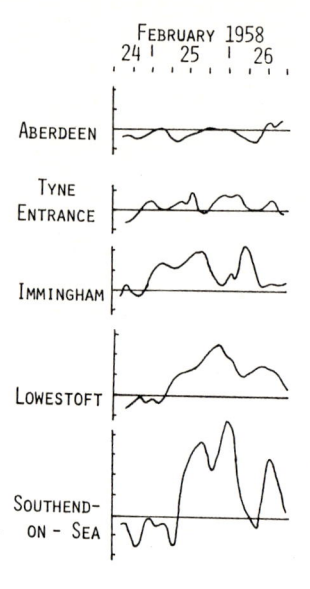

Figure 5.3 Observed Water Elevation 24-26 February 1958 (Heaps, 1969)

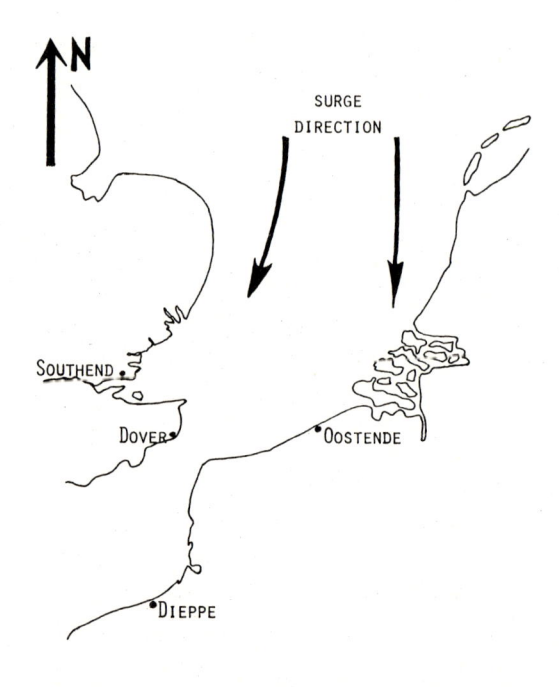

Figure 5.4 Funelling Effect

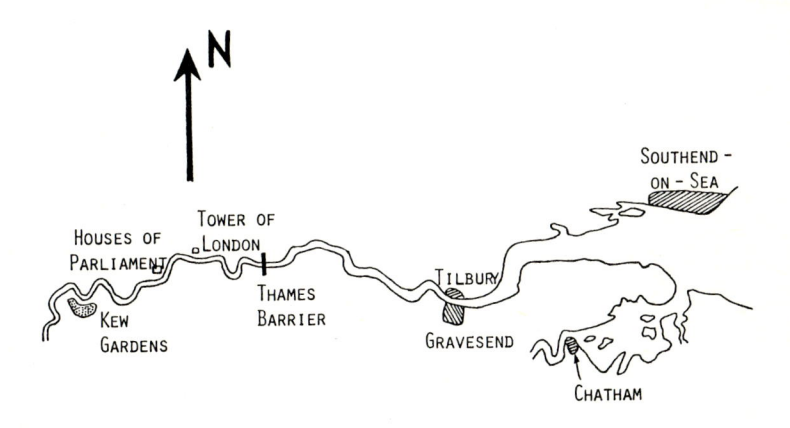

Figure 5.5   Areas of Greater London

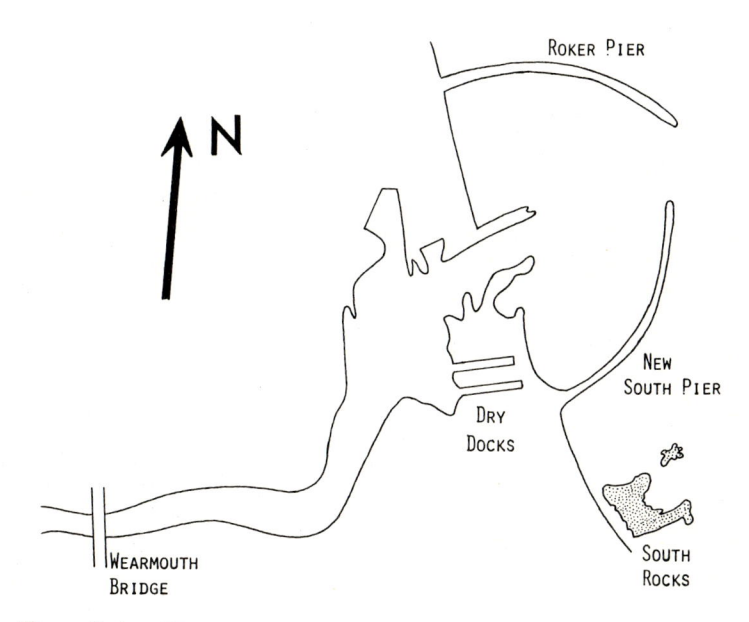

Figure 5.6   Wear Estuary

E. H. Twizell

> D = depth of bed below assumed mean water level,
>
> F,G = wind stress in x,y directions (see below),
>
> C_o = friction coefficient,
>
> ω = angular rotation of the earth,
>
> ϕ = latitude of the location,
>
> ρ = density of sea water,
>
> g = acceleration due to gravity,

The wind stresses in equations (5.4.3) and (5.4.4) are given by

$$F = 0.002\rho_a u^2 \quad , \quad G = 0.002\rho_a v^2 \qquad (5.4.6)$$

respectively, where ρ_a is the density of air and u,v are the wind speeds in the x,y directions, respectively.

5.4.3 Interface at River Mouth

There are two important points to note in interfacing a river model with a sea model. The first is that there must be no discontinuity in the elevation of the water surface above the assumed mean water level. The second is that there must be no discontinuity in the velocity of the water at the interface. Prandle (1974), has given a full description of how to ensure that these discontinuities do not arise in the model.

5.4.4 Initial Conditions

Initial values of all dependent variables must be specified in order for the model to be able to predict values of the dependent variables at any subsequent time t.

5.4.5 Boundary Conditions

Elevations of the water level above the assumed mean water level must be specified at the open boundaries of the model. Attention was drawn to the desirability of using time-dependent boundary conditions in section 5.1

5.5 Computational Aspects

In this section of the paper concentration will be focused on the southern North Sea and the River Thames and its estuary. The shorelines of the adjacent land masses will be approximated as shown in Figure 5.7; the

Numerical Modelling of Storm Surges

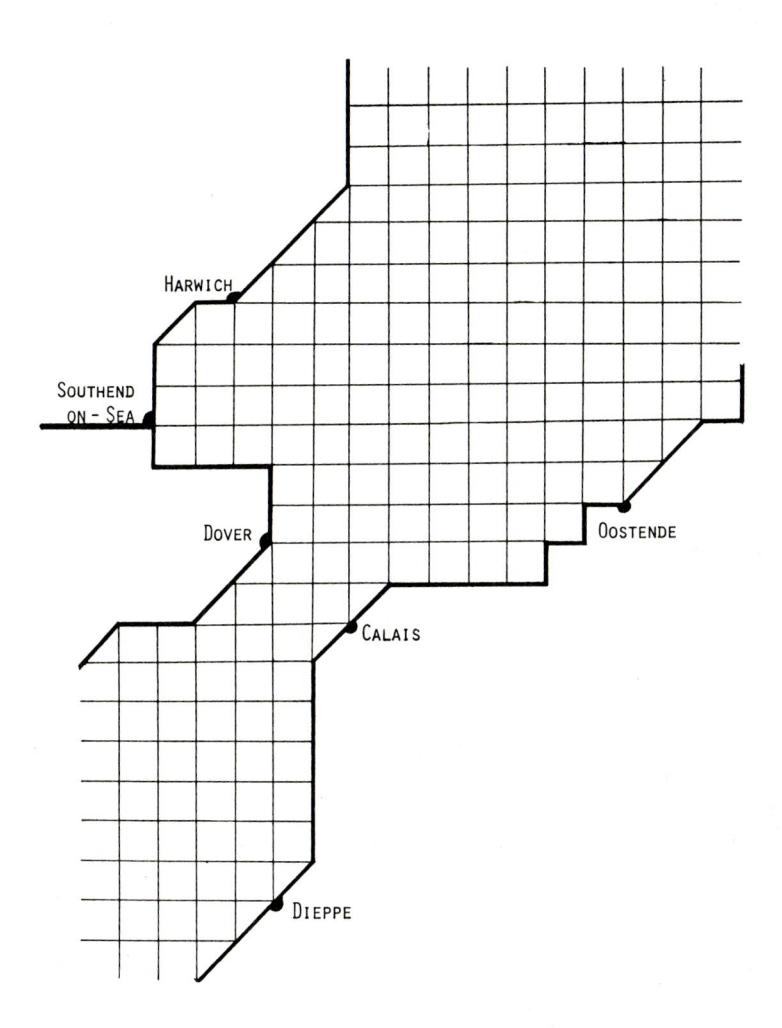

Figure 5.7   Approximation of Southern North Sea in Computation

E. H. Twizell

River Thames will be approximated by a straight line running east to west. The sea model in Figure 5.7 is discretized by a square mesh of step length h = 15km (the distance from Dover to Calais is approximately 43km) and the river model is also discretized in steps of 15km.

The three space derivatives in the differential equations of the river model, and the four space derivatives in the differential equations of the sea model, are now replaced by the usual first order finite difference approximants.

The partial differential equations, with the finite difference replacements, are now applied to all the mesh points of Figure 5.7 at which the values of U, V, Z are sought. These mesh points include those at sea, those on the coastlines and those along the length of the river model.

This procedure of approximating the space derivatives with finite difference replacements effectively reduces the number of independent variables in the differential equations to one (the time variable t). The derivatives remaining in the river model are thus

$\frac{dZ}{dt}$ and $\frac{dU}{dt}$, and in the sea model are $\frac{dU}{dt}$, $\frac{dV}{dt}$ and $\frac{dZ}{dt}$.

The numerical solutions of the systems of non-linear first order hyperbolic partial differential equations of the river and sea models are thus determined by solving a single associated system of non-linear *ordinary* differential equations. This procedure is known as the *method of lines* and is described in detail for hyperbolic and parabolic equations in the text by Twizell (1984).

The first order initial value problem associated with the river and sea models has the form

$$\frac{d\underline{W}}{dt} = \underline{f}(\underline{W}) \quad ; \quad \underline{W}(0) = \underline{W}_0 , \qquad (5.4.7)$$

where $\underline{W} = \underline{W}(t)$ contains the vectors $\underline{U}$, $\underline{V}$, $\underline{Z}$ of values of U, V, Z at the points of the mesh. If there are M mesh points in the river model and N points in the sea model, then $\underline{f}$ and $\underline{W}$ each have 2M + 3N elements.

Numerical Modelling of Storm Surges

The solution of the initial value problem (5.4.7) may be found explicitly using the Euler predictor formula

$$\underline{W}_{n+1} = \underline{W}_n + \ell \underline{f}_n \quad ; \quad n = 0,1,2,\ldots \qquad (5.4.8)$$

where ℓ is a convenient time step, $\underline{W}_n = \underline{W}(n\ell)$ and $\underline{f}_n = \underline{f}(\underline{W}_n)$. This method is first order accurate in time and has stability interval $\bar{\ell}\varepsilon(-2,0)$, where $\bar{\ell} = \ell\lambda$ and λ is that eigenvalue of the Jacobian $\partial \underline{f}/\partial \underline{W}$ which has maximum modulus (the eigenvalues of the Jacobian are all negative).

Second order accuracy in time may now be obtained by correcting this value of $\underline{W}_{n+1}$ using the Euler corrector formula

$$\underline{W}_{n+1} = \underline{W}_n + \tfrac{1}{2}\ell(\underline{f}_n + \underline{f}_{n+1}) \quad ; \quad n = 0,1,2,\ldots \qquad (5.4.9)$$

in *PECE* mode. In this mode, the Euler-modified Euler predictor-corrector combination also has stability interval $\bar{\ell}\varepsilon(-2,0)$ so that a larger time step than the maximum permitted by the Euler predictor formula alone, may not be used. Therefore, unless higher order replacements of the space derivatives in the partial differential equations of the river and sea models are used, the extra computational costs involved in applying (5.4.9) outweigh the achievement of higher accuracy in time.

5.6 Numerical Results : The Storm of 1953

The reliability of the model was tested by simulating the disastrous storm surge of 31 January and 01 February 1953. The various data needed for the computation were obtained from the works of Prandle (1974,1975) and the numerical results were obtained using the method of lines and the Euler predictor formula (5.4.8) with a time step, following Prandle (1974), of 0.05 hour.

Comparison of the computed results relating to Southend-on-Sea with the observed water elevation levels during the surge, show them to be in reasonable agreement. The discrepancy in magnitude and the phase lag can be explained by the simplified coastline of the model and the probable disagreement between the actual location of the observation point and the corresponding mesh point.

The observed and computed water elevations during the storm at Southend-on-Sea are depicted in Figure 5.8. The associated surge, defined as the

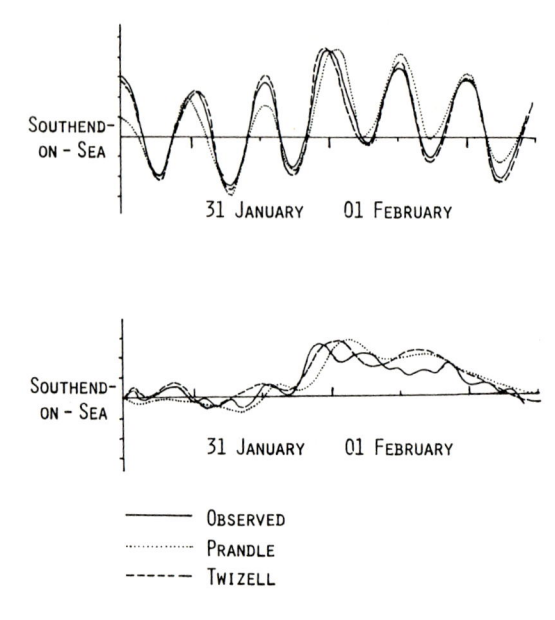

Figure 5.8 Observed and Computed Water Elevations during Storm of 1953 and Associated Surge

difference between the elevation above the assumed mean value and the predicted tide, is also depicted in Figure 5.8.

5.7 Summary

The mathematical modelling of water elevation levels above an assumed mean value and the velocity of the water during storm surges have been considered in this paper. The advantages and disadvantages of mathematical models have been outlined, and the points to be considered in compiling a model have been discussed.

The partial differential equations relating to river models and sea models were listed separately and the discontinuities to be avoided at a river/sea interface were noted. Using the method of lines, the systems of partial differential equations were reformulated as a single system of non-linear, first order, ordinary differential equations for which the initial conditions were provided by the prevailing conditions at the beginning of the simulation period. The first order initial value problem was solved using the low order Euler formula.

Numerical results relating to the storm surge of 31 January and 01 February 1953 were obtained and compared with those recorded at Southend-on-Sea.

E. H. Twizell

References

Brebbia, C. A. & Partridge, P. W., 1974
Appl.Math.Mod. 1, 101-107.

Dronkers, J. J., 1969
J.Hydraul.Div.Am.Soc.Civ.Engrs., 95, 29-77.

Heaps, N. S., 1969
Phil.Trans.R.Soc.London. A265, 93-137.

Prandle, D., 1974
I.O.S. Report No. 4, Institute of Oceanographic Sciences, Bidston Observatory, Birkenhead, Merseyside, England. L43 7RA.

Prandle, D., 1975
Proc.R.Soc.Lond. A344, 509-530.

Rossiter, J. R. & Lennon, G. W., 1965
Proc.Instn.Civ.Engrs. 31, 25-36.

Twizell, E. H., 1984
Computational Methods for Partial Differential Equations, Ellis Harwood/John Wiley and Sons, Chichester, England.

6. COASTAL SEDIMENT MODELLING

B.A. O'Connor,
Simon Engineering Laboratories,
University of Manchester, Oxford Road,
Manchester. M13 OPL.

6.1 Introduction

The movement of sediment in the nearshore zone is a direct result of hydrodynamic forces exerted by tide, wave and wind action. Engineering works which modify nearshore water and wave movements will also modify the hydrodynamic forces applied to the sea bed sediments, and thereby modify sea bed topography. Unfortunately, the severity of the wave and current climate at a particular site is also influenced by the depth and orientation of the sea bed contours over a wide area adjacent to the site. Consequently, any attempt to predict the environmental impact of a particular engineering scheme must concern itself with how sediment is moved about under the action of waves, tides and winds.

Particular environmental studies aimed at assessing changes in beach or sea bed contours will usually involve a combination of field measurements, physical model studies and, in more recent times, mathematical (computer) model tests. This lattermost area is the subject of the present paper. However, the complexity of both the environment and the various modelling techniques presently being developed, together with the short time available to present ideas, means that only an outline of current practice can be given. Attention is also focused on the sediment side of coastal models since other papers discuss the nearshore wave and current climate. Further details of current modelling practice can also be obtained by consulting the references given at the back of the present paper.

6.2 Need for Computer Models

Typical engineering problems involving the need to estimate sediment movements and changes in beach and sea bed topography include:-

i)   construction of harbour breakwaters, leading to shoreline changes and harbour siltation,

ii) dredging of new or deeper port approach channels, leading to changes

B.A. O'Connor

in maintenance dredging requirements.

iii) construction of coastal power station cooling water intakes and outfalls, leading to changes in coastal bathymetry,

iv) construction of coastal revetments, sea walls, groynes and coastal jetties, leading to shoreline changes,

v) reclamation of coastal land from the sea for industrial or recreational purposes, leading to changes in shoreline configuration both upstream and downstream of the work,

vi) construction of nearshore islands for industrial purposes, including oil and gas exploration,

vii) closure of estuaries at their seaward limits, leading to costal erosion problems,

viii) placing of oil and gas platforms on the sea bed, leading to local scour problems,

ix) cutting of sea bed trenches for installation of oil and gas pipelines.

Many of the above problems involve changes of sea bed contours stretching over many kilometres. For example, the construction of the new harbour breakwaters at Zeebrugge in Belgium has extended the shoreline seawards by some 1-2 kilometres and means that beach and nearshore changes can be expected over an area of some 4x6 kms, extending both upstream and downstream of the works. Reproduction of such large areas in a physical model requires the use of very large scale distortions, if large expensive models are to be avoided - a 20km length of coastline would occupy 200m at a scale of 1:100. It is also difficult in physical models to reproduce geostrophic forces, wind-driven currents, the correct bed roughness and scale of sediment transport.

Computer models on the other hand have the potential to include geostrophic and wind action and to minimise sediment ransport scale effects by using relationships which relate sediment movements to local hydrodynamic forces. They can also be operated to give the worst conditions

Coastal Sediment Modelling

over the lifetime of a structure; occupy little storage space; and can be modified rapidly to study new conditions.

6.3 Model Types

The majority of coastal zones are composed of sandy sediments with the coarsest fractions found on the beach. The following discussion is therefore, confined to the discussion of models of sand movement. Some details of cohesive material is, however, given in Lean (1980), O'Connor & Tuxford (1980).

i) Coastline Models

In many parts of the world, movement of beach and cliff material occurs due to the direct action of breaking waves. The study of beach problems involving groynes, reclamations and jetties may then be undertaken by treating the surf zone as a single entity without regard to the detailed processes at work when the waves break, entrain sediment and run up the beach. Such "one-line" models make use of the simple continuity equation, Figure 6.1:-

$$\partial Q/\partial x + h\partial y/\partial t = 0 \tag{6.1}$$

where Q = the longshore transport rate (m³/s),
h = depth at which sediment is first moved by wave action,
y = a coordinate measuring scour or accretion of the coastline.

Various models have been built using finite difference methods to solve equation (6.1). Often Q is given by the equation:-

$$Q = K/((1-p)(\rho_S-\rho)g)(E \cdot C_g)_B \cos\alpha_B \sin\alpha_B \tag{6.2}$$

where K = a constant, commonly 0.77
p = beach porosity, commonly 0.40
ρ_S, ρ = sediment and fluid density, respectively
g = acceleration due to gravity
E = wave energy at breaking per unit surface area ($\rho g H^2/8$)
H = wave height
C_g = group velocity of waves
α_B = the angle the wave crests make with the sea bed contours at breaking
B = a subscript indicating breaking.

B.A. O'Connor

Most models also use a wave ray computer program to route waves inshore from deep water. Wave transformations due to refraction, reflection, shoaling and friction are also included in the routing process (see Mogel & Street (1974), Komar et al (1976), Ozasa & Brampton (1979), Barber (1981), Skovgaard et al (1975)).

It should also be noted that equation 6.2 takes no account of the grain size properties of the beach or its slope or roughness. However, field measurements (Bruno et al, 1980) and detailed computer modelling, as described in section (ii) overleaf, have shown that allowance for different sized beach material can be made in equation (6.1) by making K a function of grain size, as shown in figure 6.2.

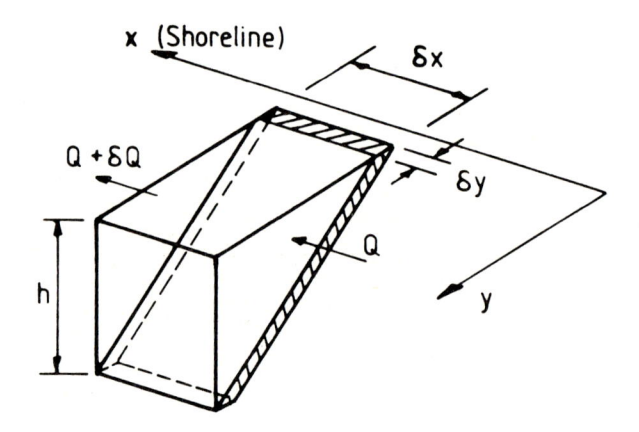

Figure 6.1 Shoreline Continuity Equation

Equation (6.2) may be further modified to include the effects of edge-waves, rip currents and longshore variations in wave heights by adding an extra term to the equations, Barber (1981). However, the effects of beach roughness, slope and grain density can only be included in shoreline models by the use of different equations for Q (Cacoutas, (1982), Swart & Fleming (1980)).

Coastal Sediment Modelling

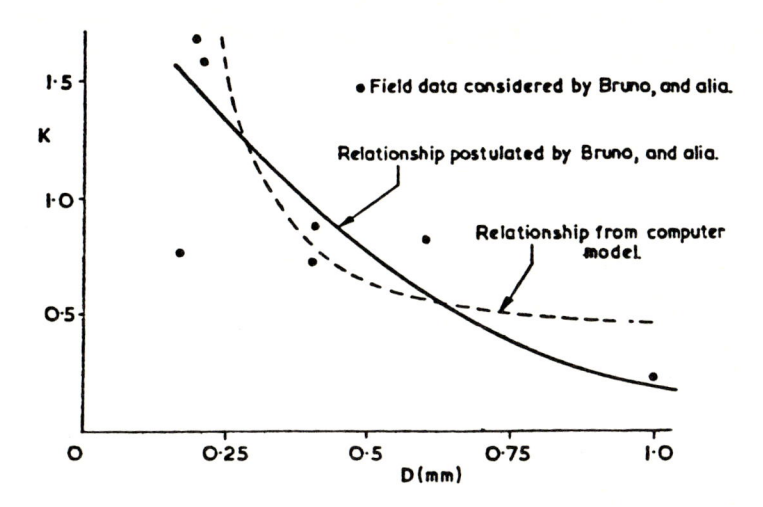

Figure 6.2 Comparison of Relationships between K (defined in text) and D (median grain size).

Equation (6.1) may also be solved analytically for simple problems, such as the build-up behind a groyne or breakwater by relating the longshore variation in Q to the plan shape of the beach $\partial y/\partial x$ (Muir-Wood & Fleming (1981)). Movement around the end of a groyne or breakwater can be included by dividing the shoreline into two or more longitudinal zones, an inner zone where longshore sediment is stopped by the structure and outer zones where sediment is allowed to pass freely; all zones being connected by equations defining some measure of onshore-offshore transport involving the slope of the sea bed in relation to some equilibrium value (Cacoutas, (1982), Bakker (1968)).

ii) Coastline and Offshore Models

Situations involving significant tidal or other currents cannot be studied effectively with shoreline models since sediment movement occurs in offshore areas. Such problems are usually tackled by solving some form of spatially-averaged suspended load equation, supplemented by an equation to describe bed load movements (Kerssens et al (1977), O'Connor (1975, 1982), O'Connor & Tuxford (1980)). Work with such models in tidal environments has shown that medium to coarse sands respond quite quickly to changes in environmental forces. Consequently, the majority of coastal models employ a de-coupling technique to avoid the expense of solving a two or three dimensional load equation covering the continuum. Usually such models involve some description of how the suspended load varies over the flow depth. This lattermost information is then combined with the varying patterns of wind, wave and tide-induced currents, which also vary in strength and direction over the flow depth, to provide information on suspended load, figure 6.3.

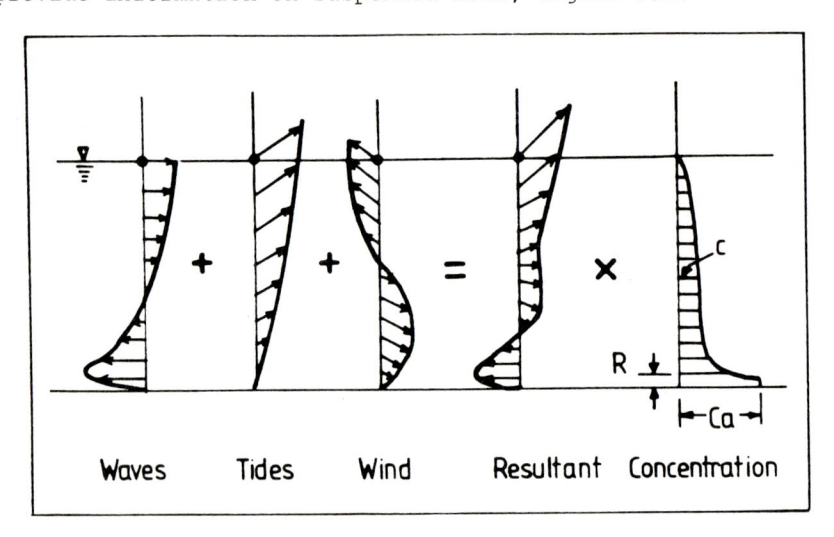

Figure 6.3 Calculation of Suspended Load

The total transport of sediment is then obtained by addition of the local bed load transport rate. This leads in turn to the change in sea bed contours by solving a simple two dimensional total load continuity equation:-

$$\partial Q_x/\partial x + \partial Q_y/\partial y + \partial z/\partial t = 0 \qquad (6.3)$$

where Q_x, Q_y = total load sediment transport rates in m³/s in the x, y co-ordinate directions respectively, figure 6.4

z = the change in elevation of the sea bed relative to a horizontal datum.

Various techniques have been involved for specifying Q_x and Q_y, (Bijker (1968), Fleming & Hunt (1976), Sorenson et al (1980), Morcos Fanos (1979), Wang & Liang (1975)). The bed load, Q_B, in Bijker's approach (1968) and its derivatives is described by a

Coastal Sediment Modelling

Frijlink (1952) equation:-

$$Q_B = K_1 D_{50} (V/C) g^{\frac{1}{2}} \cdot \exp(-\rho g \Delta K_2 D_{50} / (\mu \tau r)) \qquad (6.4)$$

where K_1, K_2 = constants with typical values of 5 and 0.27 respectively
- V = depth-mean current velocity
- D_{50} = median grain size of bed material
- C = Chezy's roughness coefficient
- $\Delta = (\rho_s/\rho) - 1$
- μ = ripple coefficient defined by equation (6.5)
- τ_r = resultant wave and current shear stress given by equation (6.6).

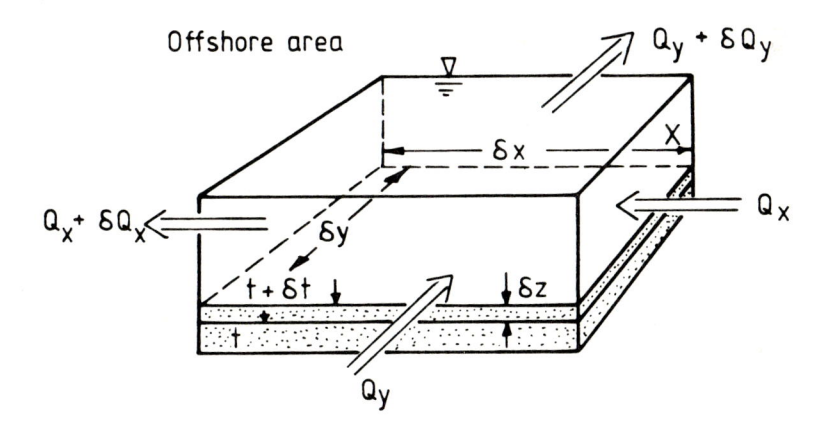

Figure 6.4   2-D Continuity Equation

Thus:-

$$\mu = (C/C_{90})^{3/2} \qquad (6.5a)$$

$$C = 18\log(12h/R) \qquad (6.5b)$$

$$C_{90} = 18\log(12h/D_{90}) \qquad (6.5c)$$

- R = effective roughness of the sea bed
- D_{90} = 90% of grain size
- h = flow depth

$$\tau_r = \tau_c + 0.5\tau_w = \rho U_{*R}^2 \qquad (6.6)$$

B. A. O'Connor

τ_c = current shear stress (= $\rho g(V/C)^2$)

τ_w = wave-induced maximum bed shear stress ($f_w U_o^2 /2$)

U_o = wave orbital velocity at the top of the wave boundary layer, usually given by first order wave theory

f_w = wave friction factor given as:-

$f_w = \exp(-5.977 + 5.213(a/R)^{-0.194})$ for a/R > 1.57 \hfill (6.7a)

$f_w = 0.28$ for a/R < 1.57 \hfill (6.7b)

$a = U_o T/(2\pi)$ \hfill (6.7c)

The suspended load (Q_s) is given by the equation:-

$$Q_s = \int_R^h vc \, dz \qquad (6.8)$$

where v = the resultant water velocity at level z above the sea bed, excluding wave orbital motion, figure 6.3.

c = the wave-period-mean suspended sediment concentration at level z above the sea bed, equation 6.9.

$C = Ca((\frac{h}{z} - 1)/(\frac{h}{R} - 1))^z$ \hfill (6.9a)

$Z = 2.5w/U_{*R}$ \hfill (6.9b)

$Ca = Q_B/(6.35 R U_*)$ \hfill (6.9c)

and   w = the fall velocity of the sediment grains in motion.

Equation 6.8 is usually solved by numerical integration, once the vertical distribution of non-oscillating currents is specified. This requires the use of one or more hydrodynamic models, either with or without interaction between the wave and current fields, as appropriate (Skovgaard & Jonsson (1976), Ebersole & Dalrymple (1980), Barber (1981)).

The computations also require the specification of sea bed roughness (R), which is a far from easy task (McDowell & O'Connor (1981)). In wave action alone, R is usually related to bed form dimensions, for example:-

$R = 25 H_R (H_R / L_R)$ \hfill (6.10)

where H_R, L_R = the height and wavelength of bed forms respectively.

Coastal Sediment Modelling

Various complex empirical formulae are available to help with predictions (Muir-Wood & Fleming (1981), Delft Hydr. Lab (1976), O'Connor (1982)).

Once wave and current roughness have been estimated, it is possible to estimate the likely roughness of the sea bed in combined wave and currents by using the criteria:-

$U_o/U_* < 3$ - use current value

$3 < U_o/U_* < 10$ - use largest value

$U_o/U_* > 10$ - use wave value

It is usual also to determine the wave and current conditions over a wide area, using boundary values well away from the area of interest, (O'Connor et al (1981), Fleming & Hunt (1976), Barber (1981)). This will usually mean the use of a number of computation grids, which may become more refined as the shoreline is reached, figure 6.5.

The inter-relating elements of a "point-by-point" model are summarised in figure 6.6, where the total model is seen to be made up of a number of separate modules. This has the advantage of ease of replacement as scientific knowledge increases.

Any model will need to be operated for a number of representative tides and tidal states with different incident wave conditions and directions. The resultant patterns would then be combined to give long term predictions (Fleming & Hunt (1976), Morcos Fanos (1979)). Alternatively, the wave climate may be described by a simplified form of directional wave spectrum and patterns of accretion/erosion combined depending on the joint probability of wave and current climates (Barber (1977)).

A very simple application of such a coastline/offshore (or "point-by-point") model for predicting the changes in sea bed contours following the construction of a long groyne is shown in figures 6.7-6.10. Figure 6.7 shows the model grid system, while figures 6.8-6.10 show the build-up behind the groyne as shown by the computer model built at HR Ltd, Wallingford. Quite good agreement is evident.

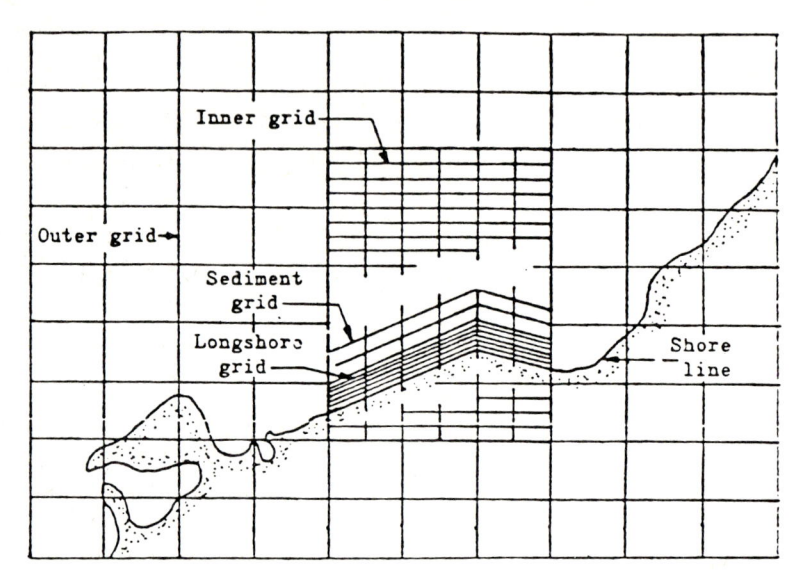

Figure 6.5   Diagrammatic Sketch showing Grids in a Coastal Area

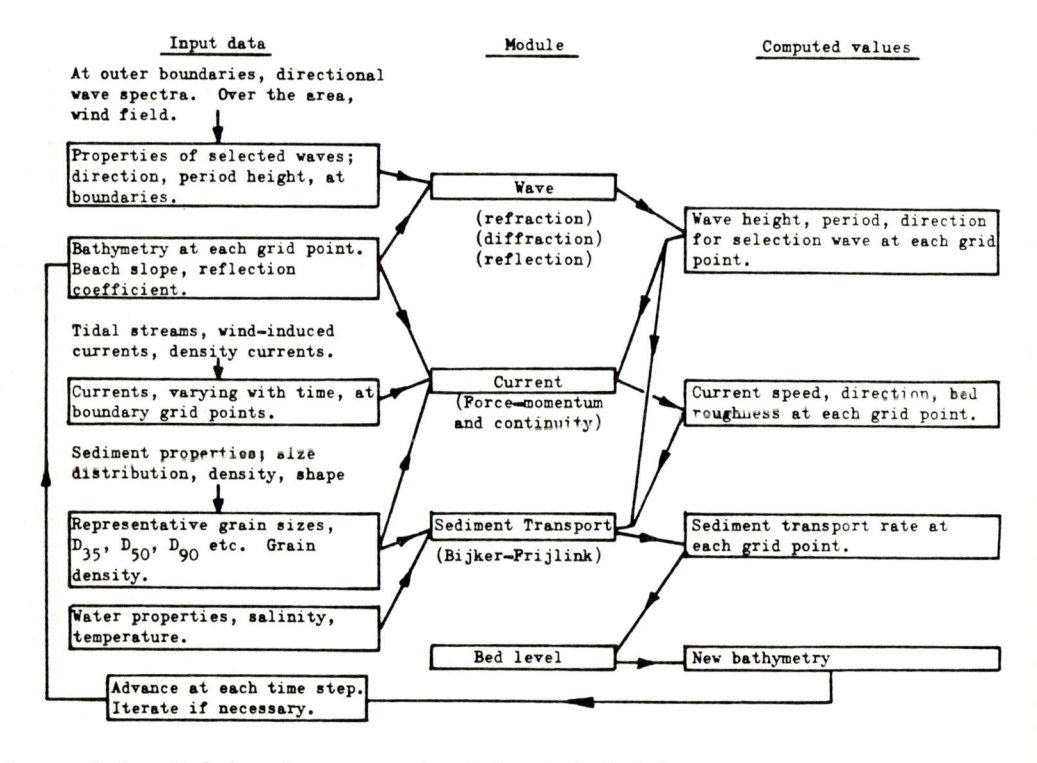

Figure 6.6   Modular Arrangement of Coastal Models

Coastal Sediment Modelling

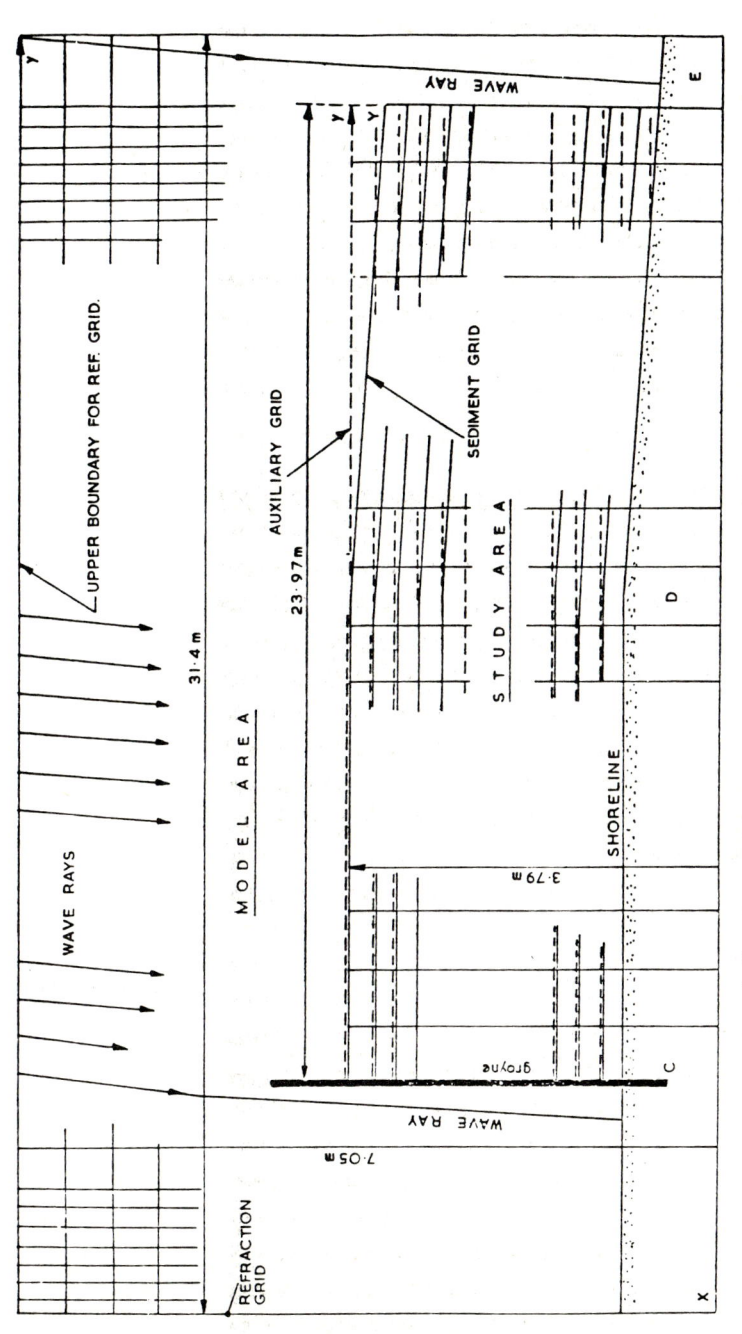

FIG. 6.7 THE COMPUTATIONAL GRID. (DISTORTED SCALE) Morcos Fanos (1979).

B. A. O'Connor

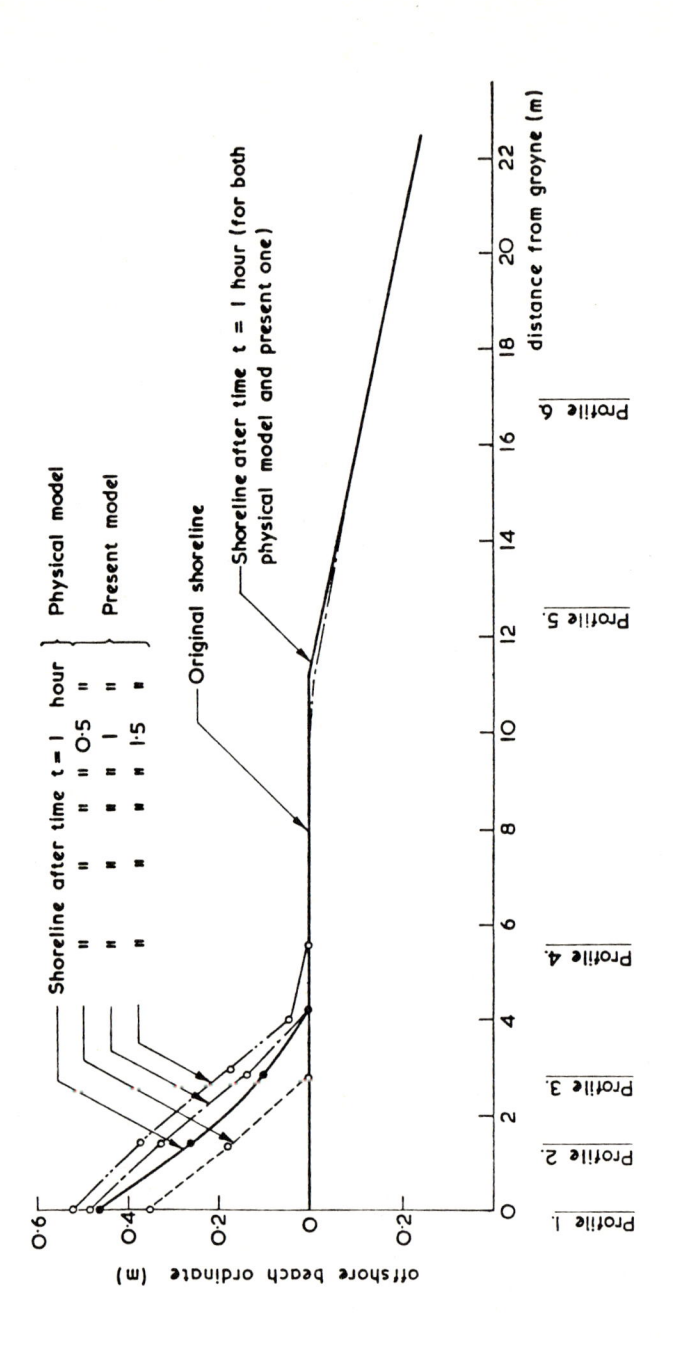

FIGURE 6.8 VARIATION OF SHORELINE DUE TO THE CONSTRUCTION OF THE GROYNE. Morcos Fanos (1979)

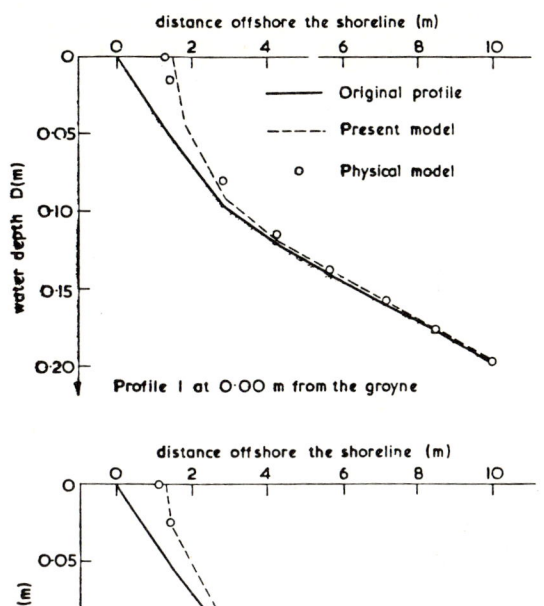

Coastal Sediment Modelling

Profile 1 at 0.00 m from the groyne

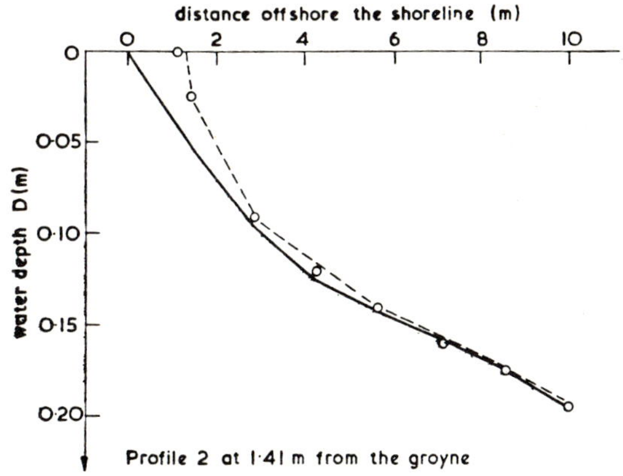

Profile 2 at 1.41 m from the groyne

Note: the numbers of the profiles are shown in Figure 6.8

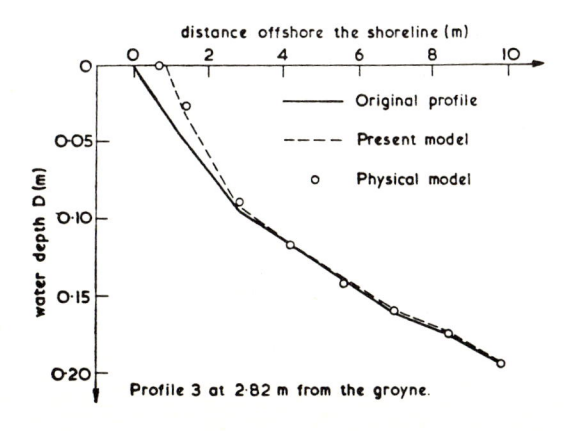

Profile 3 at 2.82 m from the groyne.

FIGURE 6.9 VARIATION OF BEACH PROFILES, 1 HOUR AFTER CONSTRUCTION OF THE GROYNE (Morcos Fanos (1979))

B. A. O'Connor

Profile 4 at 5·64 m from the groyne

Profile 5 at 12·69 m from the groyne

— Original profile
---- Present model
o Physical model

Profile 6 at 16·92 m from the groyne

FIGURE 6.10 VARIATION OF BEACH PROFILES, I HOUR AFTER CONSTRUCTION OF THE GROYNE (Morcos Fanos (1979)

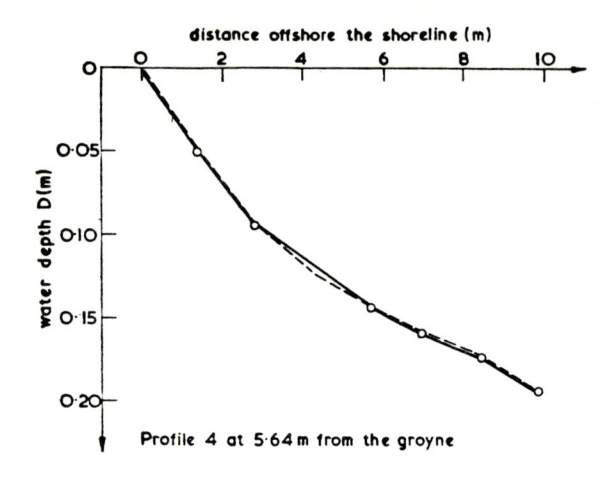

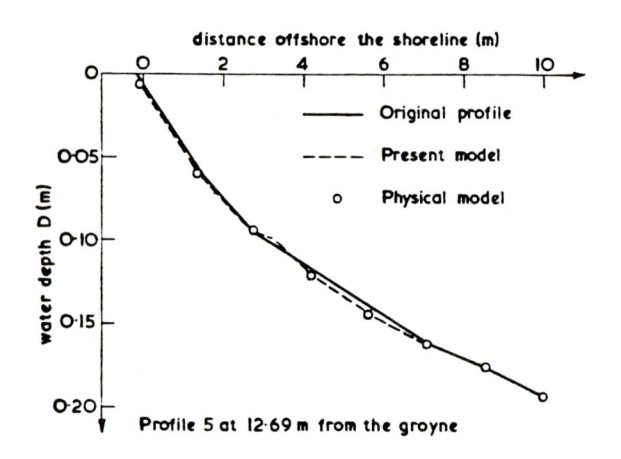

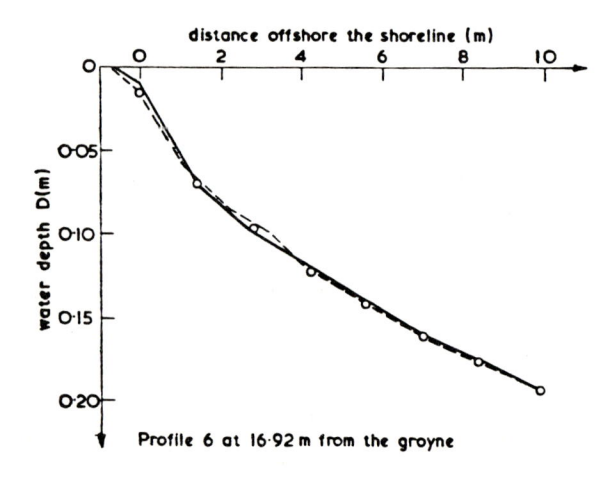

Coastal Sediment Modelling

Similar approaches have been used to study problems of siltation in a power station cooling water intake in South Africa (Fleming & Hunt (1976)); the loss of beach in front of a vertical sea wall on the Wirral coastline, Merseyside (Barber (1981, 1977)); changes in the sea bed off the Delaware coastline in the USA (Wang, 1977); and changes along the Dutch coastline, following construction of the Deltaworks (Hoopen & Bakker (1974), Svasek & Versteegh (1977)).

iii) Dredged Channels

The increasing size of bulk cargo ships has led to the deepening of many existing port approach channels and the construction of new ones of considerable length, often in excess of 10km. The maintenance dredging requirements of such channels must be predicted quite accurately in order to assess the economic viability of the new port. In some cases, costs can exceed £5M.

Channel dredging problems usually require that the channel be divided into a series of boxes and the nett siltation determined by consideration of both the sediment transport rates approaching the channel and in the channel itself, figure 6.11. If waves are the dominant factor causing infill, it is possible to use some of the equations already discussed in section (i) to estimate infill rates through the surf zone. For offshore zones, it is necessary to use additional equations describing the transport of sediment by wave-induced mass-transport currents. If tidal currents are also involved in the problem, it is then necessary to use the "point-by-point" approach developed in section (ii).

For offshore zones, without tidal currents, the wave-induced sediment transport rate is often determined by the following equation (Hydr. Res. (1972)):-

$$Q_w = 530 H^6 / ((h^2 T^5)(\sinh kh)^6) \qquad (6.11)$$

where Q_w = the wave-induced transport rate in kg/m/s.

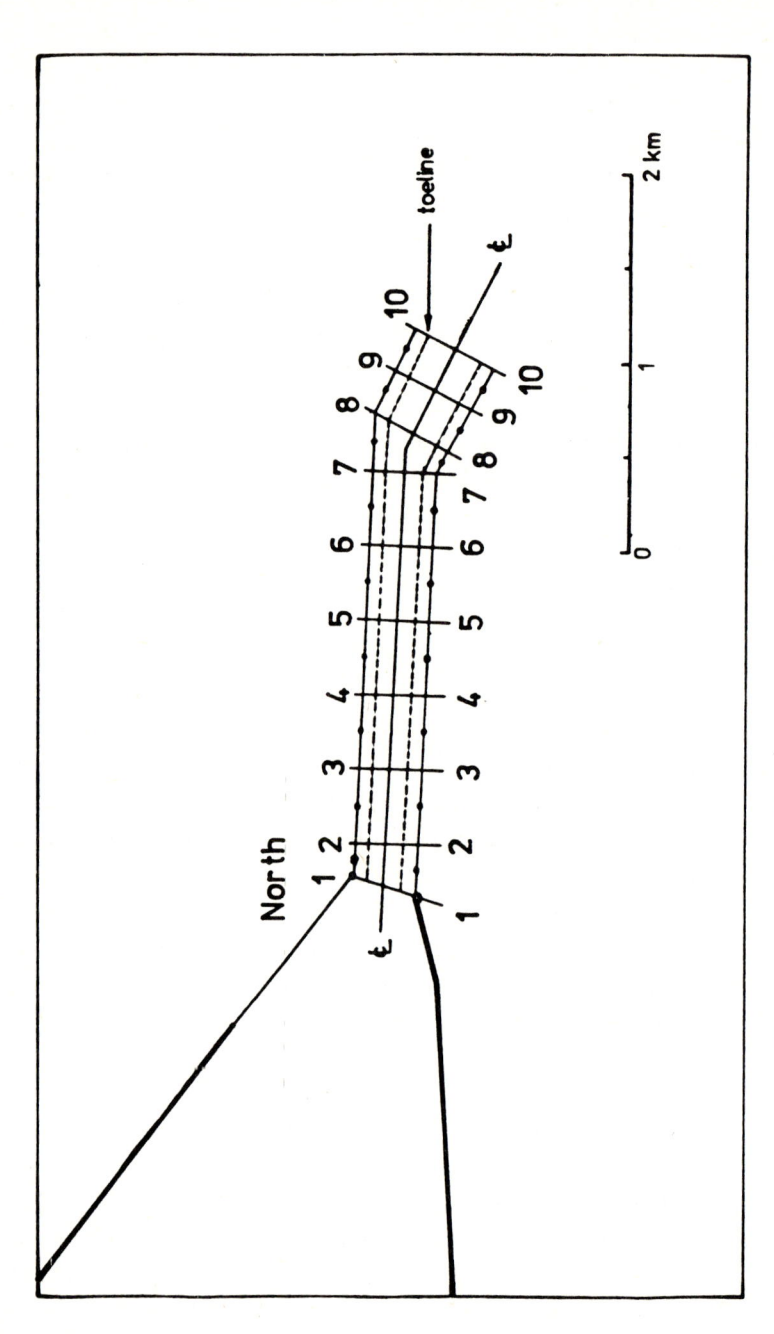

Fig. 6.11 Location of computation sections

Coastal Sediment Modelling

For inshore boxes through the surf zone, a modified form of equation (6.2) has been used, namely:-

$$Q_L = K_L \Delta E \cdot \sin 2\alpha \qquad (6.12)$$

where Q_L = the sediment transport rate in weight per unit time

K_L = a constant

ΔE = the rate at which energy is shed between two successive bed contours h_1 and h_2

α = the angle between the wave crest and the bed contours.

The rate at which energy is shed by a spectrum of waves with a Rayleigh distribution of wave heights can be calculated from the equation (Hydr. Res. (1972)):-

$$\Delta E = H_s^2 T ((\exp(-2H_2^2/H_s^2))K_{S2}^2 - (\exp(-2H_1^2/H_s^2))/K_{S1}^2)/2 \qquad (6.13)$$

where H_1, H_2 = breaker heights in water of depth h_1 and h_2 respectively

K_{S1}, K_{S2} = shoaling coefficients for depths h_1 and h_2 respectively.

Equations (6.11) and (6.12) have been used to study siltation in port approach channels at Karachi (Hydr. Res. (1972)) and Port Quasim in Pakistan. The lattermost study was particularly interesting because the direction of wave approach was such that waves were reflected from the deep channel and caused scour of the upstream banks, which led, in turn, to increased siltation in the deep channel.

Situations involving waves and tidal currents cannot be treated by the previous approach. It is necessary to include the effects of longshore and other currents in any analysis. The "point" approach outlined in section (ii) is to be preferred. The nett siltation (Q_n) in the channel may then be calculated from the equation (Lean, (1980), O'Connor & Lean (1977), O'Connor (1982)):-

$$Q_n = B_0(Q_0 - B_1 Q_1/B_0)(1 - \exp(-s/L)) \qquad (6.14)$$

where B_0, B_1 = widths between current streamlines, figure 6.12.

$$L = V_1 h_1 / w$$
$$s = W B_0 / (B_1 \sin\theta_0)$$
$$B_1 = B_0 (\sin^2\theta_0 + H_1 \cos^2\theta_0)^{\frac{1}{2}}$$
$$H_1 = h_1 / h_0$$

which also includes the refracting effect of the deep channel on current streamlines (Lean (1980)).

The results of using the above equations to study siltation in a port approach channel in the Irish Sea is shown in figure 6.13. The study also showed that during average weather years wave action was responsible for some 30% of the yearly channel siltation, although waves only existed for some 10-13% of the year. During extreme weather years, channel siltation was found to increase nearly fourfold due to the formation of a wind-driven circulation during strong, persistant E to ESE gales, figure 6.14.

Other "point" approaches have been used by Bijker (1980) and Fredsoe (1978), while Sorenson et al (1980) have used a particular "point" approach to study siltation in a port approach channel in Nigeria. It is also possible to use a continuum approach, based on a two dimensional (x,z) sediment continuity equation (Kerrsens et al (1977), O'Connor (1975), O'Connor & Lean (1977), Smith & O'Connor (1977)).

iv) Beach Situations

A knowledge of sediment transport distribution in the surf zone is of importance in some problems. The "point" approach of section (ii) can also be used to describe the detailed picture of sediment movement in the surf zone. The technique requires information on the distribution of longshore wave-induced current across the surf zone. This information can be obtained by solving relevant hydrodynamic and fluid continuity equations with added terms to allow for the dynamic forces (radiation stresses) introduced by wave action. Various analytical and numerical solutions exist for the case of a plane beach with parallel contours (O'Connor et al (1981), Marcos Fanos (1979), Longuet-Higgins (1970), Swart & Fleming (1980)).

Coastal Sediment Modelling

These plane beach velocity distributions can be combined with equations 6.4-6.9 to provide details of the distribution of sediment in the surf zone.

Figures 6.15 and 6.16 show results for conditions on the east coast of India at Paradip, where the information was used in the re-design of a sand by-passing plant and sand trap, figure 6.16.

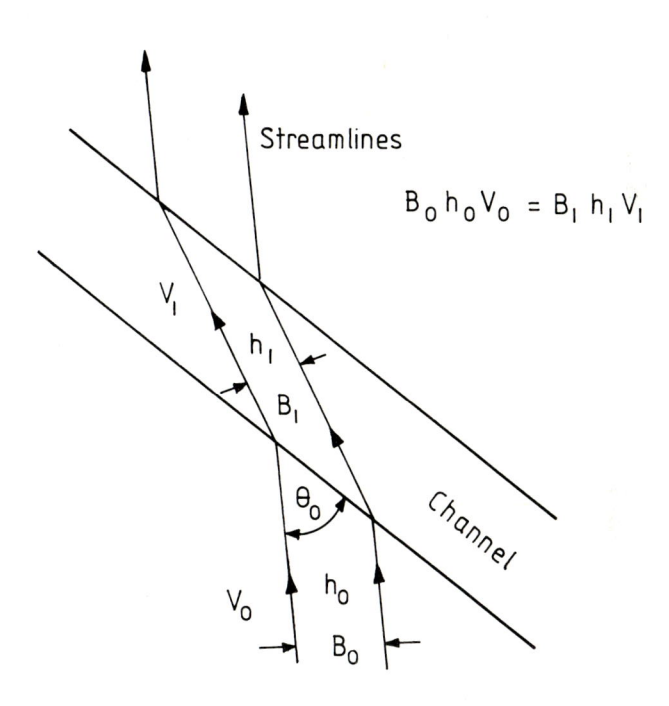

$$B_o h_o V_o = B_1 h_1 V_1$$

Figure 6.12   Flow over a Dredged Channel

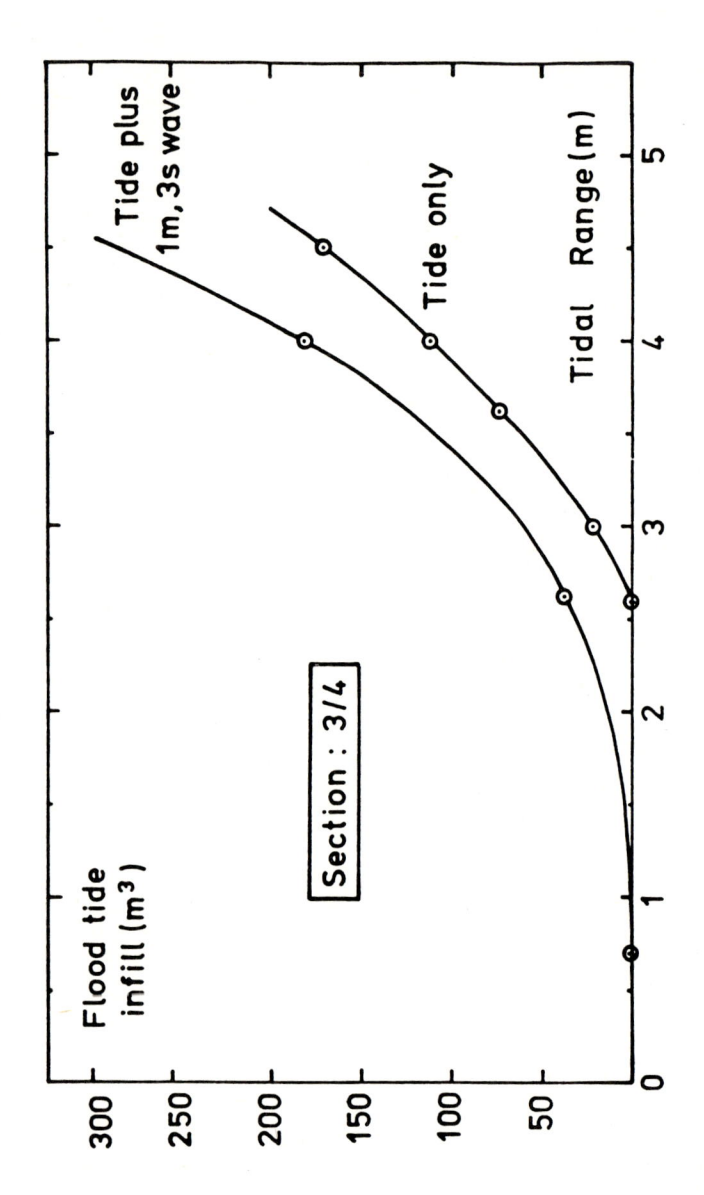

Fig. 6.13 Variation of flood tide infill with tidal range and waves

Coastal Sediment Modelling

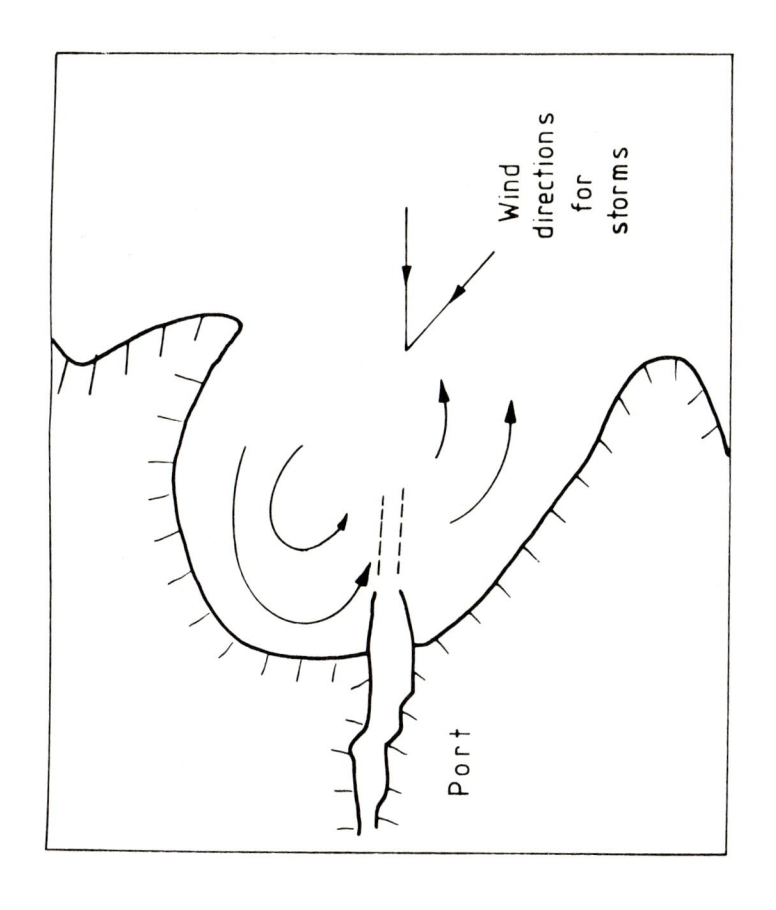

Figure 6.14 Wind-Driven Circulation

B. A. O'Connor

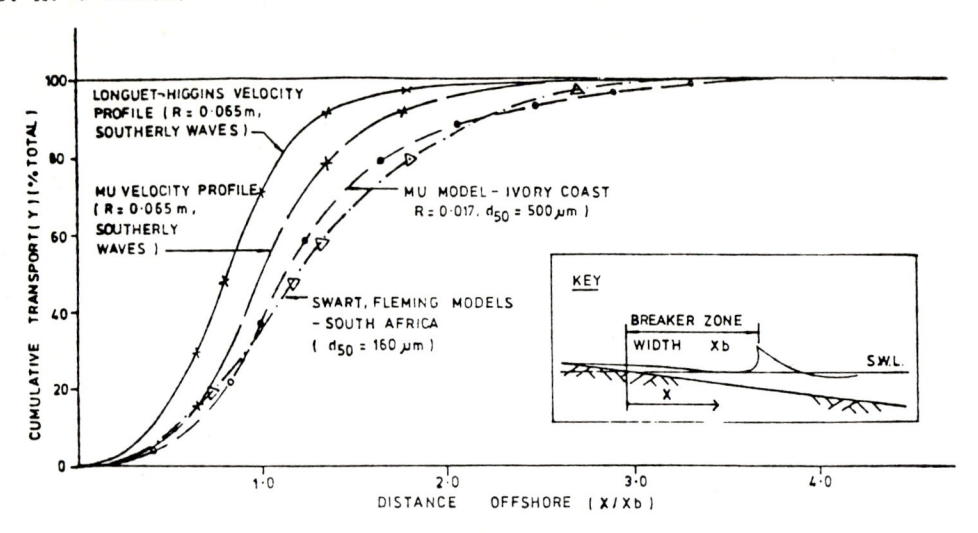

Figure 6.15a CUMULATIVE DISTRIBUTION OF LONGSHORE TRANSPORT ACROSS BEACH.

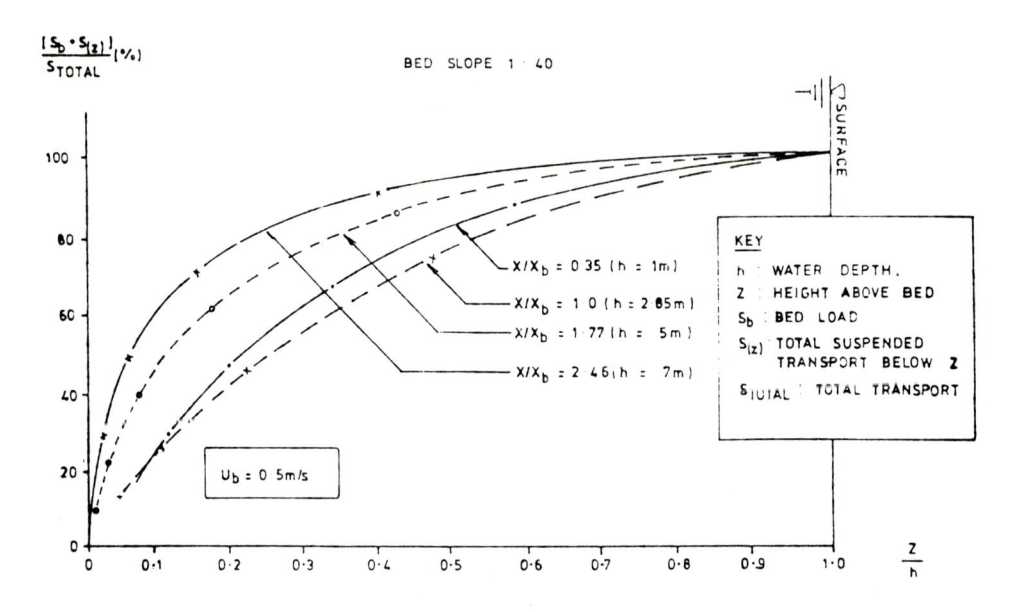

Figure 6.15b VERTICAL DISTRIBUTION OF TRANSPORT FOR VARIOUS DISTANCES OFFSHORE
(LONGUET HIGGINS VELOCITY PROFILE , NO OFFSHORE CURRENTS)

FIGURE 6.16 ADVANCE & RECESSION OF SHORELINE

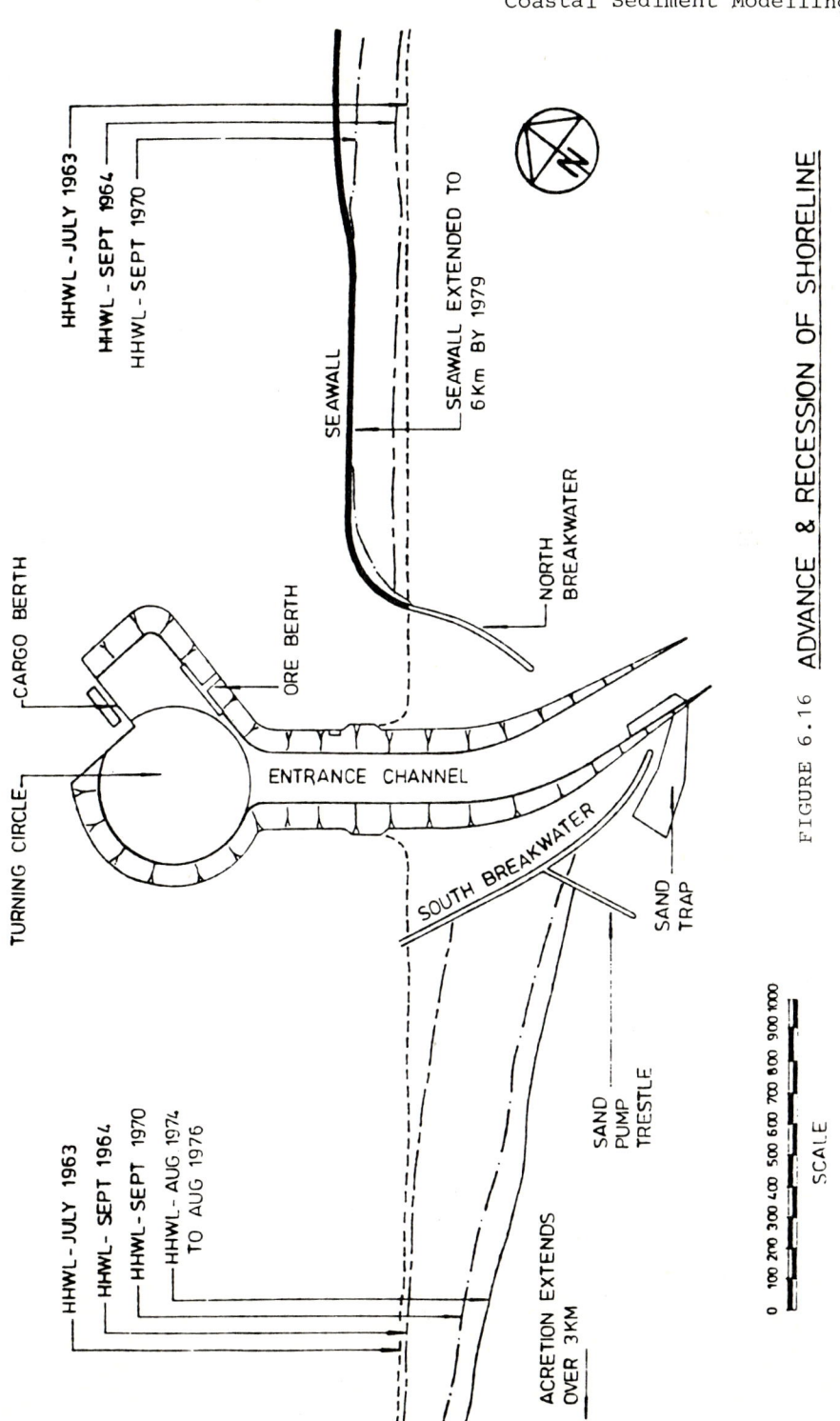

B. A. O'Connor

6.4 Conclusions

Various types of coastal sediment model have been outlined. Such modelling is in progress in many countries, but much remains to be done before such models can be used on a routine basis, as occurs at present with physical models. The complexity of the coastal environment means that computer models are likely to evolve slowly, with a continuous updating every few years as engineers improve their understanding of the environment. It should also be remembered that research into sediment movements in rivers has been in progress for more than 100 years in many institutions throughout the world and yet there is still no universal theory available, which is capable of describing transport rates to an accuracy of better than a factor of two. Present methods must, therefore, rely heavily on field data to improve their accuracy.

There is also a great need to develop new instruments so that data can be collected during storms and thus provide information to improve existing models. Only when models can be shown to reproduce correctly environmental changes with the minimum of a field data will engineers adopt them for general use.

References

Bakker, W. T., 1968
The dynamics of a coast with a groyne system. Proc. 11th Coastal Engs. Conf.

Barber, P. C., 1977
A preliminary investigation into the causes of sand erosion along the foreshore at King's Parade on the North Wirral coast. M.Eng. Thesis, Liverpool University.

Barber, P. C., 1981
A further investigation into the causes of beach erosion at King's Parade on the North Wirral coast. Ph.D. Thesis, Liverpool University.

Bijker, E. W., 1968
Littoral drift as a function of waves and currents. Proc. 11th Coastal Eng. Conf.

Bijker, E. W., 1980
Sedimentation in channels and trenches. Proc. 17th Coastal Eng. Conf.

Bruno, R. O., Dean, R. G. and Gable, C. G., 1980
Longshore transport at a detached breakwater. Proc. 17 Coastal Eng. Conf.

Cacoutas, A., 1982
A microcomputer model for shoreline evolution. M.Sc. Thesis, Manchester University.

Delft Hydraulics Lab., 1976
Coastal Sediment Transport, computations of longshore transport. Report R468, pt. 1.

Ebersole, B. A. & Dalrymple, R. A., 1980
Numerical modelling of nearshore circulation. Proc. 17th Coastal Eng. Conf.

Fleming, C. A. & Hunt, J. N., 1976
Application of a sediment transport model. Proc. 15th, Coastal Eng. Conf.

Fredsoe, A., 1978
Natural backfilling of pipeline trenches, 10th Offshore Technology Conf.

B. A. O'Connor

Frijlink, H. C., 1952
Discussion des formules de debit solide de Kalinske, Einstein et Mayer-Peter et Mueller compte terme des mesures recentes de transport dans les rivieres Neerlandaises. 2nd J.Hyd.Soc.Hyd. de France.

ten Hoopen, H. G. & Bakker, W. T., 1974
Erosion problems of the Dutch island of Goeree. Proc. 14th Coastal Eng. Conf.

Hydraulics Research Ltd., 1972
Calculation of infill in dredged channel. Hydraulics Research Newsletter.

Kerrsens, A., Van Rijn, L. & Wyngaaden, N., 1977
Model for non-steady suspended sediment transport. Proc. 17th I.A.H.R. Congress.

Komar, P. D., Lizzaraga-Aremiaga, J. R. & Tenich, T. A., 1976
Oregon coast shoreline changes due to jetties. Proc. A.S.C.E. J.Water Harb. & Coast. Eng.

Lean, G., 1980
Estimation of maintenance dredging for navigation channels. H.R. Ltd.

Longuet-Higgins, M. S., 1970
Longshore currents generated by obliquely incident sea waves. J.Geophys. Res.

McDowell, D. M. & O'Connor, B. A., 1981
Numerical analysis of sediment transport in coastal engineering problems. I.C.E. Publ. Hydraulic modelling applied to maritime engineering problems.

Mogel, T. R. & Street, R. L., 1974
Computation of alongshore energy and littoral transport. 12th Coastal Eng. Conf. 1974.

Morcos Fanos, A., 1979
A mathematical computer model for coastal sediment transport. Ph.D. Thesis, Manchester University.

Muir-Wood, A. & Fleming, C. A., 1981
Coastal Hydraulics, Macmillan.

O'Connor, B. A., 1975
Siltation in dredged cuts. 1st BHRA Int.Sym. Dredging Technology.

O'Connor, B. A., 1982
Coastal sediment transport short course notes. Simon Engineering Labs. Manchester University.

O'Connor, B. A. and Lean, G., 1977
Estimation of siltation in dredged channels in open situations.
24th PIANC Congress.

O'Connor, B. A., Morcos Fanos, A. & Cathers, B., 1981
Simulation of coastal sediment movements by computer model. Proc.
2nd Int. Conf. Eng. Software, Imperial College.

O'Connor, B. A. & Tuxford, C., 1980
Modelling siltation at dock entrances. Proc. 3rd Int. Sym. Dredging
Technology.

Ozasa, J. & Brampton, A., 1979
Models for predicting the shoreline evolution of beaches backed by sea-walls. HR Ltd., Rep. No. IT 191.

Skovgaard, O. & Jonsson, I. G., 1976
Current depth refraction using finite elements. Proc. 15th Coastal Eng.
Conf.

Skovgaard, O., Jonsson, I. G. & Bertelsen, J. A., 1975
Computation of wave heights due to refraction and friction, Proc.
ASCE. Water Harb. and Coastal Eng. Div.

Smith, T. J. & O'Connor, B. A., 1977
A two-dimensional model for suspended sediment transport. 17th IAHR
Conf.

Sorensen, T. et al, 1980
Sedimentation in dredged navigation channels in the open sea. Proc.
17th Coastal Eng. Conf.

Svasek, J. N. & Versteegh, J., 1977
Mathematical model for quantitative computation of morphological changes
caused by man-made structures along coasts and in tidal estuaries. 17th
IAHR Cong.

Swart, D. H. & Fleming, C. A., 1980
Longshore water and sediment movements. Proc. 17th Coastal Eng. Conf.

Wakeling, L., Cox, N., Gosh, A. and O'Connor, B. A., 1983
Study of littoral drift at Paradip, India. Int. Conf. Coastal, Port Eng.
in Develop. Countries, Sri Lanka.

B. A. O'Connor

Wang, H., 1977
Modelling of short term beach processes. Rep. No. 2 NERC, Japan.

Wang, H. & Liang, S. S., 1975
Mechanics of suspended sediment in random waves. J.Geophys.Res.

7. THE APPLICATION OF RAY METHODS TO WAVE REFRACTION STUDIES

I.M. Townend and I.A. Savell,
Halcrow Maritime Offshore and Energy,
Burderop Park, Swindon, Wiltshire. SN4 0QD.

7.1 Introduction

In offshore and coastal engineering studies one of the prime data requirements is an accurate description of the wave climate at the site of the study. Such data is rarely available, as the collection of wave data is a lengthy and expensive business and requires at least one year's advance warning of the data requirement. If data is available, it will only refer to the location of the recorder, which may be located at some distance from the site for operational reasons. Where data is not available, data from a remote site or hindcast data must be used. In any of these cases there is a requirement for the available data to be modified to reflect the differences between conditions at the data site and those at the study site.

In modifying the data a number of phenomena must be taken into account. The most important of these are:-

1) Refraction by bottom topography. Wave crests tend to align themselves with the bottom contours as they approach the shore. This is a result of the dependence of wave celerity on water depth.

2) Refraction by currents. Currents, as well as depth changes can cause changes in wave celerity, resulting in changes in the wave direction.

3) Diffraction. This may be either internal diffraction, resulting from large localised changes in wave height caused by refraction or similar phenomena, or external diffraction. The latter form results from a discontinuity in the wave field caused by the presence of a structure.

4) Shoaling. The response of the wave profile to changes in water depth. Waves become higher and steeper as they approach their breaking point.

5) Energy dissipation. While propagating over shallow water, waves may lose energy through a wide range of mechanisms. The most widely considered of these are bottom friction and breaking, but percolation and deformation of the bed may also be significant. When considering a spectral description of a wave field, transfer of energy across the spectrum is often significant.

6) Generation. Whenever the distance from the data site to the study site is larger than a few kilometres, the possibility of generation of additional wave energy from the wind must be considered.

A mathematical model incorporating some or all of the above phenomena is a vital part of the engineer's tool kit. The accuracy of the wave climate predicted at the study site is dependent on the accuracy with which the model transfers the remote wave characteristics to the site.

In developing and applying the model, the nature of sea waves must be taken into account. Many early wave models were capable of considering only linear, monochromatic, unidirectional waves. Today, such models in which the sea state is represented as a two dimensional frequency/ direction spectrum give more realistic results and are preferred for all quantitative work. These models are usually relatively insensitive to the exact functional description of the wave spectra, so lack of knowledge of the prototype spectra will not invalidate the model results.

Wave propagation models in use today may be divided into three categories based on the computational approach used. Finite element models have received a lot of attention in the technical press, due to the complexity of the mathematical description and the solution routines used. While the finite element method is well established in many fields, it is not yet an economically viable method for wave modelling. Finite difference models are more widespread, and may give good results in some cases. These models are very demanding of computer power, and this places considerable restrictions on the size of the problem that can be studied. A practical limit using today's technology is an area some 20 wavelengths, say 1 or 2 kilometres, square. This is usually insufficient for studying anything larger than a small harbour. A further problem with these models is the difficulty of defining the input wave boundary conditions. This limits many models to unidirectional waves, which in some cases must be parallel to the model boundary.

Application of Ray Methods

The third type of model is the ray model, so-called because it calculates the paths of "wave rays", also known as wave orthogonals, across the model area. These models have been in use for many years and are known to be reliable and robust when used within the limits of their applicability. The difficulty with these models is that the limits are relatively narrow. This paper is concerned with ray models, and outlines the way in which they may be used, and improvements that may be made to widen the limits of applicability.

7.2   Ray Models

Two types of ray model can be used to investigate the effects of refraction and shoaling. The first of these evaluates the effects of these phenomena as the offshore wave propagates landwards by constructing diagrams showing orthogonal paths as they move towards the area of interest, generally known as forward tracking.

The fundamental relationship describing the curvature of a wave orthogonal, expressed in cartesian co-ordinates x and y, is:

$$\frac{d\alpha}{ds} = \frac{1}{C}\frac{dC}{dx}\sin(\alpha) - \frac{dC}{dy}\cos(\alpha) \tag{7.1}$$

where s is directed along the orthogonal and α is the angle between the wave orthogonal and the x-axis. The wave speed C for sinusoidal waves of small amplitude is given by the dispersion equation:

$$C = \frac{gT}{2\pi}\tanh\left(\frac{2\pi h}{CT}\right) \tag{7.2}$$

where T is the wave period, h the water depth and g the acceleration due to gravity.

The computer model defines the sea bed topography by a number of rectangular arrays of depth values, which together form a grid covering the entire area of interest. The resolution in each cell of the grid may be chosen for an optimum trade-off between accuracy and computational efficiency. Starting from a known point and a known direction, equation 7.1 may be solved repeatedly to calculate the path of the wave ray across the element. The solution technique used is a development of the "circular arc" method originally developed at the Hydraulics Research Station (UK). Briefly, the method assumes that within a small triangular element the variation of wave speed is linear in both x and y coordinate

I. M. Townend and I. A. Savell

directions. Using this assumption, the ray path within an element is described by a circular arc. This arc is centred on the zero celerity line in the local celerity plane and is tangential to the corresponding arcs in adjacent elements.

The solution requires definition of the initial values so that wave rays may be tracked from any point to some unknown destination. Wave ray diagrams may be constructed by tracking a number of rays from the offshore boundary to cover the inshore area of interest. Whilst this method provides a good picture of the general wave refraction patterns in an area, associated methods of calculating wave refraction coefficients are unsatisfactory for a number of reasons which are discussed by Abernethy et al (1975).

This forward tracking technique is used by the authors only where a qualitative idea of refraction patterns is required. An example of a forward tracked ray diagram and a picture of wave crests derived from the ray diagram appear in figures 7.1 and 7.2.

An alternative to this method is to track wave rays "backwards" from an inshore point of particular interest to the offshore boundary of the model. This is done for a number of wave directions and periods from each point of interest. This allows complete two-dimensional frequency - direction wave spectra to be transferred to the inshore points and has been found to give more stable and realistic results. The spectrum refraction equation may be stated as:-

$$S(f,\phi) = \frac{C_0 C_{g_0}}{C C_g} S_0(f,\theta) \qquad (7.3)$$

where S is the spectrum function, ϕ is the wave direction in shallow water, θ the wave direction in deep water, f is the wave frequency (= 1/T), C_g is the wave group velocity and subscripts $_0$ refer to deep water.

The wave refraction model output associates a deepwater wave direction with each inshore wave direction for each period considered. Using a suitable spectral description of the deepwater wave conditions, the energy density associated with the deepwater end of each wave ray may be transferred to the inshore spectrum at the appropriate direction and period. Equation 7.3 is used to modify the energy density to account

Application of Ray Methods

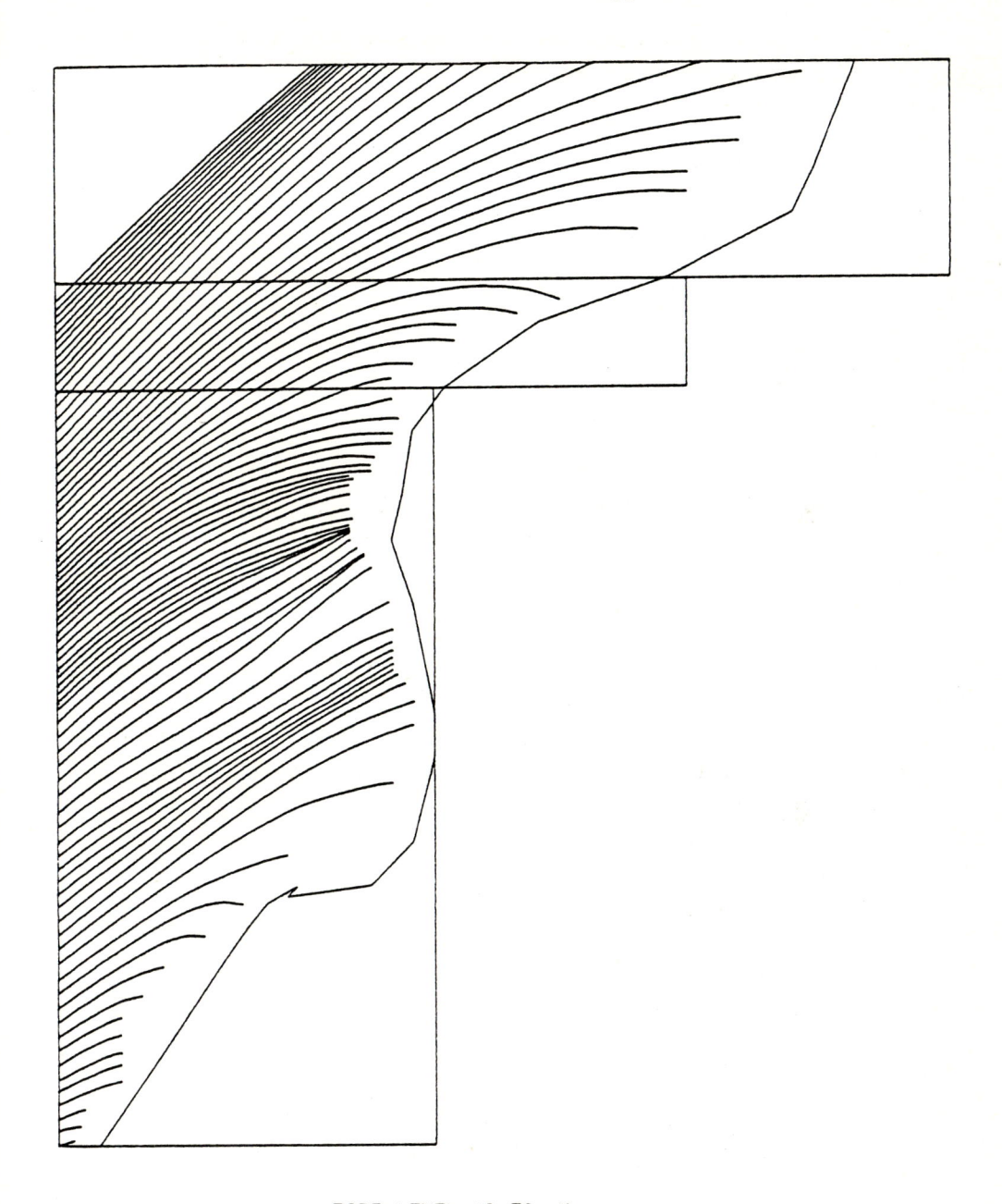

TIDE LEVEL 10.70 M

WAVE PERIOD 7.1 SECS

DIRECTION 315.0 DEGREES TRUE

Figure 7.1　Wave Refraction - Forward Tracking

I. M. Townend and I. A. Savell

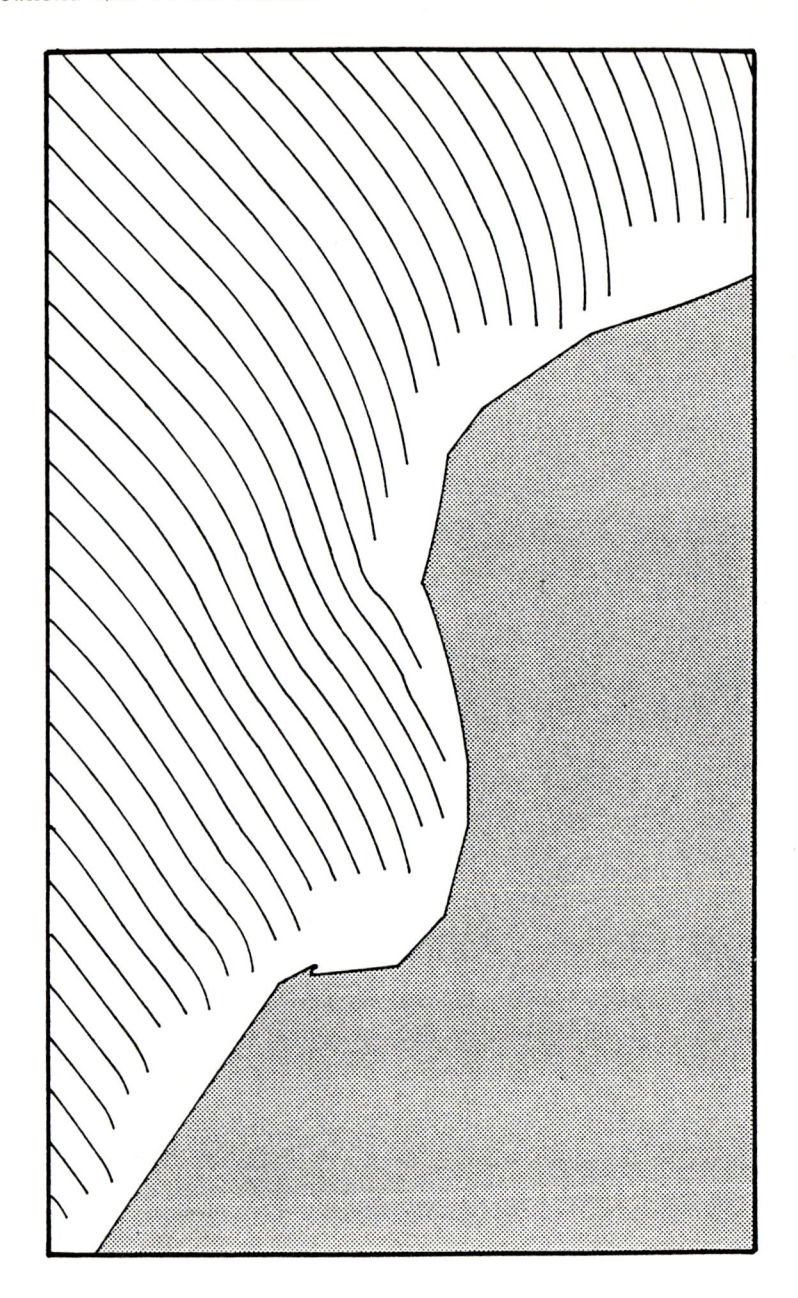

Figure 7.2   Wave crest diagram

Application of Ray Methods

for refraction effects. From the inshore wave spectrum, relevant quantities such as wave height, mean direction and frequency may be calculated.

The offshore two-dimensional frequency-direction spectrum is usually assumed to be to the product of two mutually independent components S'(f) and G(ϕ) for which the former is a one dimensional frequency spectrum and the latter is a suitable direction distribution function. In most cases some form of the JONSWAP formulation is used to represent the deepwater spectrum (Carter, 1982). This describes a developing sea, which is considered to be appropriate for most coastal sites around the UK, and has the functional form:-

$$S'(f) = \frac{g}{(2\pi)^4 f^5} \exp\left(\frac{-5}{4}\left(\frac{f}{f_p}\right)^{-4}\right) \gamma^A \qquad (7.4)$$

where $A = \exp\left(\frac{-(f-f_p)^2}{2\lambda^2 f_p^2}\right)$

λ is 0.07 when $f < f_p$ and 0.009 when $f > f_p$, α and γ are constants obtained either from measured offshore spectra, or by parameterisation (Carter, 1982). f_p is the peak frequency of the spectrum. The functional form of the direction distribution is assumed to be:-

$$G(\theta) = \frac{\cos^n(\theta-\bar{\theta})}{\int_{-\pi/2}^{\pi/2} \cos^n(t) dt} \qquad (7.5)$$

where $\bar{\theta}$ is the mean direction of the spectrum and the exponent n is typically 2 or 10. An exponent of 2 corresponds to a broad banded directional distribution which can be associated with locally wind-generated waves whilst a value of 10 corresponds to a relatively narrow-banded distribution which can be associated with seas generated over a long fetch.

The functional shapes of typical two-dimensional spectra with direction indices of 2 and 10 are shown in figures 7.3 and 7.4. Comparison of the results of models using differing forms of the above spectral shapes, or using prototype spectral data, has shown that the results are relatively insensitive to variations in the spectra used. Standard analyses always use two different values of the direction index as a check on the sensitivity of the results to varations in this parameter.

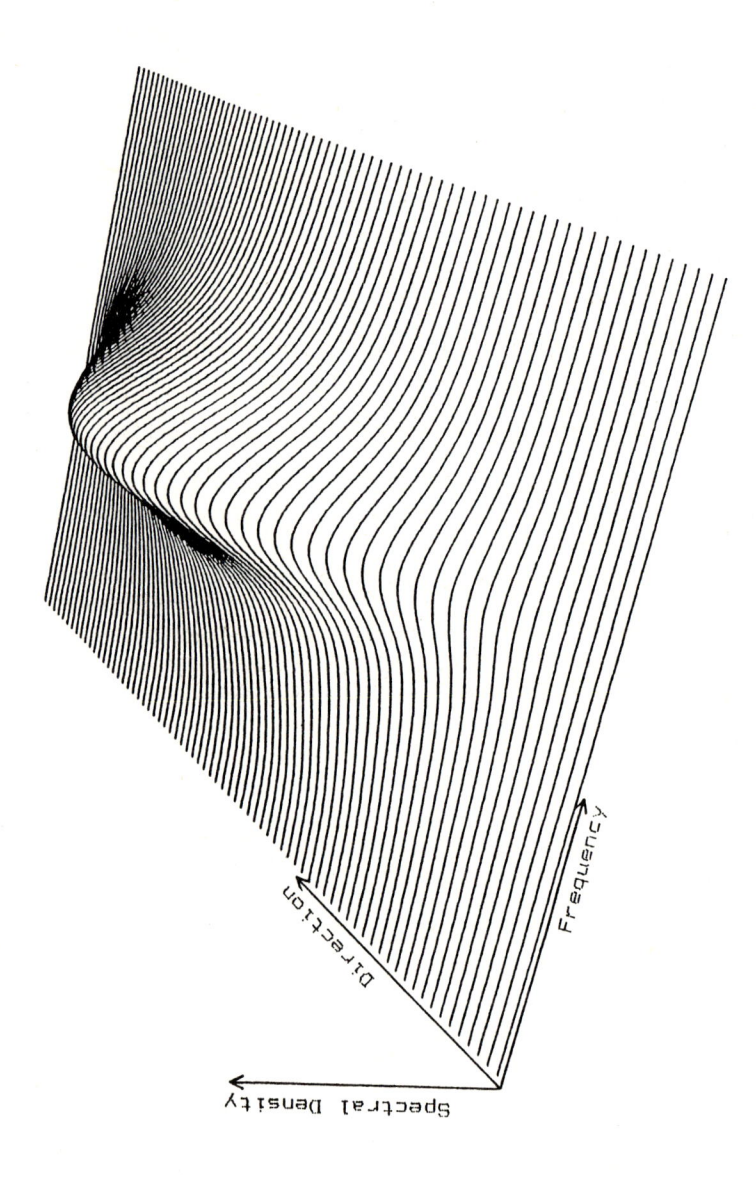

Figure 7.3   Two dimensional frequency/direction spectrum, direction index 2

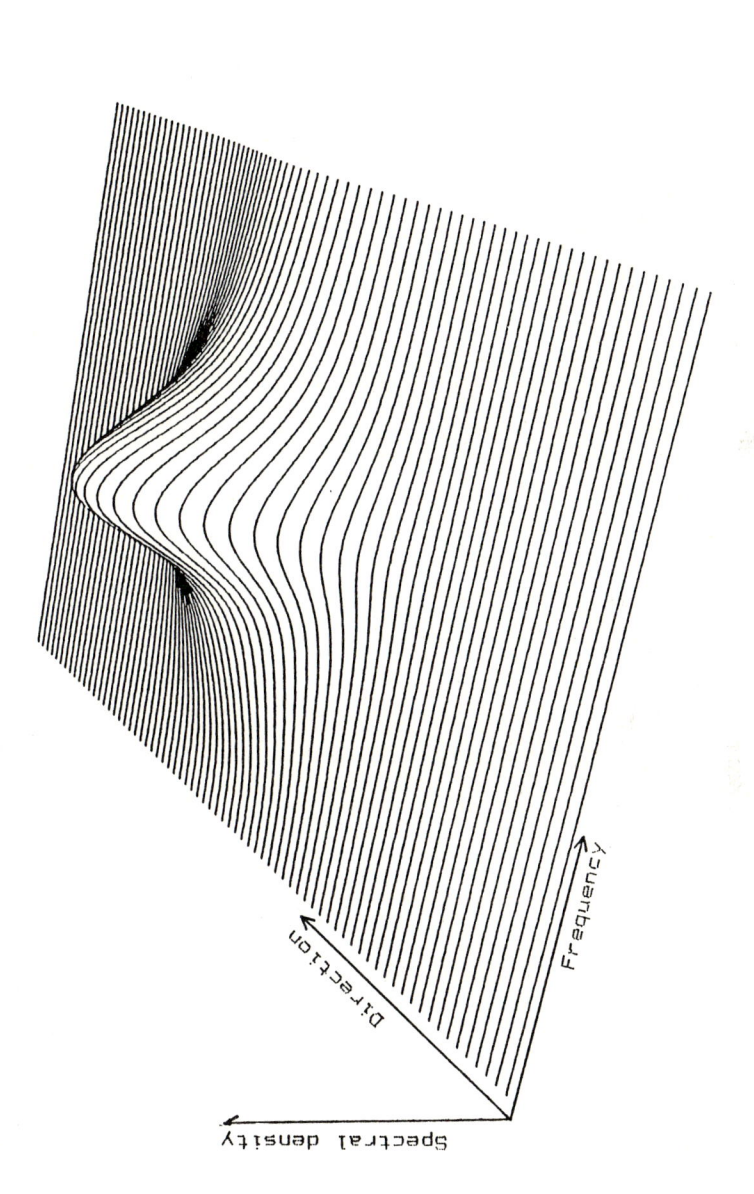

Figure 7.4 Two dimensional frequency/direction spectrum, direction index 10

I. M. Townend and I. A. Savell

The inshore wave height, mean direction and period may be determined by integrating to obtain the moments of the inshore spectrum. The above parameters are then obtained from these, as follows:-

$$K_w = \left(\frac{\int\int S(f,\phi) \, df d\phi}{\int\int S_0(f,\theta) \, df d\theta} \right)^{\frac{1}{2}} \tag{7.6}$$

$$\bar{\phi} = \frac{\int\int \phi \, S(f,\phi) \, df d\phi}{\int\int S_0(f,\phi) \, df d\phi} \tag{7.7}$$

$$T = \left(\frac{\int\int S(f,\phi) \, df d\phi}{\int\int f^2 S(f,\phi) \, df d\phi} \right)^{\frac{1}{2}} \tag{7.8}$$

K_w expresses the inshore wave height as a proportion of the offshore wave height, making the refraction study independent of the deepwater wave height in this simple case. $\bar{\phi}$ is the mean direction of the inshore spectrum and T is the mean period of the inshore spectrum.

7.3 Application of the Ray Model, a Simple Case

The simple ray model described above has been used in many studies where a wave refraction analysis was required. As a typical study we have chosen that carried out for the Wessex Water Authority at Minehead, on the North Devon coast. Figure 7.5 shows the location of the site. The purpose of the study was to recommend measures to alleviate flooding of the low-lying land to the east of the town. Part of the work involved determining design waves for a rock revetment to protect over 2 kilometres of the coast.

Wave data was available from three offshore sites in the Bristol Channel and also from a wind-wave hindcasting study. The refraction model was used to transfer this data to inshore locations along the area of the coastline being studied. Figure 7.5 shows the location of these points. As the basic model is not capable of taking wave breaking into account, the inshore points were chosen to be outside the breaker zone for the waves being studied.

A refraction grid consisting of 8 cells was set up using bathymetric data from Admiralty charts. These are not always ideal for this purpose, as they are produced with the mariner in mind, not the coastal engineer.

Application of Ray Methods

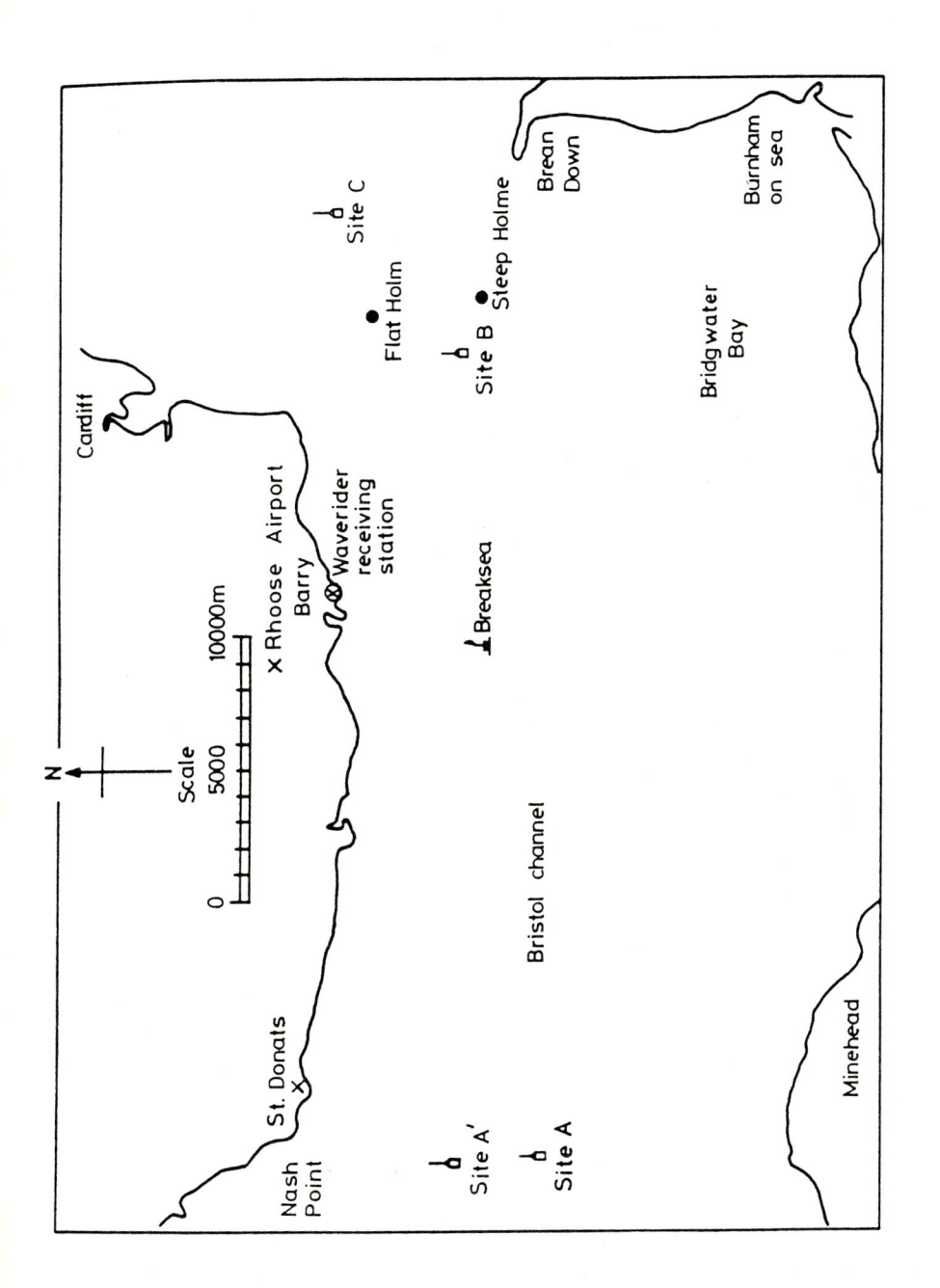

Reproduced from H.R.S. report no EX 994 (ref 3)

Figure 7.5   Location of Waverider Buoys

I. M. Townend and I. A. Savell

Use of collector charts, where available, rather than the published charts, allows a more accurate representation of the bathymetry. Until recently, preparation of depth grids was a tedious and lengthy manual task. Use of an A0 digitising tablet, together with surface fitting computer software, has now considerably reduced the effort involved.

The area covered by the grid was determined by inspection of the bathymetry of the area, the location of the wave data, and possible wave paths to the site. This suggested a grid 22km by 19km, extending to the middle of the Bristol Channel and 10km either side of the area of interest. This is adequate for all locally generated waves but cannot include the effects of refraction on waves propagating up-channel from the south west. Deciding where to terminate the refraction model under these circumstances is difficult, and many conflicting factors must be taken into account. Ultimately the project engineer must use his experience to settle the matter.

As a check on the accuracy of the grid, a contour plot is usually produced and compared with the chart. Any gross errors generally appear as patterns of concentric diamonds or rectangles on the contour plot and are rapidly dealt with.

Before using the refraction grid, it was smoothed. This was done with care, as smoothing progressively removes the very topographical features whose effect on the waves the model is to determine. The aims of the smoothing process is to remove any minor irregularities that may affect the ray model more severely than the actual waves. Consequently, only the deeper parts of the grid are smoothed.

The smoothing algorithm used conserves the volume of the seabed, distributing material between each point and its neighbours. When smoothing the edges of a grid cell, the depths in the neighbouring cell are taken into account to ensure that the model seabed remains continuous across the cell boundary. Discontinuities here are not recognised by the ray model, and will therefore distort the results of the refraction analysis.

Current practice is not to smooth at all if possible, as the surface fitting software now employed generates a much less lumpy seabed than the old manual techniques.

Application and Ray Methods

SEA DEFENCES

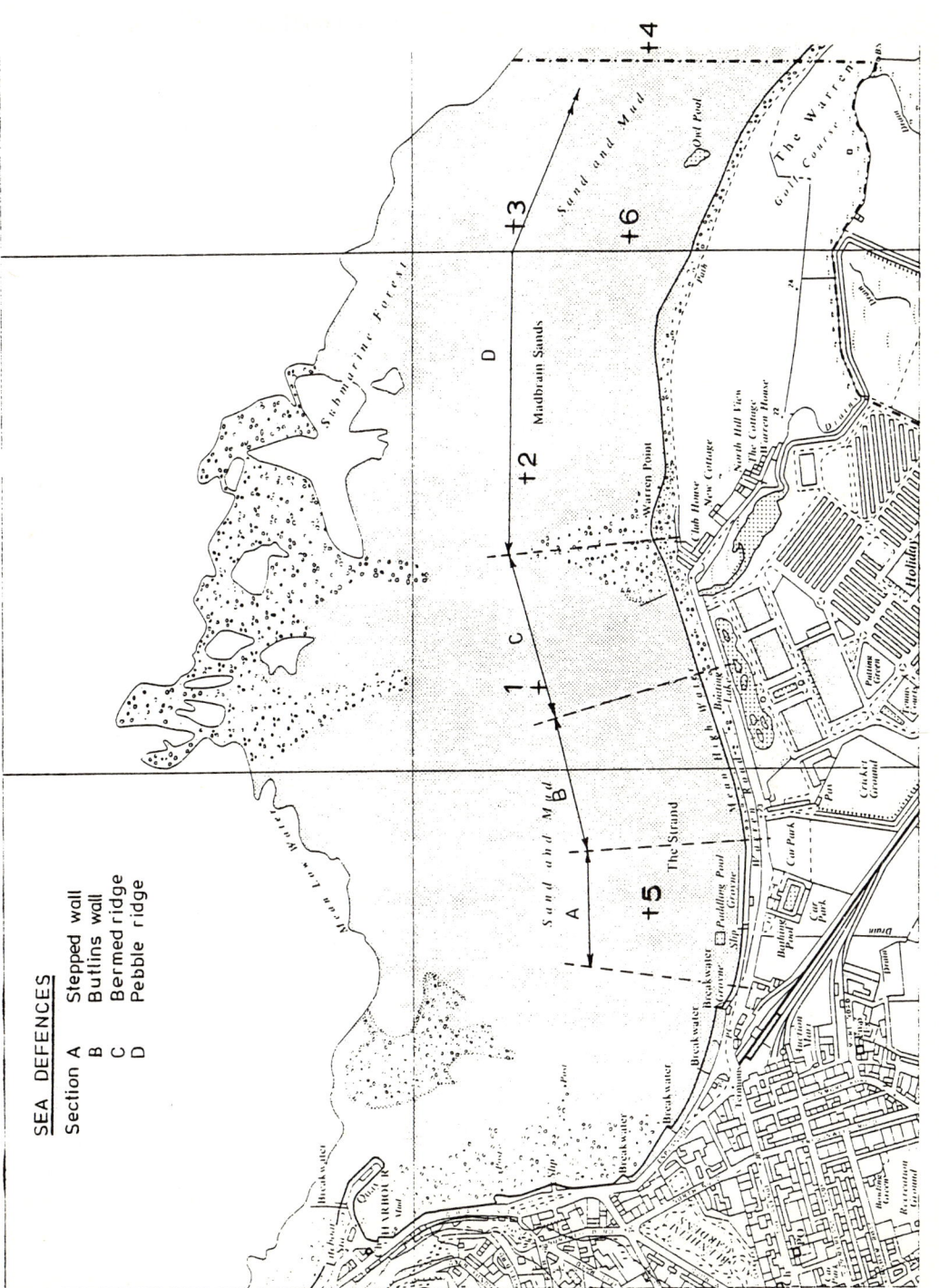

Section A Stepped wall
        B Butlins wall
        C Bermed ridge
        D Pebble ridge

Figure 7.6   Plan of Minehead Sea Defences

I. M. Townend and I. A. Savell

At one time, a study such as this, where wave conditions must be determined over a length of coastline, would have used a forward tracking ray model. Reference to figure 7.1 which shows forward tracked rays over part of the Minehead grid, indicates the type of output which would be obtained. Comparing figure 7.1 with figure 7.6 shows that the forward tracking study predicts a concentration of wave energy near points 1 and 3, while points 2, 4 and 5 are in areas of diverging rays where wave heights will be reduced. Diagrams similar to figure 7.1 are frequently published in the technical press as evidence that some area of coastline or length of breakwater is subject to particularly severe wave attack. Later we will see that in this case at least the forward tracking results are grossly in error when the spectral nature of the waves is taken into account.

For this study the backtracking technique was used for all the quantitative results. Fans of wave rays were sent out from each inshore point at 1 degree spacing for each of six wave periods. This much computation would have been prohibitively expensive only a few years ago, but today the cost is only a few hundred pounds. Figure 7.7 shows a typical fan of rays, in this case for a period of 7.9 seconds at position 3. These fan diagrams are produced for checking purposes only, the essential data is transferred automatically between programs. Figure 7.7 shows that the area to the east of the site is inadequately covered by rays, leading to the possibility of spurious results when considering north-easterly storms. This area was filled in by a further run of the refraction program and the results merged automatically.

Results from the spectrum transfer program are shown in figures 7.8 and 7.9. The wave height factor is the ratio of the inshore wave height to the offshore height. Figure 7.8 shows how the wave height factor and the mean inshore wave direction vary with the peak period of the offshore storm for an offshore wave direction of 45 degrees, i.e. north east. The two lines relate to different values of the index "n" in the directional spectrum. The results are relatively insensitive to the value of this index, as may be seen in this figure. Note, however, that as the directional spectrum becomes narrower, the wave height factor increases. This indicates that a unidirectional wave train, as assumed by the forward tracking model, would predict a marked increase in the inshore wave height at this point.

Application of Ray Methods

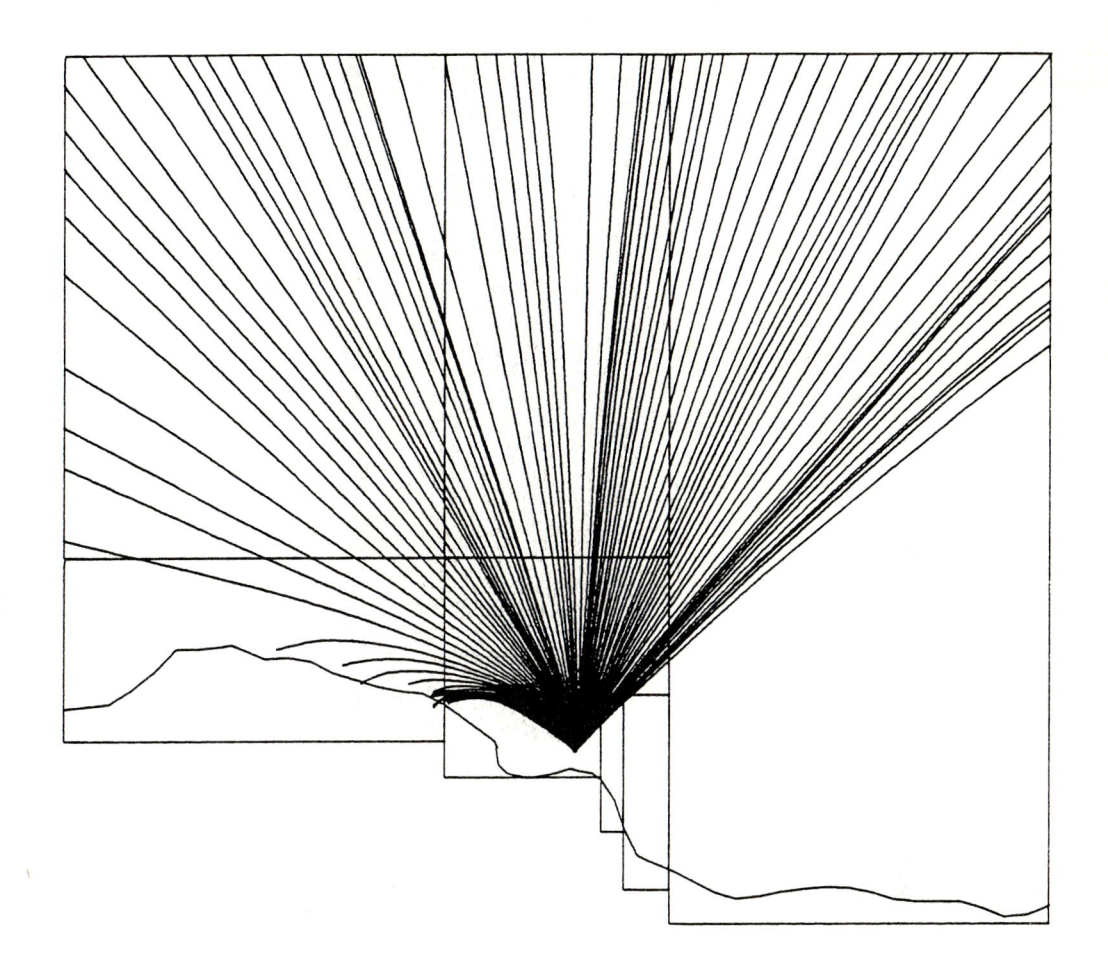

TIDE LEVEL 10.70 M

WAVE PERIOD 7.9 SECS

POSITION 3

Figure 7.7   Wave Refraction - Backward Tracking

I. M. Townend and I. A. Savell

POSITION NO: 03   DIRECTION INDEX: 2
TIDE LEVEL: 10.70M   10
OFFSHORE WAVE DIRECTION: 45·0 DEGREES TRUE

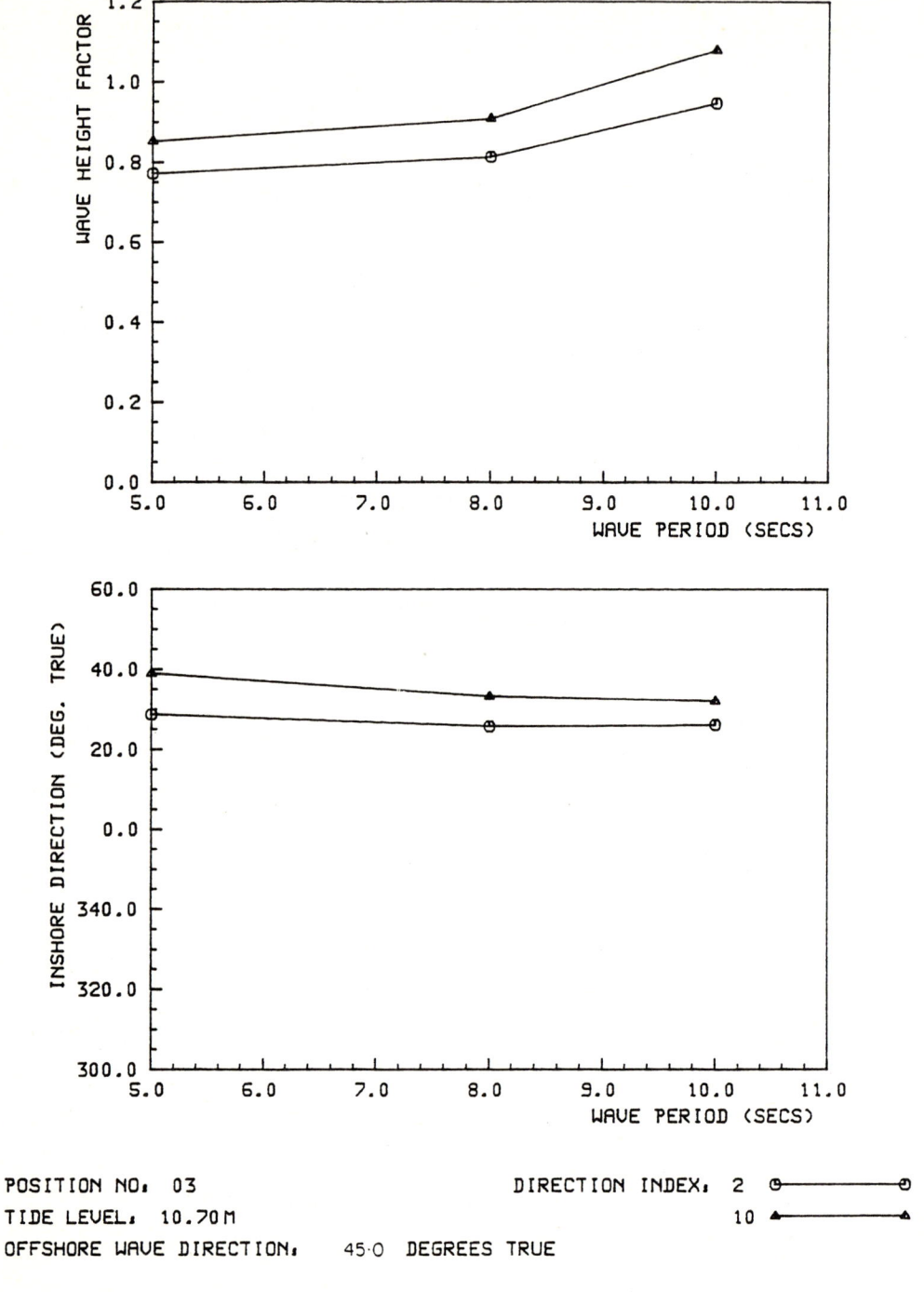

Figure 7.8   Wave Height Factors - Constant Direction

Application of Ray Methods

Figure 7.9 Wave Height Factors - Constant Period

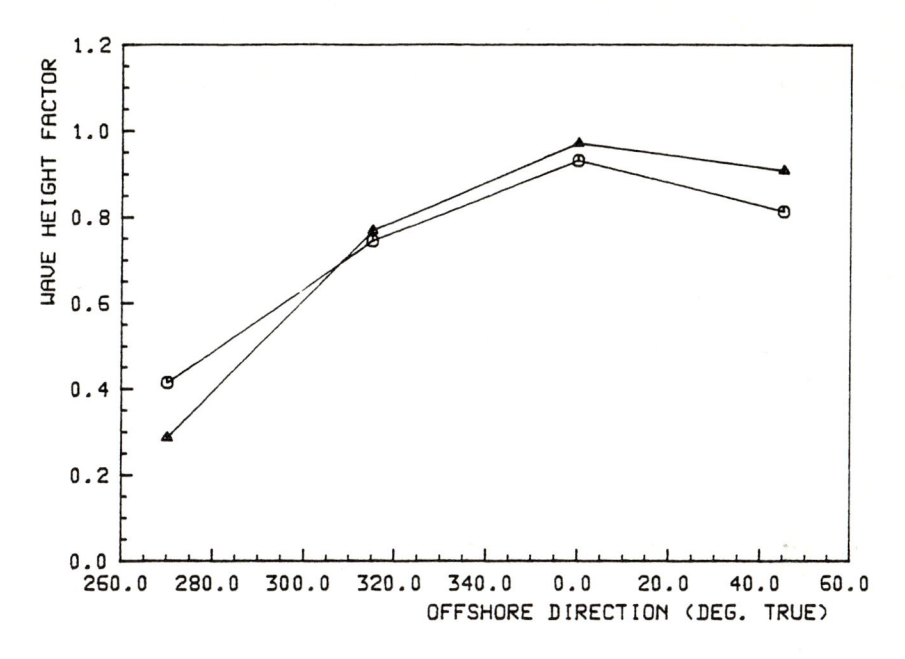

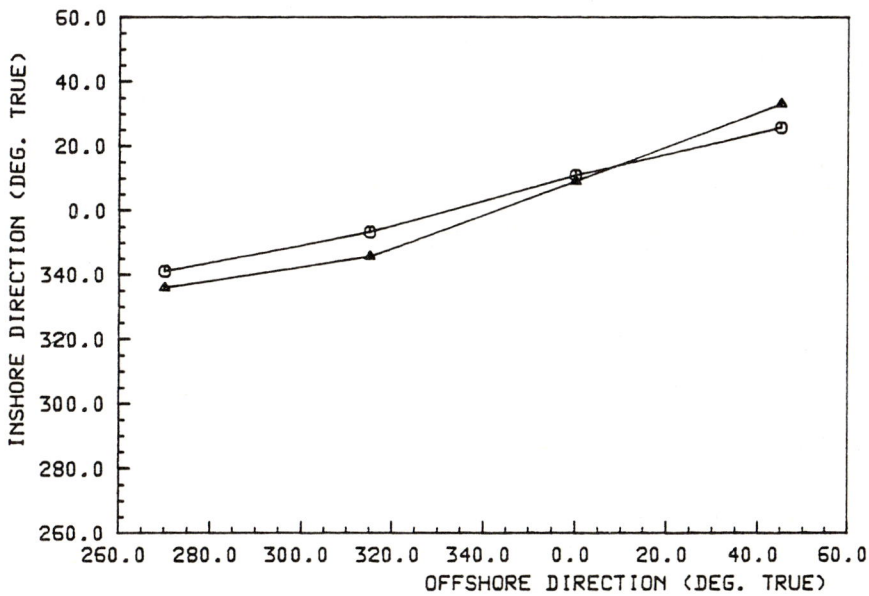

POSITION NO: 03
TIDE LEVEL: 10.70 M
PERIOD: 8.0 SECS

DIRECTION INDEX: 2
                  10

I. M. Townend and I. A. Savell

The model used for the Minehead study took account of refraction by bottom topography only. The model was run for the high tide case only, avoiding the need to consider energy losses over the Madbrain Sands and Submarine Forest, which present a large area of shallow water at low tide levels. If for some reason, the model were required to be run at a lower tide level, energy losses due to bottom friction and breaking would become the dominant mechanism modifying the nearshore waves, and the model would have to take these effects into account. We next consider a study where wave breaking was judged to be the dominant effect.

7.4 A Study Including Wave Breaking

Hurst Spit extends eastwards into the western arm of the Solent from Christchurch Bay (Figure 7.10). The natural development of the spit results in erosion of the south facing shore and the extension of the end of the spit towards the north and northeast. The erosion of the south shore has accelerated in recent years following the construction of coast protection works at Barton-on-Sea, Milford and Christchurch. The authors were commissioned to prepare a design for works to ensure the stability of the spit in the area of Hurst Castle.

It is evident from inspection of the plan shape of the spit that it is sheltered from wave action by the Shingles Bank and the North Head. Designing the protection works using offshore wave heights would therefore be needlessly conservative, so an analysis of wave refraction, shoaling and breaking over the offshore banks was undertaken. An earlier study by the then Hydraulics Research Station had identified the problems associated with modelling wave action across the banks, and made an esimate of the inshore wave climate at the spit. The authors' aim was to see if a method could be developed to account for the energy losses over the banks with acceptable accuracy.

The basic model used was the same as that used at Minehead, and it was set up in much the same way. Figure 7.11 shows the behaviour of a monochromatic sea, forward tracked across the bank, with all the "safety stops" turned off. The modelling approach adopted was to use the wave refraction program in backtracking mode, with switches set to stop the ray path if the refraction approximations broke down. Thus rays running into shallow water or turning tightly over the banks were terminated.

Application of Ray Methods

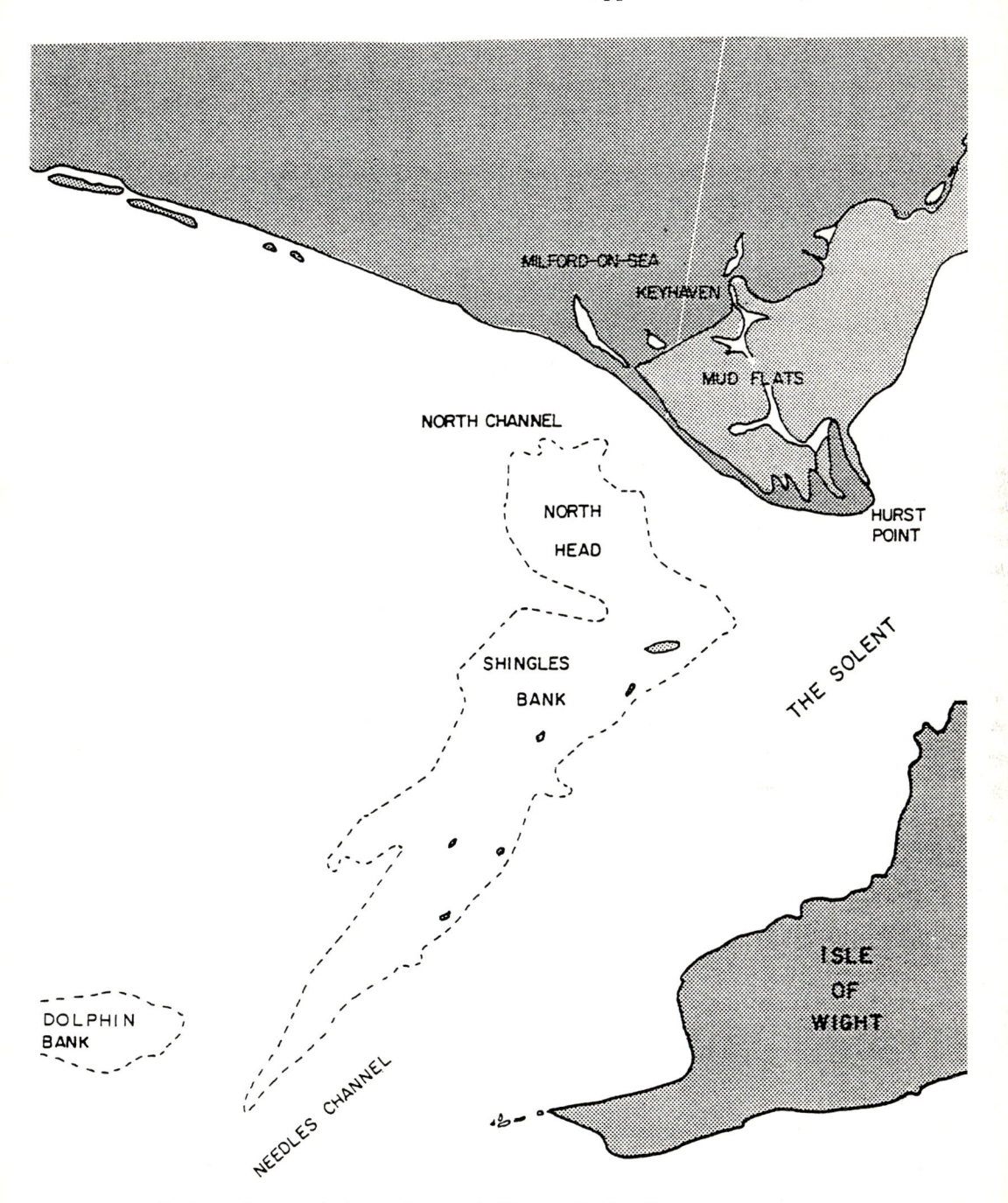

Figure 7.10 Hurst Spit - General Plan of the Area

I. M. Townend and I. A. Savell

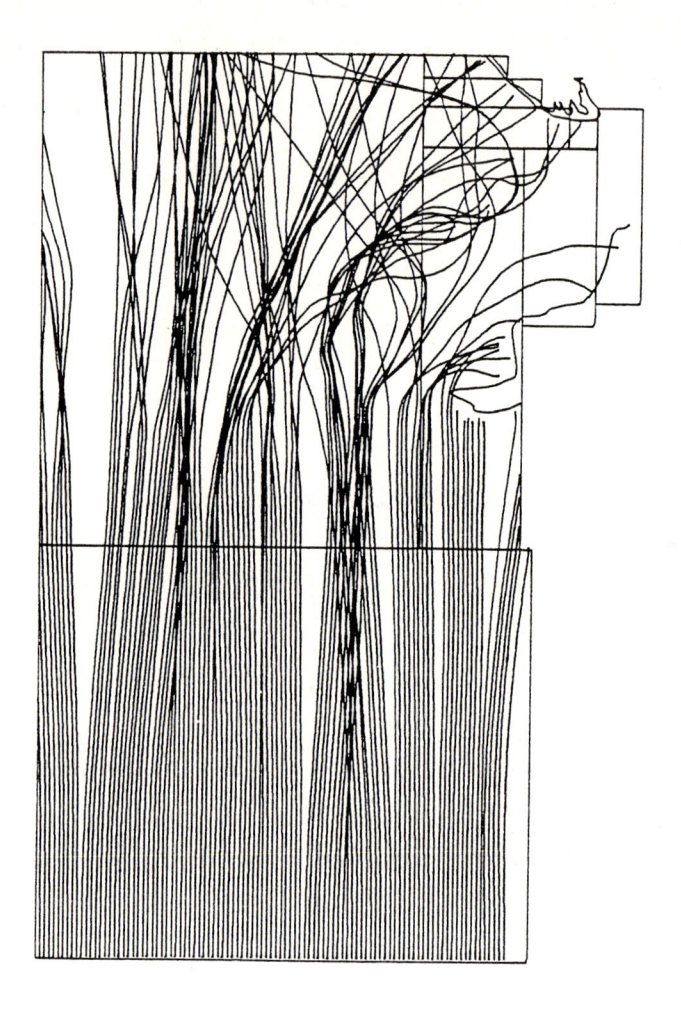

WAVE PERIOD 10.0 SECS

DIRECTION 180.0 DEGREES TRUE

Figure 7.11   Hurst Castle Protection - Wave Refraction - Forward Tracking from South

Application of Ray Methods

A typical fan diagram is shown in Figure 7.12 and it is easy to see that most of the rays terminate on the North Head, indicating that most waves will break before reaching the site. Those rays which terminated on the banks were replaced by straight rays, along which the wave energy was progressively reduced using a model of wave energy loss due to breaking. The spectral refraction equation (7.3) was then modified to take these energy losses into account when transferring the offshore spectrum to the inshore point.

The wave breaking model uses the method proposed by Gerritsen (1979) who suggested the following expression for the wave energy loss due to breaking:

$$\varepsilon_b = \frac{\zeta}{4\sqrt{2}} \rho g f H^2 \qquad (7.9)$$

where ε_b is the rate of energy dissipation, ζ is a dimensionless energy dissipation coefficient with a value between 0.3 and 0.6, f is the wave frequency and H is the wave height immediately prior to breaking.

By equating the energy flux with the rate of energy dissipation, an expression for the resultant wave height can be derived:-

$$H_2 = \left(\frac{Cg_1}{Cg_2} (H_1)^2 - \Delta F - \Delta B \right)^{\frac{1}{2}} \qquad (7.10)$$

where Cg_1/Cg_2 represents the shoaling factor, ΔF the loss due to bottom friction and ΔB the loss due to wave breaking.

The rate of energy dissipation due to bottom friction (wave decay), based on linear assumptions (Skovgaard et al, 1975) is given by:-

$$\varepsilon_f = \frac{2}{3} f_w \rho \frac{U_0^3}{\pi} \qquad (7.11)$$

where f_w is the Jonsson friction factor (Jonsson & Carlson (1976)) and U_0 is the maximum wave orbital velocity at the bed. By summing the losses over finite intervals along a given wave ray path the net loss due to bottom friction is obtained.

In a similar manner the loss due to wave breaking is derived from equation 7.9 and for a particular wave ray path is computed each time the wave height exceeds the allowable depth limited value (Lick, 1978)

I. M. Townend and I. A. Savell

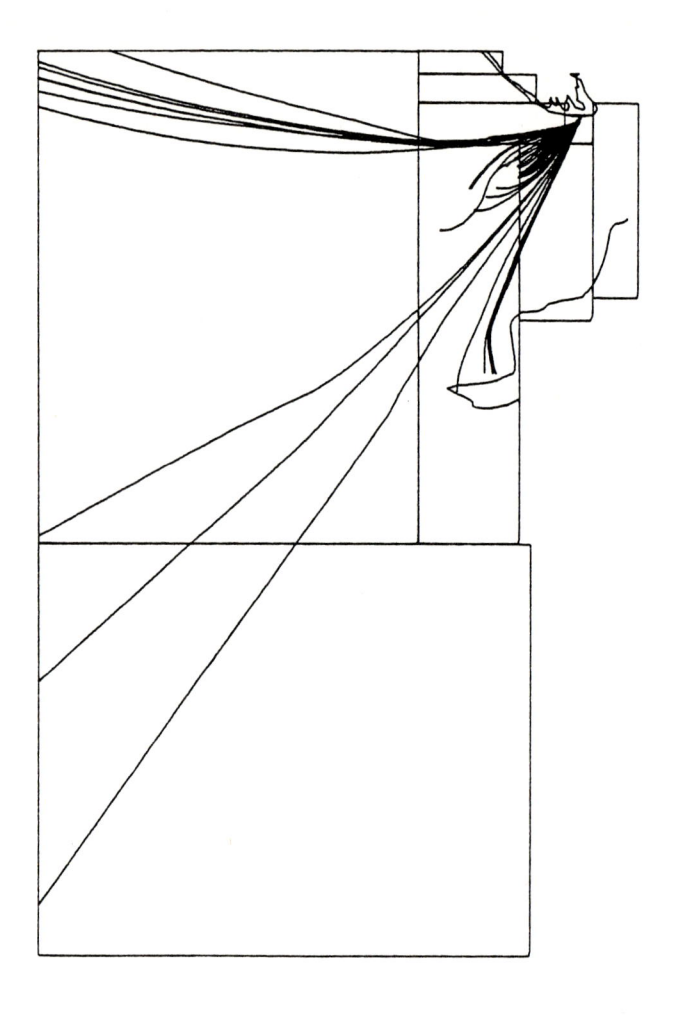

Position No 1
Wave period 5.6 secs.

Figure 7.12   Hurst Castle Protection - Wave Refraction - Back Tracking

Application of Ray Methods

given by:-

$$\frac{H_b}{h} = b - a \frac{H_b}{gT^2} \qquad (7.12)$$

where h is the depth, H_b the breaking wave height and a and b are functions of the bed slope.

When using the simple wave refraction model to transfer a spectrum from offshore to the inshore location there is no need to consider a specific wave height. If wave breaking or bottom friction are to be examined, then a specific wave height has to be considered. The representative offshore wave height (H_1) used for each frequency is obtained from the offshore spectrum by using the relationship:-

$$H \propto \sqrt{S(f)} \qquad (7.13)$$

The expression is valid for low amplitude deep water waves and is used in this case to give an appropriate modulation for each frequency used.

Having obtained an inshore wave height from equation (7.10), the ratio of wave heights, for a given frequency, $H_2(f)/H_1(f)$ is used to modify the spectrum such that:-

$$S_2(f) = S_1(f) \left(\frac{H_2}{H_1}\right)^2 \qquad (7.14)$$

In this manner it is possible to take account of the contribution to the inshore spectrum made by waves that break while travelling to the site, as well as that made by waves which simply undergo shoaling and refraction to reach the inshore site.

This model may be unsatisfactory for several reasons, but gave results which agreed well with the HRS estimates and showed the expected variations of wave height with the period and direction of the offshore storm. The shortcomings of the model all tend towards the production of conservative estimates of wave height, and were considered acceptable in an engineering study of this kind. This is not the case, however, in all such studies and more research is necessary to try to improve these aspects of the model.

I. M. Townend and I. A. Savell

The main shortcoming of the wave breaking model is the representation of each component of the wave spectrum by a single representative wave when considering breaking. In reality, breaking occurs when the actual waves present become too steep or too high for the water depth. At this point the individual spectral components, however these are determined, may be well inside their breaking limit. As a result of this, the model may fail to reduce the wave energy sufficiently during breaking, or fail to break the waves at all.

This problem leads many people to abandon the spectral approach and return to the use of monochromatic waves and a forward tracking model. Given the severe problems involved in the interpretation of forward tracked wave rays, the authors see this approach as a step backwards and are reluctant to adopt it. A better approach would be to transform the frequency/direction spectrum into a height/period spectrum, perhaps using a development of the Rayleigh wave height distribution. This would permit a much better representation of the component waves to be made. Meanwhile it is necessary to apply a certain amount of judgement when using the spectral model to ensure that the model does not predict excessive amounts of energy passing over submerged banks.

Other shortcomings of the breaking model include the inability of the model to redistribute the energy of the breaking waves to other frequencies or directions in the spectrum, and the restricted description of the breaking process. The first of these phenomena could be incorporated relatively easily, but there is no consensus on where the energy goes. Some of the energy certainly moves to lower frequencies, and a small amount moves to high frequencies, but the authors are not aware of any reliable quantitative models of the processes involved.

The description of the breaking process has been shown by Gerritsen to be potentially very accurate in the conditions for which it was derived. These conditions were plunging breakers on a Hawaian reef. The energy dissipation factor ζ in equation 7.9 has been used to adjust the energy loss to account for other forms of breaking, e.g. spilling, but there is little quantitative data on which to base this value.

There is considerable scope for research into the mechanics of wave breaking. Mathematical models of plunging breakers have been developed which will predict the form of the breaking wave up to the point at

Application of Ray Methods

which the falling nappe touches the water in front of the wave, but this is not sufficient to predict energy losses. Other models are based on the energy loss in a travelling bore, but cannot show how or when the bore develops from the breaking wave. The authors favour a field based research program and it is hoped that research projects being planned at present by several universities will provide useful results in the next few years.

7.5 A Study Including Diffraction and Reflection

To date, the most ambitious development of the basic ray model is that used to predict wave conditions around a proposed caisson retained island to be constructed in the Beaufort sea. The proximity of a vertical walled structure means that the phenomena of diffraction and reflection assume some importance. The details of the modifications made to the basic model are very complex, but the principle was relatively simple.

Reflection was included by causing wave rays striking the structure to reflect with incident angle equal to reflected angle. As the model is a spectral one, the phase change on reflection was discounted. This represents the worst case, as the initial phase of each way ray is unknown. Reflected rays were given an additional offshore to inshore energy coefficient to account for the loss of energy by diffraction around the edges of the structure. This coefficient was calculated using equations for diffraction of waves by a wedge developed by Lick (1978). The same equations were used to calculate diffraction coefficients for rays diffracting to the point of interest from the far side of the island.

Figure 7.13 shows the ray pattern for one point with reflecting and diffracting rays. The radial pattern used for the diffracted rays is not a good approximation, but the contribution of these rays to the spectrum at the point of interest is small, so the error is acceptable. The ray model was used to provide wave data for a full sediment transport model of the area around the island. Predicted scour patterns agreed well with those observed in a physical model when similar conditions were input to both models (Fleming, 1983). This gave confidence in the accuracy of all parts of the computer model, particularly the wave model which is little affected by scaling errors.

I. M. Townend and I. A. Savell

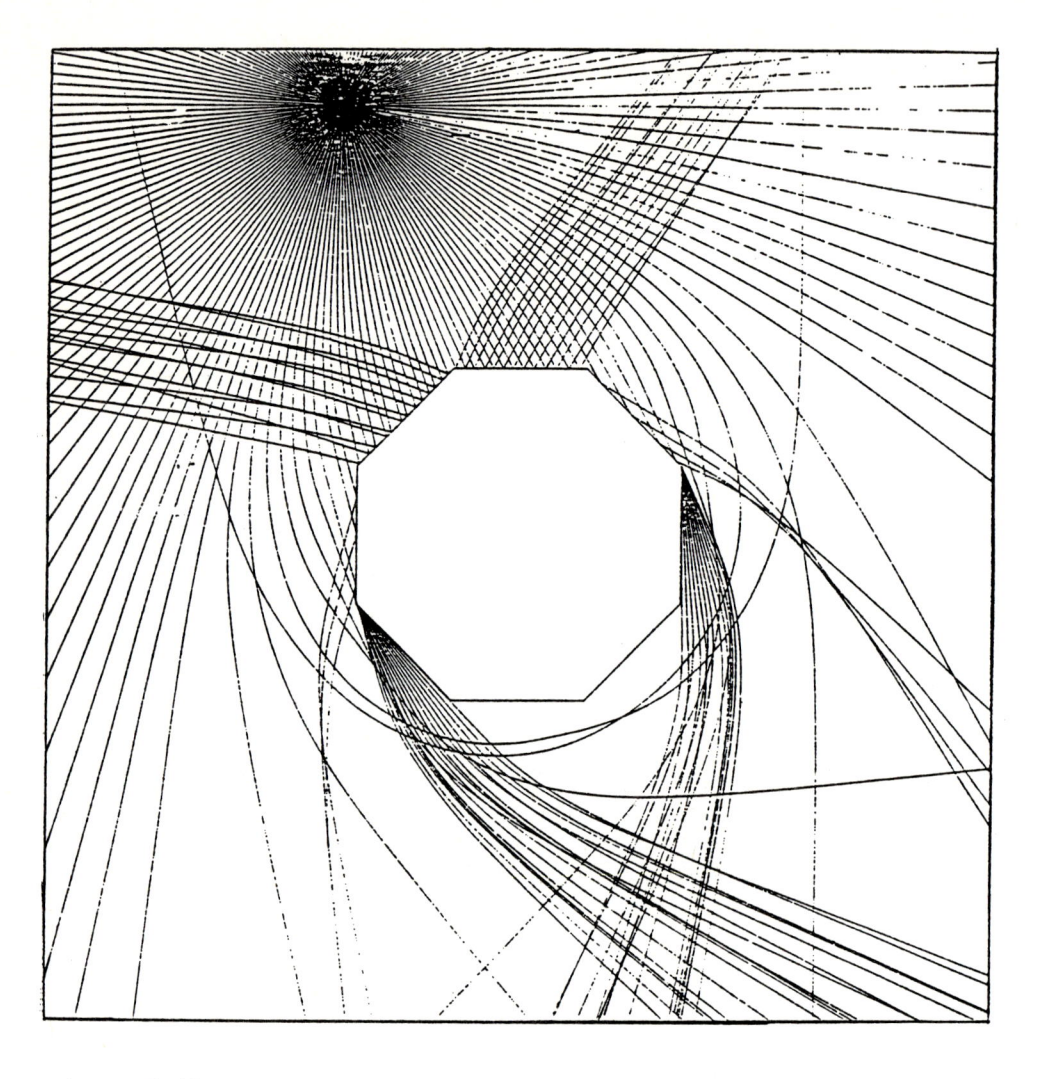

Figure 7.13 Ray pattern for one point with reflecting and diffracting rays

Application of Ray Methods

7.6 Future Developments

The range of applicability of ray models continues to grow. A non-linear version of the model has been developed in association with D. H. Swart, using his Vocoidal wave theory (Crowley et al, 1982). Results to date suggest that this model may be an economic proposition with existing computers, and it may eventually displace the linear version in many studies. For the moment, however, the wave height dependency of the model limits it to forward tracking studies for most practical applications. Routine use of the model must wait until the problems with the interpretation of these results are overcome.

Ray methods have been successfully applied to the problem of current/depth refraction by several workers. Christoffersen et al (1981) describes a very comprehensive analysis which the authors expect to include in the present ray model in the near future. This will open up a wide range of new possibilities.

Southgate (1981) has developed a very good solution to the problem of incorporating diffraction into a ray model. He uses the Sommerfeld flat bed solution to find a ray pattern for the diffracted and diffracted reflected components of the wave field near the breakwater tip. These rays are then tracked forwards, together with the incident and reflected rays. A sophisticated averaging procedure, which includes the initial diffraction and reflection coefficients as well as the refraction coefficients, is then used to derive wave height figures over the areas of interest. This averaging procedure also takes account of the phase differences between rays and between wave trains.

These developments all help to ensure that ray methods will maintain their position as a relatively cheap and accurate part of the coastal engineers tool kit for many more years.

I. M. Townend and I. A. Savell

References

Abernathy, C. L. & Gilbert, G., 1975
Refraction of wave spectra. Hydraulics Research Station, Int.Rep. INT 117.

Carter, D. J. T., 1982
Estimation of wave spectra from wave height and period. MIAS Rep.No. 4, I.O.S. Warmley.

Christoffersen, J. B., Skovgaard, D. & Jonsson, I. G., 1981
A numerical model for current depth refraction of dissipative waves. Proc.Int.Symp. on Hydrodyn. in Ocean Eng., Trondheim, VI.

Crowley, J. B., Fleming, C. A. & Cooper, C. N., 1982
A computer model for the refraction of non-linear waves. Proc 18th Conf. On Coastal Engr.

Fleming, C. A., 1983
A numerical model for scour round a caisson island. ACROSES Arctic Regional Workshop, Calgary, Canada.

Gerritsen, F., 1979
Energy dissipation in breaking waves. Proc 5th Conf. P.O.A.C.

Jonsson, I. G. & Carlsen, N. A., 1976
Experimental and theoretical investigation in an oscillating turbulent boundary layer. J.Hydr.Res $\underline{14}$.

Lick, W., 1978
Diffraction of waves by a wedge. Proc. ASCE WW2.

Skovgaard, O., Jonsson, I. G. & Bertelsen, J. A., 1975
Computations of wave heights due to refraction and friction. Proc. ASCE WW1 $\underline{101}$.

Southgate, H. N., 1981
Ray methods for combined refraction and diffraction problems. H.R.S. Report IT214.

8. A MODEL FOR SURFACE WAVE GROWTH

A.J. Croft,
Department of Mathematics,
Coventry (Lanchester) Polytechnic.

8.1 Introduction

Most theories, for example, Kelvin-Helmholtz (1968), Miles (1957) and Phillips (1957) are concerned with the instability which gives rise to the formation of surface water waves due to the action of the wind. These theories fail to predict by a substantial amount the observed growth in the surface wave elevation. The model to be considered goes to the other extreme in that we consider some effects of a high speed wind on an already existing train of surface waves.

The central feature of the model is that it is assumed that the wind has generated the waves and is supplying energy so that they will grow, and furthermore, as the mean wind speed increases particles of fluid will be removed from the wave surface to generate a spray. As the mean wind speed increases further this process becomes more accentuated so that there will be a dense spray full of water droplets above, and in contact with, the surface of the water. In addition, the wind shear stress will act at the free surface of the layer and the object of the investigation is to determine whether the two taken in conjunction will have any significant effect as far as wavegrowth is concerned.

Experimental work undertaken by Monahan, Spiel and Davidson (1982) indicates that for wind speeds greater than $9ms^{-1}$ a mechanism distinct from those associated with bursting whitecap bubbles is a significant contributor to large spray droplets. Thus, for a high speed wind ($\gg 9ms^{-1}$) this mechanism will be even more pronounced, therefore a layer of such droplets incorporated in the model adds realism to it. Since we are considering a layer of dense spray its mean density is only slightly lower than that of the water below. In order to make progress the simplest type of layer is considered namely a homogenous one whose density is constant and only slightly less than that of water.

Turning next to the wind shear stress distribution, there appears to be disagreement as to the importance of this. Stewart (1967) argues that the shear stress distribution yields a wave generation mechanism. Also there will be little stress in the troughs but there will be a signifi-

A. J. Croft

cant amount at the crest, the distribution being in phase with the waves. This broadly agrees with Longuet-Higgins (1977). Barnett and Kenyon (1975) point out that the theory of Phillips intuitively suggests that the shear stress could not be very effective in setting up the irrotational water motion of gravity waves. In the model being considered here the flow in the upper layer will be rotational when viscosity is taken into account. On balance, the evidence would suggest that the wind shear stress could be important. The variable wind shear stress distribution specified at the free surface in phase with it is compatible with Jeffrey's sheltering hypothesis (1924). In fact it would appear intuitively that this effect will be much more pronounced in the high speed wind situation than as a mechanism for generating waves as originally conceived.

8.2 Formulation of the Problem

The model being considered is that of two immiscible liquids, the heavier one being bounded below by a fixed horizontal plane and the lighter one above by the atmosphere. In both liquids the velocity field is induced by a one-dimensional train of progressive sinusoidal waves for which the ratio of the amplitude to the wavelength is small.

A Stokes expansion up to the second order is undertaken assuming the fluids to be inviscid and the flow irrotational. For the third order approximation all the non-ideal behaviour is confined to the layer, whereas the ideal nature of the lower liquid is unaffected.

Up to the second order we have to solve Laplace's equation for the two liquids subject to the appropriate boundary conditions. Thus, we seek solutions of

$$\nabla^2 \phi_1 = 0 \quad \text{and} \quad \nabla^2 \phi = 0,$$

where

$$\nabla^2 \equiv \frac{\partial^2}{\partial x^2} + \frac{\partial^2}{\partial z^2}$$

the undistributed interface being the xy plane and z is measured vertically upwards. The suffix 1 refers to the upper liquid, ϕ and ϕ_1 being the respective velocity potentials. The elevation of the free surface from its undisturbed position is denoted by ζ_1 while ζ is the elevation of the interface relative to its undisturbed position.

A Model for Surface Wave Growth

The kinematic condition at the free surface is

$$\frac{\partial \zeta_1}{\partial t} + \frac{\partial \phi_1}{\partial z} - \frac{\partial \phi_1}{\partial x}\frac{\partial \zeta_1}{\partial x} = 0 \quad \text{on} \quad z = h_1 + \zeta_1 \quad (8.2.1)$$

where h_1 is the undisturbed thickness of the upper liquid.

The dynamic condition at the free surface is

$$\tfrac{1}{2}q_1^2 + gz - \frac{\partial \phi_1}{\partial t} = 0 \quad \text{on} \quad z = h_1 + \zeta_1 \quad (8.2.2)$$

where $q_1 = -\nabla \phi_1$. The pressure at the free surface is equated to zero because only the shear stress due to the wind will be eventually considered. For the third order solution (8.2.2) is replaced by the continuity of the shear and normal stresses.

A Taylor expansion about $z = h_1$ applied to (8.2.1) gives

$$\frac{\partial \zeta_1}{\partial t} + \left(\frac{\partial \phi_1}{\partial z} - \frac{\partial \phi_1}{\partial x}\frac{\partial \zeta_1}{\partial x}\right)_{z=h_1} + \left[\frac{\partial}{\partial z}\left(\frac{\partial \phi_1}{\partial z} - \frac{\partial \phi_1}{\partial x}\frac{\partial \zeta_1}{\partial x}\right)\right]_{z=h_1} + \ldots = 0 \quad (8.2.3)$$

and (8.2.2) becomes

$$g\zeta_1 + \left(\tfrac{1}{2}q_1^2 - \frac{\partial \phi_1}{\partial t}\right)_{z=h_1} + \zeta_1\left[\frac{\partial}{\partial z}\left(\tfrac{1}{2}q_1^2 - \frac{\partial \phi_1}{\partial t}\right)\right]_{z=h_1} + \ldots = 0 \quad (8.2.4)$$

absorbing gh_1 into ϕ_1.

For the lighter liquid we write

$$\begin{aligned} q_1 &= \varepsilon q_1^{(1)} + \varepsilon^2 q_1^{(2)} + \ldots \\ \phi_1 &= \varepsilon \phi_1^{(1)} + \varepsilon^2 \phi_1^{(2)} + \ldots \\ \zeta_1 &= \varepsilon \zeta_1^{(1)} + \varepsilon^2 \zeta_1^{(2)} + \ldots \\ p_1 &= \varepsilon p_1^{(1)} + \varepsilon^2 p_1^{(2)} + \ldots \end{aligned} \quad (8.2.5)$$

where p_1 is the pressure at any point in the upper liquid and ε is a small parameter of order of the wave steepness.

Similarly for the heavier liquid,

A. J. Croft

$$q = \varepsilon q^{(1)} + \varepsilon^2 q^{(2)} + \ldots$$
$$\phi = \varepsilon \phi^{(1)} + \varepsilon^2 \phi^{(2)} + \ldots$$
$$\zeta = \varepsilon \zeta^{(1)} + \varepsilon^2 \zeta^{(2)} + \ldots \qquad (8.2.6)$$
$$p = \varepsilon p^{(1)} + \varepsilon^2 p^{(2)} + \ldots$$

For the interface progressive wave train

$$\zeta^{(1)} = a \sin m(x-ct) \qquad (8.2.7)$$

Substituting (8.2.5) into (8.2.3) the free surface kinematic condition to the first order is

$$\frac{\partial \zeta_1^{(1)}}{\partial t} + \left(\frac{\partial \phi_1^{(1)}}{\partial z}\right)_{z=h_1} = 0 \qquad (8.2.8)$$

and to the second order is

$$\frac{\partial \zeta_1^{(2)}}{\partial t} + \left[\frac{\partial \phi_1^{(2)}}{\partial z} - \frac{\partial \phi_1^{(1)}}{\partial x}\frac{\partial \zeta_1^{(1)}}{\partial x} + \zeta_1^{(1)}\frac{\partial^2 \phi_1^{(1)}}{\partial z^2}\right]_{z=h_1} = 0 \qquad (8.2.9)$$

Substituting (8.2.5) into (8.2.4) the free surface dynamic condition to the first order

$$g\zeta_1^{(1)} - \left(\frac{\partial \phi_1^{(1)}}{\partial t}\right)_{z=h_1} = 0 \qquad (8.2.10)$$

and to the second order is

$$g\zeta_1^{(2)} + \left[\tfrac{1}{2}q_1^{(1)} - \frac{\partial \phi_1^{(2)}}{\partial t} - \zeta_1^{(1)}\frac{\partial^2 \phi_1^{(1)}}{\partial z \partial t}\right]_{z=h_1} = 0 \qquad (8.2.11)$$

The continuity of pressure at the interface gives

$$\rho\left[\frac{\partial \phi}{\partial t} - \tfrac{1}{2}q^2\right]_{z=\zeta} - g\rho\zeta = \rho_1\left[\frac{\partial \phi_1}{\partial t} - \tfrac{1}{2}q_1^2\right]_{z=\zeta} - g\rho_1\zeta \qquad (8.2.12)$$

where ρ and ρ_1 are the densities of the lower and upper liquids respectively.

A Model for Surface Wave Growth

A Taylor expansion about z = 0 gives

$$\rho[\frac{\partial \phi}{\partial t} - \tfrac{1}{2}q^2]_{z=0} + \rho\zeta[\frac{\partial}{\partial z}(\frac{\partial \phi}{\partial t} - \tfrac{1}{2}q^2)]_{z=0} + \ldots$$

$$= g\zeta(\rho-\rho_1) + \rho_1[\frac{\partial \phi_1}{\partial t} - \tfrac{1}{2}q_1^2]_{z=0} + \rho_1{}^3[\frac{\partial}{\partial z}(\frac{\partial \phi_1}{\partial t} - \tfrac{1}{2}q_1^2)]_{z=0} \qquad (8.2.13)$$

An alternative approach used by Stokes for small amplitude surface waves, (Witham, 1974) is adapted to the present problem and essentially means expanding ζ, ϕ, ζ_1 and ϕ_1 in a series of cosines using the amplitude 'a' of the interface wave train as the scaling parameter. By fitting these expansions to the boundary conditions it can be shown that for long waves the second and third harmonics for the interface waves can be neglected. The same analysis also shows that the speed of propagation c for the first order solution is sufficiently accurate up to the third order solution. For the model being considered, the third order solution takes into account the viscosity of the upper liquid, therefore, the velocity potential does not exist. Nevertheless, this work does furnish some evidence for neglecting the third harmonic and the third order correction term to the speed of propagation. Thus, using (8.2.7) and (8.2.6)

$$\zeta = \varepsilon\zeta^{(1)} = \varepsilon a \sin m(x-ct) \qquad (8.2.14)$$

Therefore, the continuity of the pressure at the interface to the first order is

$$[\rho \frac{\partial \phi^{(1)}}{\partial t} - \rho_1 \frac{\partial \phi_1^{(1)}}{\partial t}]_{z=0} = g(\rho-\rho_1) a \sin m(x-ct) \qquad (8.2.15)$$

and to the second order is

$$[\rho\{\frac{\partial \phi^{(2)}}{\partial t} - \tfrac{1}{2}q^{(1)^2}\} - \rho_1\{\frac{\partial \phi_1^{(2)}}{\partial t} - \tfrac{1}{2}q_1^{(1)^2}\}]_{z=0}$$

$$= a[\rho_1 \frac{\partial^2 \phi_1^{(1)}}{\partial z \partial t} - \rho \frac{\partial^2 \phi^{(1)}}{\partial z \partial t}]_{z=0} \sin m(x-ct) \qquad (8.2.16)$$

A. J. Croft

At the interface the kinematic condition is

$$\frac{\partial \phi_1}{\partial z} - \frac{\partial \phi_1}{\partial x}\frac{\partial \zeta}{\partial x} = \frac{\partial \phi}{\partial z} - \frac{\partial \phi}{\partial x}\frac{\partial \zeta}{\partial x} \quad \text{at} \quad z = \zeta$$

Carrying out a Taylor expansion about $z = 0$ and substituting (8.2.5), (8.2.6) and (8.2.14) the interface kinematic condition to the first order is

$$\left(\frac{\partial \phi_1^{(1)}}{\partial z}\right)_{z=0} = \left(\frac{\partial \phi^{(1)}}{\partial z}\right)_{z=0} = amc \; cosm \; (x-ct) \tag{8.2.17}$$

and to the second order is

$$\left[\frac{\partial \phi_1^{(2)}}{\partial z} - \frac{\partial \phi_1^{(1)}}{\partial x}\frac{\partial \zeta^{(1)}}{\partial x} - \frac{\partial \phi^{(2)}}{\partial z} + \frac{\partial \phi^{(1)}}{\partial x}\frac{\partial \zeta^{(1)}}{\partial x}\right]_{z=0}$$
$$= \left[\frac{\partial^2 \phi^{(1)}}{\partial z^2} - \frac{\partial^2 \phi_1^{(1)}}{\partial z^2}\right]_{z=0} a \; sinm(x-ct) \tag{8.2.18}$$

Lastly, we have the boundary condition

$$\left(\frac{\partial \phi}{\partial z}\right)_{z=-h} = 0 \tag{8.2.19}$$

8.3 First Order Solution

Substituting (8.2.5) and (8.2.6) into $\nabla^2 \phi = 0$ and $\nabla^2 \phi_1 = 0$ as well as (8.2.19), we have to solve

$$\nabla^2 \phi^{(1)} = 0 \quad \text{and} \quad \nabla^2 \phi_1^{(1)} = 0$$

subject to the condition

$$\left(\frac{\partial \phi^{(1)}}{\partial z}\right)_{z=-h} = 0$$

and the interface conditions (8.2.15) and (8.2.17).

The required solutions are

$$\phi^{(1)} = \frac{ac}{sinhmh} \; cosm(x-ct) \; coshm(z+h) \tag{8.3.1}$$

and

A Model for Surface Wave Growth

$$\phi_1^{(1)} = a\cos m(x-ct)\,[c\sinh mz + \{c\rho/\rho_1 \coth mh + \frac{g}{mc}(1-\rho/\rho_1)\}\cosh mx]$$

(8.3.2)

The boundary conditions at the free surface still have to be satisfied. Eliminating $\zeta_1^{(1)}$ between (8.2.8) and (8.2.10) we have

$$\left(\frac{\partial^2 \phi_1^{(1)}}{\partial t} + g\frac{\partial \phi_1^{(1)}}{\partial z}\right)_{z=h_1} = 0$$

Substituting (8.3.2) into this expression we have a quartic equation for c. Since the upper liquid is modelling a layer of air above surface water waves which is full of heavy drops of spray, it would appear a reasonable approximation to take the difference between the two densities to be small. Thus,

$$c^2 \approx \frac{g}{m}\tanh m(h_1+h_1)$$

or

$$c^2 \approx \frac{g(\rho-\rho_1)}{m(\rho_1\coth mh_1 + \rho\coth mh)}$$

The former is by far the larger and will be more in keeping with a high-speed wind model. The depth of the undisturbed lower liquid h will be taken to be sufficiently large so that

$$c^2 \approx g/m$$

From (8.2.10)

$$\zeta_1^{(1)} = \frac{mc^2 a}{g}[\sinh mh + \{\frac{\rho}{\rho_1}\coth mh + \frac{g}{mc}(1-\frac{\rho}{\rho_1})\}\cosh mh_1]\sin m(x-ct)$$

Using the above approximations

$$\zeta_1^{(1)} \approx a(1+mh_1)\sin m(x-ct)$$

The linear model is clearly inadequate as far as wave growth is concerned and it is necessary to obtain higher order solutions.

A. J. Croft

8.4 Second Order Solution

In this case the governing differential equations are

$$\nabla^2 \phi^{(2)} = 0 \quad \text{and} \quad \nabla^2 \phi_1^{(2)} = 0$$

One boundary condition is

$$\frac{\partial \phi^{(2)}}{\partial z} = 0 \quad \text{when} \quad z = -h$$

The continuity of pressure at the interface contains the terms $[q^{(1)}{}^2]_{z=0}$ and $[q_1^{(1)}{}^2]_{z=0}$ when taken to the second order. These are respectively

$$\frac{m^2 a^2 c^2}{2 \sinh^2 mh} [\cosh 2mh - \cos 2m(x-ct)]$$

and

$$\tfrac{1}{2} m^2 a^2 c^2 \{1 + [\frac{\rho}{\rho_1} \coth mh + \frac{g}{mc^2}(1 - \frac{\rho}{\rho_1})^2]$$

$$+ \tfrac{1}{2} m^2 a^2 c^2 \{1 - [\frac{\rho}{\rho_1} \coth mh + \frac{g}{mc^2}(1 - \frac{\rho}{\rho_1})]^2\} \cos 2m(x-ct)$$

In view of the above observations appropriate solutions of the governing differential equations are of the form

$$\phi^{(2)} = E \sin 2m(x-ct) \cosh 2m(z+h) + kt$$

and

$$\phi_1^{(2)} = \sin 2m(x-ct)(\alpha_1 e^{2mz} + \beta_1 e^{-2mx}) + k_1 t$$

Substituting these into (8.2.16), the second order term associated with the continuity of pressure at the interface, and using the first order solutions, we have

$$K = \tfrac{1}{4} m^2 a^2 c^2 \operatorname{cosech}^2 mh,$$

$$K_1 = \tfrac{1}{4} m^2 a^2 c^2 \{[\frac{\rho}{\rho_1} \coth mh + \frac{g}{mc^2}(1 - \rho/\rho_1)]^2 - 1\}$$

and

A Model for Surface Wave Growth

$$\rho[\frac{ma^2c}{2}(\tfrac{1}{2}\text{cosech}^2 mh - 1) - 2E \cosh 2mh]$$

$$= -\rho_1[2(\alpha_1+\beta_1) + \frac{ma^2c}{4}\{3 - [\frac{\rho}{\rho_1}\coth mh + \frac{g}{mc^2}(1 - \rho/\rho_1)]^2\}]$$

Similarly, from the second order kinematic condition at the interface,

$$\alpha_1 - \beta_1 - E \sinh 2mh = \tfrac{1}{2}ma^2c\,(1 - \frac{\rho}{\rho_1})(\coth mh - \frac{g}{mc^2})$$

Thus, α_1 and β_1 can be obtained in terms of E, which is the only remaining unknown. Eliminating $\zeta_1^{(2)}$ between the kinematic and dynamic conditions at the free surface to the second order, namely (8.2.9) and (8.2.11), we have a single equation which enables E to be calculated. This is extremely complicated but simplifications can be effected. The calculation which shows that the second and third harmonics of the interface wave can be neglected also shows that the speed of propagation $c^2 = g/m$ is acceptable up to the third order. Also, m and h, are small compared with the wavelength, therefore we obtain

$$\zeta_1^{(2)} = -\frac{ma^2\rho}{2\rho_1(2\rho/\rho_1 - 1)}\cos 2m(x-ct)$$

8.5 Third Order Solution

For the third order solution the lower fluid is still assumed to be ideal, but now the upper liquid is considered to be viscous. This effect has been omitted until the third order solution because Banner and Phillips (1974) argue that the wind drift layer is established by viscous action over a time scale which is much greater than the wave period, therefore the effect of viscosity on the already vertical fluid can be neglected. This implies that the effect of the viscosity is small. On the other hand it is necessary for the model being considered as it allows the wind shear stress to be built into it.

For the lower liquid we seek the solution of

$$\nabla^2 \phi^{(3)} = 0$$

subject to

$$\frac{\partial \phi^{(3)}}{\partial z} = 0 \quad \text{when} \quad z = -h_2$$

A. J. Croft

and the boundary conditions at the interface. A solution of the form

$$\phi^{(3)} = \{E_2 \cos m_2(x-c_2t) + F_2 \sin m_2(x-c_2t)\} \cosh m_2(z+h)$$
$$+ \{E_3 \cos m_3(x-c_3t) + F_3 \sin m_3(x-c_3t)\} \cosh m_3(z+h)$$

For the upper liquid it is convenient to use an orthogonal coordinate system (ζ, η) which is such that $\eta = 0$ is the interface wave, see figure 8.1.

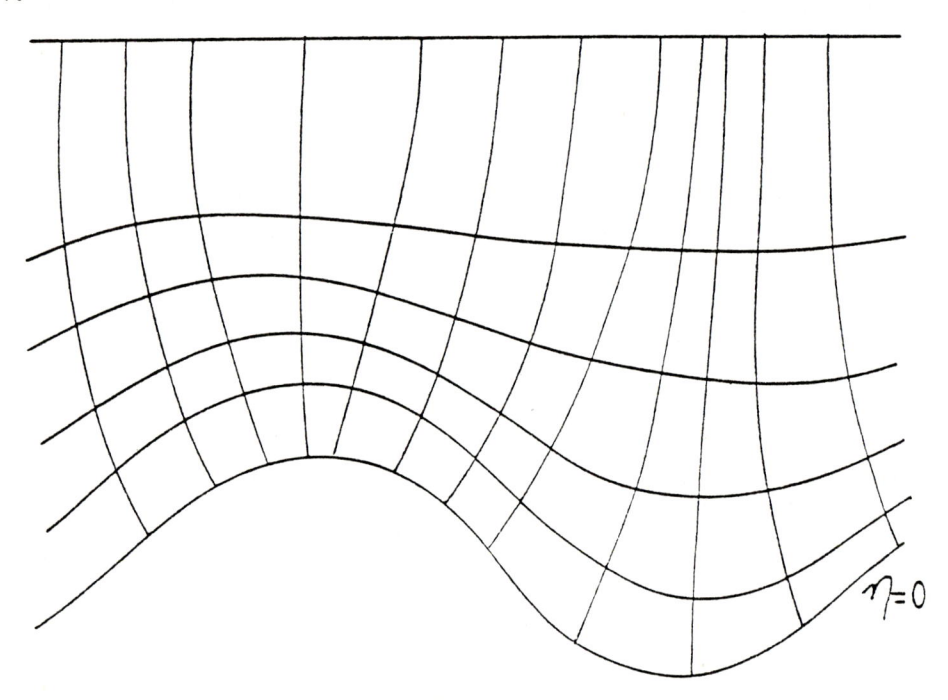

Figure 8.1   Orthogonal Co-ordinate System for Third Order Solution for the Upper Liquid (not to scale)

At the free surface the various expressions will be evaluated at $\eta = h_1$ rather than $z = h_1$, therefore being more accurate. The boundary conditions for the upper liquid can be fitted exactly at the interface. Also, the arguments for neglecting the non-linear terms in the Navier-Stokes equations are more acceptable in the new coordinate system.

Since the curvature, k, is $O(am^2)$ and $\partial k/\partial x$ is $O(am^3)$ the exact Navier-Stokes equations for flow along a curved boundary reduce to

A Model for Surface Wave Growth

$$\frac{\partial u}{\partial t} + u \frac{\partial u}{\partial \zeta} + v \frac{\partial u}{\partial \eta} = - \frac{1}{\rho_1} \frac{\partial p_1}{\partial \zeta} + \nu \nabla^2 u$$

and

$$\frac{\partial v}{\partial t} + u \frac{\partial v}{\partial \zeta} + v \frac{\partial v}{\partial \eta} = - \frac{1}{\rho_1} \frac{\partial p_1}{\partial \eta} + \nu \nabla^2 v,$$

where

$$\nabla^2 \equiv \frac{\partial^2}{\partial \zeta^2} + \frac{\partial^2}{\partial \eta^2}$$

Also the equation of continuity reduces to

$$\frac{\partial u}{\partial \zeta} + \frac{\partial v}{\partial \eta} = 0$$

where u and v are the velocity components parallel and perpendicular to the boundary respectively.

Since we are considering a very thin layer (non-dimensional) and interface waves having a wavelength much greater than the amplitude, v and its variation across the layer can be expected to be small. Therefore, from the equation of continuity $\partial u/\partial \zeta$ is small. Since we are neglecting the viscosity of the fluid beneath the layer, and also the magnitude of wind shear stress at the free surface is small, we do not have the normal boundary layer situation in which u is zero at the boundary (assumed at rest) and then rises to the free stream value over a very short distance so that $\partial u/\partial \eta$ is very large with the attendant large shear stress at the boundary. In the model being considered we would not expect u to vary by a great amount across the layer, whereupon $\partial u/\partial \eta$ can be taken to be small. Since the shear stress at the free surface is small, and the layer is thin, we would expect $\partial v/\partial \zeta$ to be small. For the first and second order solutions the two components of velocity are of the same order and there is no reason to believe that for the third order solution u will be significantly larger than v. It would, therefore, appear to be a realistic approximation to neglect the non-linear terms on the left-hand side of the Navier-Stokes equations.

From the equations relating (ζ,η) to (x,y) it is seen that for long interface waves $\zeta \approx x$.

Thus, the third order correction for the upper liquid requires the solution of

A. J. Croft

$$\frac{\partial u}{\partial t} = -\frac{1}{\rho_1}\frac{\partial p_1}{\partial x} + \nu\left(\frac{\partial^2 u}{\partial x^2} + \frac{\partial^2 u}{\partial \eta^2}\right),$$

$$\frac{\partial v}{\partial t} = -\frac{1}{\rho_1}\frac{\partial p_1}{\partial \eta} + \nu\left(\frac{\partial^2 v}{\partial x^2} + \frac{\partial^2 v}{\partial \eta^2}\right)$$

and

$$\frac{\partial u}{\partial x} + \frac{\partial v}{\partial \eta} = 0$$

Let

$$u = -\frac{\partial \phi}{\partial x} - \frac{\partial \psi}{\partial \eta} \quad \text{and} \quad v = -\frac{\partial \phi}{\partial \eta} + \frac{\partial \psi}{\partial x},$$

where ϕ and ψ are functions to be found.

From the equation of continuity

$$\frac{\partial^2 \phi}{\partial x^2} + \frac{\partial^2 \phi}{\partial \eta^2} = 0$$

and from the Navier-Stokes equations

$$\frac{\partial \psi}{\partial t} = \nu\left(\frac{\partial^2 \psi}{\partial x^2} + \frac{\partial^2 \psi}{\partial \eta^2}\right)$$

Also, it can be shown that the pressure in the viscous layer is

$$\frac{p_1}{\rho_1} = \frac{\partial \phi}{\partial t} - g\eta$$

An appropriate solution of $\nabla^2 \phi = 0$ is

$$\begin{aligned}\phi = &\ (A \cosh k\eta + B \sinh k\eta) \cos k(x-ct) \\ &+ (C \cosh k\eta + D \sinh k\eta) \sin k(x-ct) \\ &+ (E \cosh k_1\eta + F \sinh k_1\eta) \cos k_1(x-ct) \\ &+ (G \cosh k_1\eta + H \sinh k_1\eta) \sin k_1(x-ct)\end{aligned} \quad (8.5.1)$$

Also, a solution of

$$\frac{\partial \psi}{\partial t} = \nu \nabla^2 \psi$$

is

A Model for Surface Wave Growth

$$\psi = [(A_1 e^{a'\eta} + B_1 e^{-a'\eta})\cos b'\eta + (B_2 e^{-a'\eta} - A_2 e^{a'\eta})\sin b'\eta]\cos w$$
$$- [(A_2 e^{a'\eta} + B_2 e^{-a'\eta})\cos b'\eta + (A_1 e^{a'\eta} - B_1 e^{-a'\eta})\sin b'\eta]\sin w$$
$$+ [(A_3 e^{a_1'\eta} + B_3 e^{-a_1'\eta})\cos b_1'\eta + (B_4 e^{-a_1'\eta} - A_4 e^{a_1'\eta})\sin b_1'\eta]\cos w_1$$
$$- [(A_4 e^{a_1'\eta} + B_4 e^{-a_1'\eta})\cos b_1'\eta + (A_3 e^{a_1'\eta} - B_3 e^{-a_1'\eta})\sin b_1'\eta]\sin w_1$$

where

$w = k(x-ct)$, $w_1 = k_1(x-ct)$, $a' = R\cos\theta/2$, $b' = R\sin\theta/2$, $R^2 \simeq kc/\nu$ and $\theta = -\tan^{-1}(c/k\nu)$.

Also, a_1' and b_1' are as above with k replaced by k_1.

Extending the expansion in (8.2.3) and substituting (8.2.5), the kinematic condition at the free surface to third order is

$$\frac{\partial \zeta_1^{(3)}}{\partial t} - [v]_{\eta=h_1} - [\frac{\partial \phi_1^{(1)}}{\partial x}\frac{\partial \zeta_1^{(2)}}{\partial x} + \frac{\partial \phi_1^{(2)}}{\partial x}\frac{\partial \zeta_1^{(1)}}{\partial x}]_{z=h_1}$$

$$+ \zeta_1^{(1)}[\frac{\partial^2 \phi_1^{(2)}}{\partial z^2} - \frac{\partial}{\partial z}(\frac{\partial \phi_1^{(1)}}{\partial x}\frac{\partial \zeta_1^{(1)}}{\partial x})]_{z=h_1} \qquad (8.5.2)$$

$$+ \zeta_1^{(2)}[\frac{\partial^2 \phi_1^{(1)}}{\partial z^2}]_{z=h_1} + \tfrac{1}{2}\zeta_1^{(1)^2}[\frac{\partial^3 \phi_1^{(1)}}{\partial z^3}]_{z=h_1} = 0$$

Using the lower order solutions and the previous approximations the above becomes:-

$$\frac{\partial \zeta_1^{(3)}}{\partial t} - [v]_{\eta=h}$$

$$+ \frac{m^2 g a^3 \rho}{c\rho_1}[m\, h_1 + \frac{1}{8(2\rho/\rho_1-1)}\{2 - 3\frac{\rho}{\rho_1} + 6(\frac{\rho}{\rho_1})^2\}]\cos w$$

$$- \frac{3m^2 g a^3 \rho}{8c(2\rho/\rho_1-1)\rho_1}[2 - \frac{\rho}{\rho_1} + 2(\frac{\rho}{\rho_1})^2]\cos 3w = 0$$

where $w = m(x-ct)$.

The pressure due to the air above the upper liquid is being neglected therefore,

(normal stress)$_{\text{free surface}} = 0$

A. J. Croft

Correct to the third order this can be shown to give

$$g\zeta_1^{(3)} = \left(\frac{\partial \phi}{\partial t} - 2\nu \frac{\partial v}{\partial \eta}\right)_{\eta = h_1} \tag{8.5.3}$$

Substituting this into the above third order free surface kinematic condition we have

$$\frac{1}{g}\left[\frac{\partial^2 \phi}{\partial t^2} + 2\nu \frac{\partial^3 \phi}{\partial t \partial \eta^2} - 2\nu \frac{\partial^3 \psi}{\partial \eta \partial x \partial t} + g\left(\frac{\partial \phi}{\partial \eta} - \frac{\partial \psi}{\partial x}\right)\right]_{\eta = h_1}$$

$$+ \frac{m^2 g a^3 \rho}{c \rho_1}\left[mh_1 + \frac{1}{8(2\rho/\rho_1 - 1)}\left\{2 - 3\frac{\rho}{\rho_1} + 6\left(\frac{\rho}{\rho_1}\right)^2\right\}\right]\cos m(x - ct)$$

$$- \frac{3m^2 g a^3 \rho}{8c(2\rho/\rho_1 - 1)\rho_1}\left\{2 - \frac{\rho}{\rho_1} + 2\left(\frac{\rho}{\rho_1}\right)^2\right\}\cos 3m(x - ct) = 0$$

Substituting the previous expressions for ϕ and ψ we have that $k = m$ and $k_1 = 3m$.

For the continuity of the shear stress at the free surface it can be shown that

$$\left(2\frac{\partial^2 \psi}{\partial x^2} - 2\frac{\partial^2 \phi}{\partial x \partial \eta} - \frac{1}{\nu}\frac{\partial \psi}{\partial t}\right)_{\eta = h_1} = \tau, \tag{8.5.4}$$

where τ is the shear stress due to the wind. At the interface the continuity of shear stress gives

$$\left(2\frac{\partial^2 \psi}{\partial x^2} - 2\frac{\partial^2 \phi}{\partial x \partial \eta} - \frac{1}{\nu}\frac{\partial \psi}{\partial t}\right)_{\eta = 0} = 0$$

It can be proved that the continuity of the normal stress at the interface gives

$$\rho\left[\frac{\partial \phi^{(3)}}{\partial t} - q^{(1)} \cdot q^{(2)}\right]_{z = s} = \rho_1\left(\frac{\partial \phi}{\partial t}\right)_{\eta = 0} - 2\mu\left(\frac{\partial v}{\partial \eta}\right)_{\eta = 0}$$

Finally, the kinematic condition at the interface gives

$$-[v]_{\eta = 0} - \frac{\partial \phi^{(3)}}{\partial z} - \left(\frac{\partial \phi_1^{(2)}}{\partial x} - \frac{\partial \phi^{(2)}}{\partial x}\right) ma\cos m(x - ct)$$

A Model for Surface Wave Growth

$$+ \left(\frac{\partial^2 \phi_1^{(2)}}{\partial z^2} - \frac{\partial^2 \phi^{(2)}}{\partial z^2}\right) a\, \sin m(x-ct)$$

$$- \left(\frac{\partial^2 \phi_1^{(1)}}{\partial z \partial x} - \frac{\partial^2 \phi^{(1)}}{\partial z \partial x}\right) a^2 m\, \sin m(x-ct)\, \cos m(x-ct)$$

$$+ \tfrac{1}{2}\left(\frac{\partial^3 \phi_1^{(1)}}{\partial z^3} - \frac{\partial^3 \phi^{(1)}}{\partial z^3}\right) a^2 \sin^2 m(x-ct) = 0, \; z = 0$$

The third order solution has now been formulated. Basically all that now remains to be done is to evaluate the unknown constants in ϕ, ψ and $\phi^{(3)}$ by substituting into the equations arising from the boundary conditions. We have a large number of linear equations which by judicious manipulation decouple into two sets of four equations in four unknowns, each having the same left hand sides. By using λ and U matrices these can be easily solved.

8.6 Wind Shear Stress

This is modelled on the hypothesis of Jeffreys (1924). Although he suggested his so called sheltering theory as a mechanism for the growth of very small surface waves, it would seem by intuition to be more appropriate to the high speed wind case being discussed.

Longuet-Higgins (1977) suggested that shear stress rather than normal stress is the important agent for producing wave growth. He also suggested it would be a maximum at the crest and a minimum at the trough.

Also, as a result of the continuity of shear stress at the free surface given by equation (8.5.4) and the expressions for ϕ and ψ given by (8.5.1) and (8.5.2) respectively, with $k = m$ and $k_1 = 3m$, the shear stress will only contain first and third harmonics.

In the model being considered there is another important factor to be taken into account, namely the separation of the air flow. Intuitively when a high speed wind blows over the surface of a water-wave it would be expected that the air flow will not follow the wave profile, but will separate at a point near the crest. This idea is the basis of the sheltering theory postulated by Jeffries (1924) which suggests that air flowing over a wave separates somewhere on the leeward side of the crest

A. J. Croft

and reattaches itself on the windward side of the next wave at some point between the crest and the trough. The flow in the region sheltered from the main air current will be turbulent, but will be mainly composed of an eddy with a horizontal azis. The flow is represented diagramatically in Figure 8.2.

Banner and Melville (1976) argue that for separation to take place there must be a stagnation point on the surface relative to axes moving with the wave train. This condition is crucial to the mathematical solution of the problem.

In view of the above observations the wind shear stress τ is assumed to be of the form

$$\frac{\tau}{\tau_0} = \lambda(3\cos\psi + \cos 3\psi) + \frac{1}{9}(10\sin\psi + \sin 3\psi),$$

where $\psi = m(x-ct)$ and λ is a non-dimensional parameter which depends upon the point where the air flow separates from the surface.

For a high speed wind the separation point will be just on the leeward side of the crest. For example, $\lambda = 377$ when the separation point corresponds to $\psi = 95°$. From figure 8.3 it is seen that λ decreases rapidly as the separating point advances.

From the physical point of view h_1, a, m and λ must be related therefore, specifying these arbitrarily is open to criticism. Also, since λ is large there are numerical problems unless a and h_1 are chosen carefully. The method adopted is to choose λ and m and then calculate a and h_1 such that we have a relative stagnation point where the air separates and that $am \ll 1$ in addition to mh_1 being small.

With $\lambda = 377$ and $m = 1/6$ ft^{-1} we have $a = 3.6$ft and $h_1 = \frac{1}{2}$ft. From (8.5.3) we obtain

$$\zeta_1^{(3)} = 1.4\text{ft}$$

Figure 8.4 shows the free surface profile over one wave period. For the second and third order solutions it is to be noted that the crests are steeper and the troughs are shallower.

A Model for Surface Wave Growth

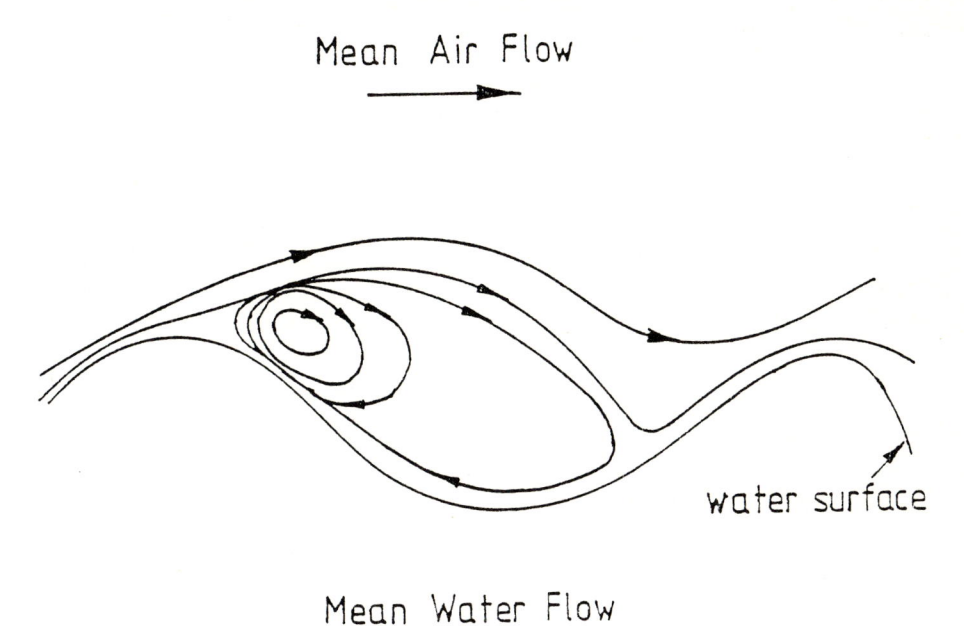

Figure 8.2   Diagrammatic Representation of Air Flow over a Water Wave when Separation takes place

A. J. Croft

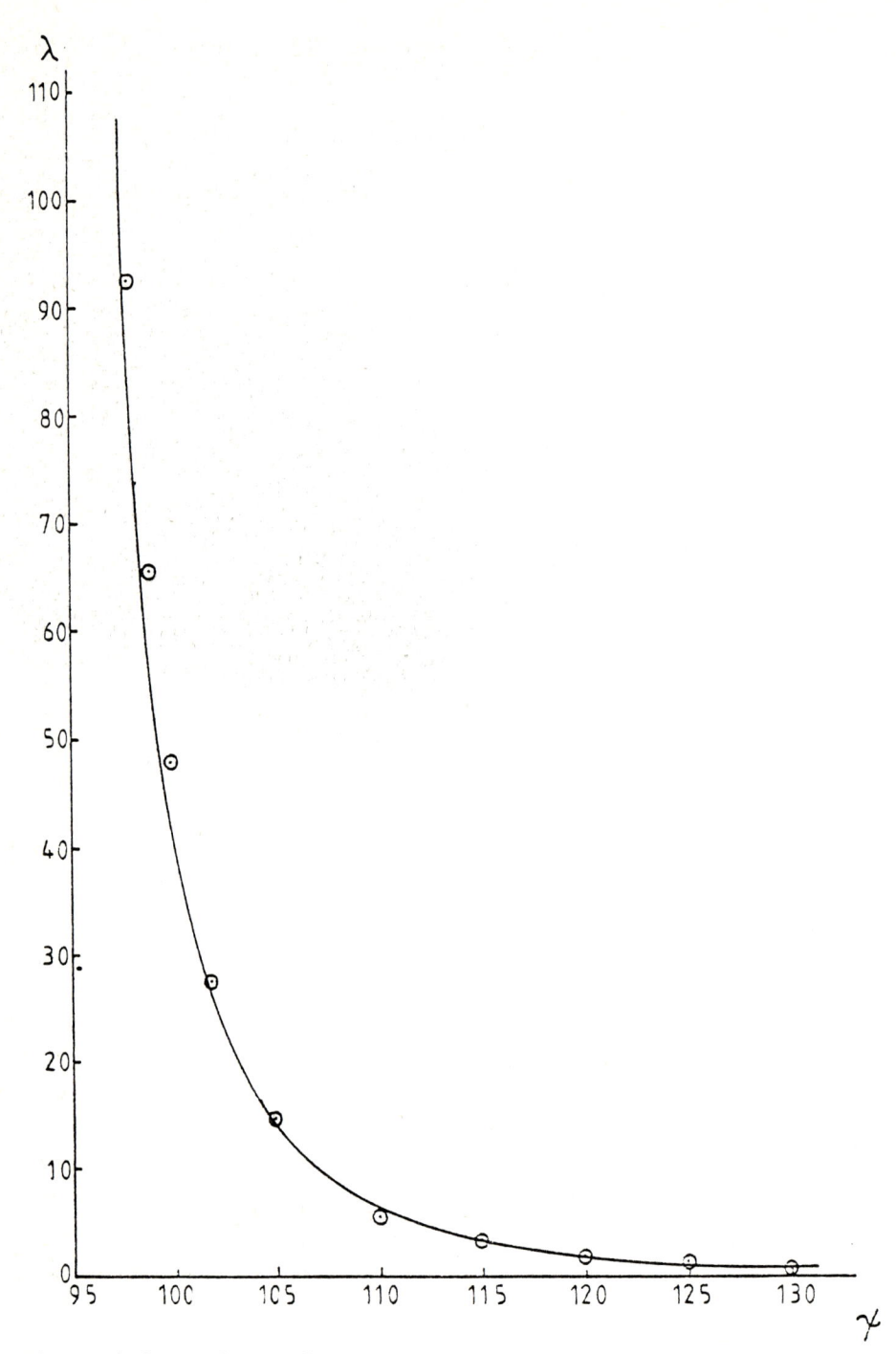

Figure 8.3   Values of λ against ψ when Separation takes place

A Model for Surface Wave Growth

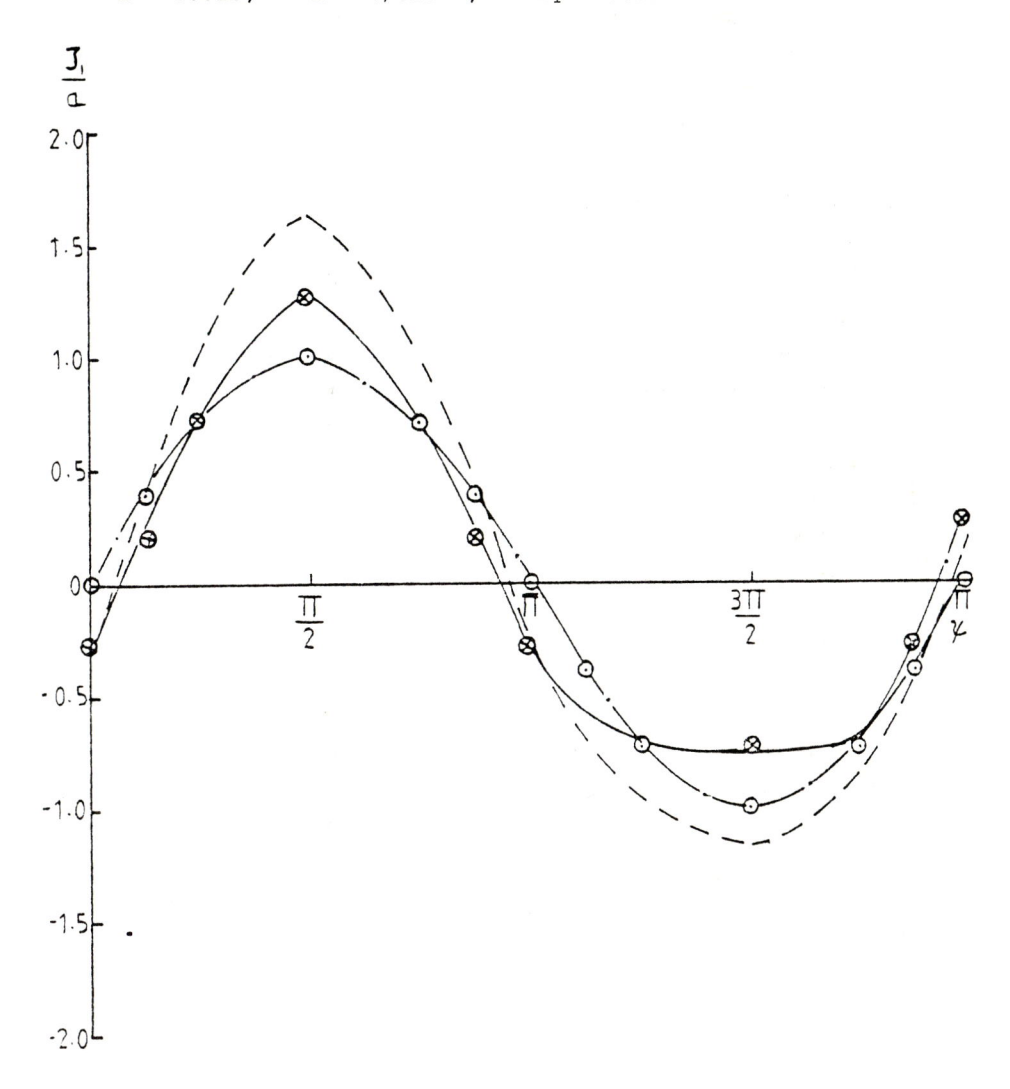

$a = 3.6 \text{ft}, \quad m = 1/6 \text{ft}^{-1}, \quad h_1 = 0.5 \text{ft}$

⊙ First Order

⊗ First Order Plus Second Order

--- First Order Plus Second Order Plus Third Order
(with Wind Shear Stress)

Figure 8.4 Non-dimensional Free Surface Elevation

A. J. Croft

8.7   Conclusions

The proposed long wave model indicates that the relatively thin layer of droplets in the presence of a wind shear stress in accordance with Jeffries' sheltering hypothesis could be a significant mechanism in the growth of the free surface elevation in the presence of a high speed wind. An acceptable substantial increase in the free surface elevation has been achieved by "stretching" the model so that it can only be just considered to be a long wave. This also indicates that the higher harmonics are coming into play, which is to be expected.

It must be stressed that the model is only a qualitative one because of the arbitrary shear stress distribution based on the sheltering theory being imposed on the system as well as the point of air separation and the wavelength of the interface wave. In practice these are related and depend upon the wind field. The structure of the wind flow would have to be studied before it could become a quantitative model.

References

Banner, M. L. and Phillips, O. M., 1974
On the incipient breaking of small scale waves. J.Fluid.Mech. 65, 647.

Banner, M. L. and Melville, W. K., 1976
On the separation of air flow over water waves. J.Fluid.Mech., 77, 825.

Barnett, T. P. and Kenyon, K. E., 1975
Recent advances in the study of wind waves. Rep.Prog.Phys. 38, 667.

Chandrasekhar, S., 1968
Hydrodynamic and hydromagnetic stability. Oxford University Press.

Jeffreys, H., 1924
On the formation of water waves by wind. Proc.Roy.Soc. A, 107, 189.

Longuet-Higgins, M. S., 1977
Some effects of finite steepness on the generation of waves by wind.
A voyage of discovery: George Deacon 10th Anniversary volume. Pergamon.

Miles, J. W., 1957
On the generation of surface waves by shear flows. J.Fluid.Mech., 3, 185.

Monahan, E. C., Spiel, E. S. and Davidson, K. L., 1982
Model of marine aerosol generation via whitecaps and wave disruption.
Private communication.

Phillips, O. M., 1957
On the generation of waves by turbulent wind. J.Fluid.Mech., 2, 417.

Stewart, R. W., 1967
Mechanics of the air-sea interface. Phys. Fluids (Suppl.), 10, S47.

Whitham, G. B., 1974
Linear and non-linear waves. Wiley.

9. POWER TAKE-OFF AND OUTPUT FROM THE SEA-LANCHESTER CLAM WAVE ENERGY DEVICE

F.P. Lockett,
Coventry (Lanchester) Polytechnic, England.

9.1 Introduction

The Clam wave energy converter has been under development by a team at Coventry (Lanchester) Polytechnic for the past six years and is now one of the few devices which show enough promise to merit continued Government and industrial support within the now much reduced British wave energy research programme. With the loss of official interest in large scale generating stations, attention is now being focused on smaller devices for power supply to isolated seaboard communities.

The device consists of partially submerged and inflated flexible air bags mounted on a rigid floating spine. Differential wave action along the spine causes the bags to pump air into and out of a common reservoir within the spine in closed system. Each bag is separately connected to the spine by a duct housing a turbogenerator. The turbines are of the Wells design which rotate in the same sense whatever the direction of the air flow, coupled directly to generators.

Current research is concentrated in a device featuring a hollow spine typically 100m long, 9m high, fitted with eight bags and capable of producing up to 500kw of electricity. The work aims to identify a design for a full scale demonstration prototype whose construction will be the next phase of development (Peatfield and colleagues, 1983).

9.2 Experimental Tests

Previous work on the 2GW scheme Clam device involved 1/50th scale experiments in the Cadnam wave tank to determine performance in 46 defined test seas as well as testing of a 1/10th scale model, some 29m long, at the team's Loch Ness test site. With the change in objective to a smaller device working in much the same high power seas, 1/14th scale was chosen for experiments at Loch Ness, to ease construction launching and testing.

Tests began in November 1983 to determine device performance in a wide range of sea states and its sensitivity to spine length, height and the number of bags. The models, up to 9m long, were constructed around a

F. P. Lockett

steel spine duct with plywood cladding and have corded latex bags of a design similar to that proposed for full scale. The fabricated steel bag-to-spine ducts house purpose built rotating dampers whose flat-bladed rotors may be driven at arbitrary speed and act as very inefficient Wells turbines, providing the linear (pressure drop proportional to flow rate) characteristic of the real thing. By varying the rotor speeds from shore, tests at various damping rates can be easily conducted to optimise this key parameter. The pressure drop across each damper is measured and from known calibrations the flow rate and hence airpower dissipated is deduced. Routinely, 32 channels of data are sampled at 5Hz for 358.4 seconds and stored digitally on disc for subsequent analysis.

9.3 Power Take-Off Simulation

The Clam bag converts wave power to airpower, the turbine airpower to mechanical shaftpower and the generator finally gives electrical output power. To complement the experimental work and complete the picture of Clam performance, digital simulation of the power take-off system has been carried out to determine sensitivity to design parameters, its mean efficiency in given seas and the character of the total output for integration studies.

The computer simulations use airpower data measured at the dampers of the experimental models which have the characteristic of the Wells turbine run at constant speed. The computations are done for all Clams on the spine simultaneously, in order to calculate the variation and statistics of total output. Provided the simulations show rotor speeds not varying greatly from the speed chosen for the turbogenerators, then the assumption that the airpower signals remain invariant is a good approximation and the simulations give a valid picture of full scale performance.

For the most part the work has focused on a system using induction generators coupled directly to the turbines. The characteristic of these machines is a strong restoring torque acting to maintain the rotor at a fixed so-called "synchronous" speed, the torque maximizing at between 5% and 15% overspeed. Any other generator system which keeps turbine speeds near constant could be simulated with constant damping test data in this way. On the other hand, gross variations in speed

Clam Wave Energy Device

would affect the damping rate and hence hydrodynamic performance, calling for simulation during model tests with interactive prescription of damping rates determined from the computed performance of real turbogenerators. The rotating dampers in the current Loch Ness models were designed with this work in mind, work which will be undertaken if the power take-off systems which emerge as desirable give wide variation in turbine speed.

9.3.1 The Wells Turbine

For these performance calculations, the Wells turbine is adequately characterized by the following parameters:

Design airpower	= DAP watts
Design speed	= ω_d rads/sec
Design damping characteristic	= ε_d m³/s/pa
Efficiency	= η, a function of instantaneous speed and airpower

These terms mean that the turbine achieves maximum efficiency in steady flow Q_d with pressure drop Δ_{pd} running at speed ω_d, where DAP = $Q_d \Delta_{pd}$ and $\varepsilon_d = Q_d/\Delta_{pd}$. Efficiency at other speeds and power levels is then given in Figure 9.1. The design point is not unique, but in conjunction with an induction generator it simplifies matters to take ω_d as the synchronous speed of the generator. The damping characteristic ε which is inversely proportional to the speed, influences hydrodynamic performance of the bag and this relationship is being determined from the model tests.

The value of DAP affects performance of the turbine at all other airpower levels and its optimum value (for a given Clam unit on a given design of spine) is dependent upon the scatter diagram of the resource (as perceived as airpower captured by the bag) and the value to the consumer of the power produced. Figure 9.2. shows mean turbine efficiency $\bar{\eta}$ running at design speed in Normally distributed flows, computed readily from η, and at varying speed in experimental data flows computed during the simulations. From performance estimates for the scatter diagram of the site chosen we may then determine the mean annual efficiency $\bar{\bar{\eta}}$ of the turbine. Typically, annual energy captured is a maximum when DAP is approximately 3 times the mean annual airpower; as DAP is reduced output regularity improves but with $\bar{\bar{\eta}}$ decreasing from a peak of around 70%.

F. P. Lockett

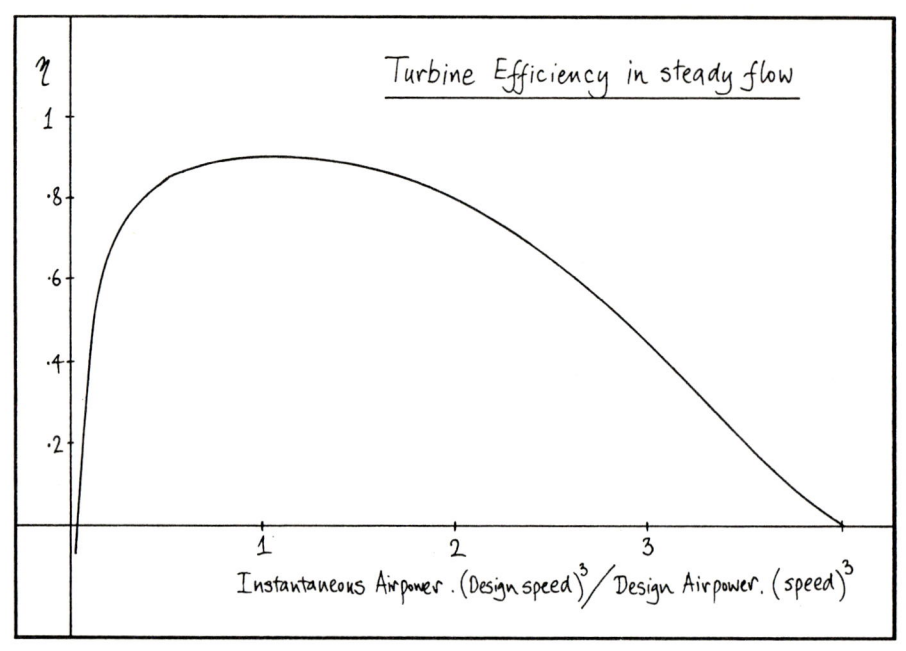

Figure 9.1

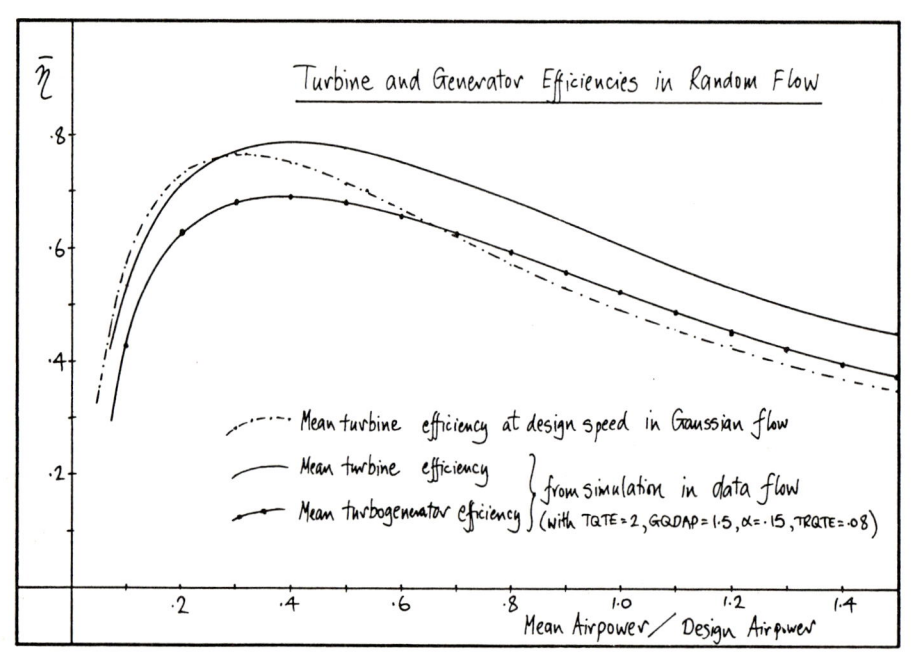

Figure 9.2

Clam Wave Energy Device

9.3.2 Rotor Dynamics

The basis for the simulation is the solution of the equation of motion of the turbogenerator rotors:-

$$d/dt(J\omega^2/2) = \text{airpower} \times \text{instantaneous turbine efficiency} - \text{generator mechanical load power}$$

where J is the turbogenerator moment of inertia (kgm²).

To model the continually transient magnetic conditions within the generator, the mechanical load it provides is here assumed to satisfy a first order lag equation with time constant τ, following the steady load which, for an induction generator, is a function of its speed and design parameters.

In order to provide comprehensive information applicable to Clam (or generically similar) systems of any size and rating in any wave climate, the defining equations and results have all been expressed in terms of various dimensionless parameters which describe the power take-off plant and airpower resource. The most significant among these are:-

$$\text{TQTE} = \frac{\text{Rotor inertia time constant}}{\text{Energy Period of Airpower signal}} = \frac{J\omega_d^2}{DAP \cdot T_\varepsilon}$$

$$\text{GQDAP} = \frac{\text{Generator maximum power}}{DAP} = \text{(approx.)} \quad \frac{2.5 \times \text{Generator rating}}{DAP}$$

$$\alpha = \text{Generator slip } (= \text{fractional overspeed}) \text{ at maximum torque}$$

$$\text{TRQTE} = \frac{\text{Generator magnetization time constant } \tau}{T_\varepsilon}$$

$$\text{MQDAP} = \frac{\text{Mean airpower of test data}}{DAP}$$

Other generator parameters enter the output calculations (and hence affect its efficiency) but dynamic response is insensitive to these and is adequately described by α, τ and the machine rating. The spectral distribution of the airpower signal, for given energy period T, also has small effect on the results for the variations observed. The model test signals show spectra well fitted by a Pierson-Moskowitz type of distribution with index six rather than index four, usually assumed for wave height in fully developed seas.

F. P. Lockett

9.3.3   Simulation Results

The airpower data used for the results presented here were recorded at Loch Ness in February 1984 during a test of a 1/14th scale model of an 8 bag, 80m long, 7m high device. The time traces shown cover a representative section equivalent to 100 seconds at full scale, the whole test, equivalent to 1341 seconds, was used for all statistical calculations. Figure 9.3 shows the airpower signals used to "drive" the simulations. Full scale equivalent mean values for the whole test were:-

Clam	1	2	3	4	5	6	7	8	average
kw	19.0	20.4	36.2	39.8	30.0	24.9	22.5	63.1	32.0

which represents roughly average annual performance for the model tested. The power distribution and spectral spread of the data was very similar in other sea states, so for the simulations other airpower levels were generated from the same data by multiplying throughout by the same factor. A feature worth pointing out is an interval of wave activity running down the spine (which was moored at an angle of about 40 degrees to the principal wave direction) starting at time 27 sec. at Clam 8 and reaching Clam 1 at time 42 sec., followed immediately by a period of about 10 seconds (1.5 waves) during which spine breathing was particularly monochromatic, displayed clearly in the spine total trace.

The format of the rest of the diagrams is the same. Each shows results computed from the airpower data for various parameter choices, Clam 5 was chosen as it usually displays average response. The time traces are:

i)   Clam 5 Rotor speed (normalized by the induction generator synchronous speed)

ii)   Clam 5 Airpower   (data)

iii)   Clam 5 Shaftpower   (the turbine output power)

iv)   Clam 5 Output power (the electrical output from the generator)

v)   Spine total output power.

All power values are expressed as multiples of DAP, the design airpower of the turbine.

Clam Wave Energy Device

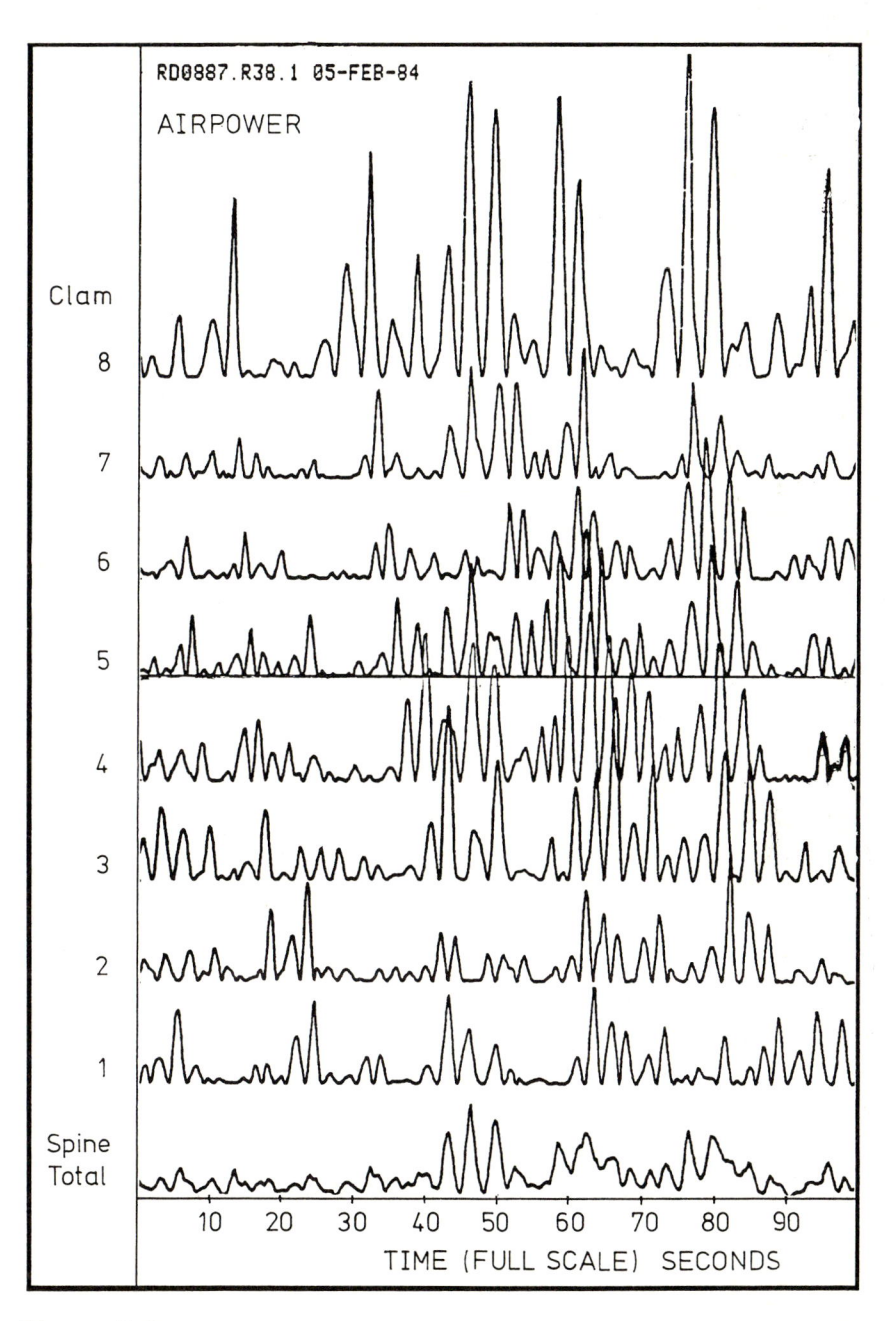

Figure 9.3

F. P. Lockett

Figures 9.4-9.9 show the effects of varying mean airpower, rotor inertia and generator "stiffness". Each diagram was obtained by changing just one of the parameters MQDAP, TQTE and α from the base case of Figure 9.4, which has TQTE = 2.0, MQDAP = 0.5, GQDAP = 1.5, α = 15 and TRQTE = .08.

The maximum mean airpower encountered through the year is approximately 3 times the annual mean, so choosing the turbine DAP to be 3 or 4 times the annual mean, the working range for MQDAP is from 0 to 1 (possibly more if Clam 8, more energetic than average, has plant identical to the rest). At the higher airpower levels energy is cast off by the turbine as flowrate increases beyond stall at the rotor blades. Figure 9.6 shows a succession of double peaks on the shaftpower trace at around 60 seconds corresponding to the high single peaks of the airpower trace. In these circumstances the entire efficiency curve of Figure 9.2 is traversed back and forth twice per wave.

Figures 9.4, 9.6 and 9.7 show the benefits of increasing rotor inertia. Indeed the lightest case (TQTE = 0.5) shows unacceptable periods of overspeed, even at moderate airpower levels, running up to 1.76 during the whole test with correspondingly low generator efficiency. Deficiency in inertia and consequent overspeed can be offset to some extent by using a more powerful generator but this increases the effective stiffness and hence the variability of the output.

The fluctuations in turbine speed in sympathy with airpower pulses, more marked at low inertias, increase mean turbine efficiency but at the expense of generator efficiency. Calculations for the whole test show a net benefit as inertia increases, mean turbine efficiency tending to approximately 74% as inertia tends to infinity, in close agreement with the efficiencies computed for Gaussian flows at constant speed shown in Figure 9.2.

For low values of slip (the fractional overspeed), the torque from an induction generator is approximately proportional to the slip. Figure 9.9 illustrates a "stiff" generator with α = .05, the torque and output varying sharply as the rotor speed fluctuates. Figure 9.4 corresponds to α = .15, most commercial machines falling within this range. Figure 9.9 indicates what can be expected from an alternator-DC system with a "ramp" control strategy as proposed for the 2GW Clam scheme (Lockett

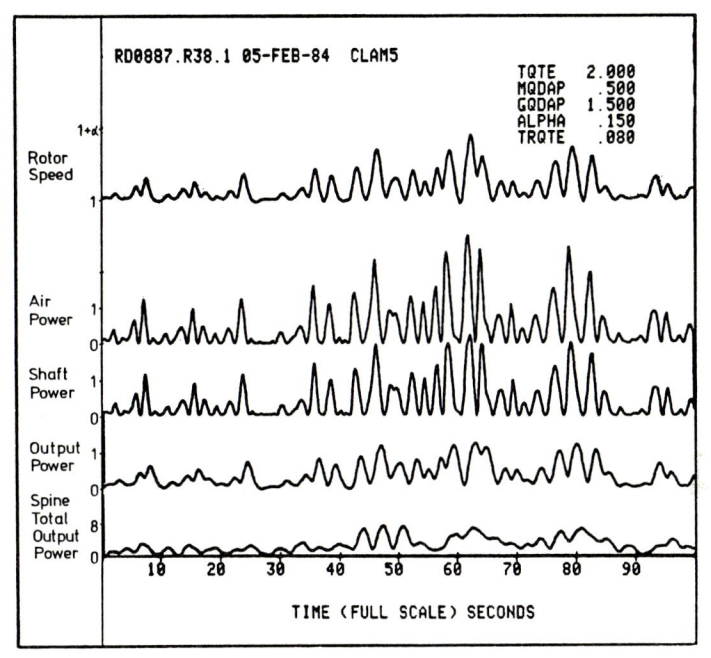

Figure 9.4

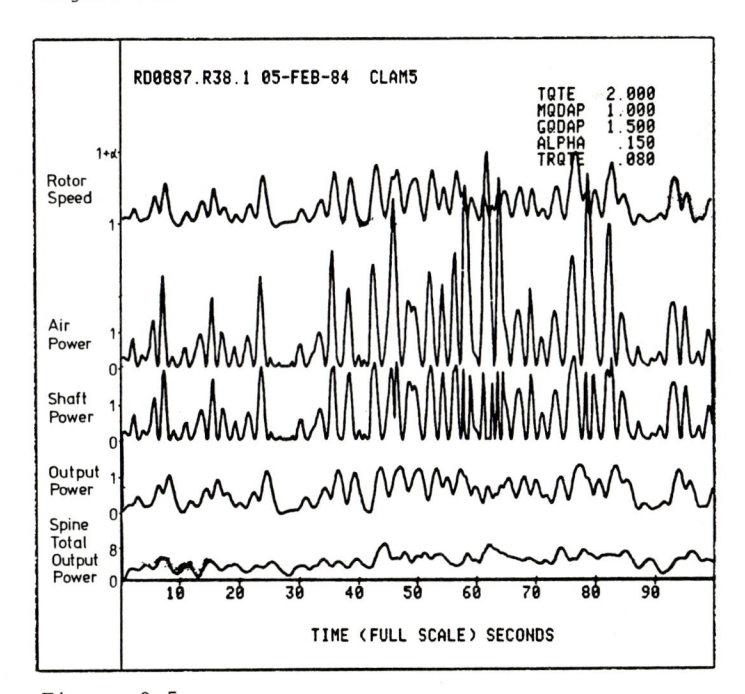

Figure 9.5

F. P. Lockett

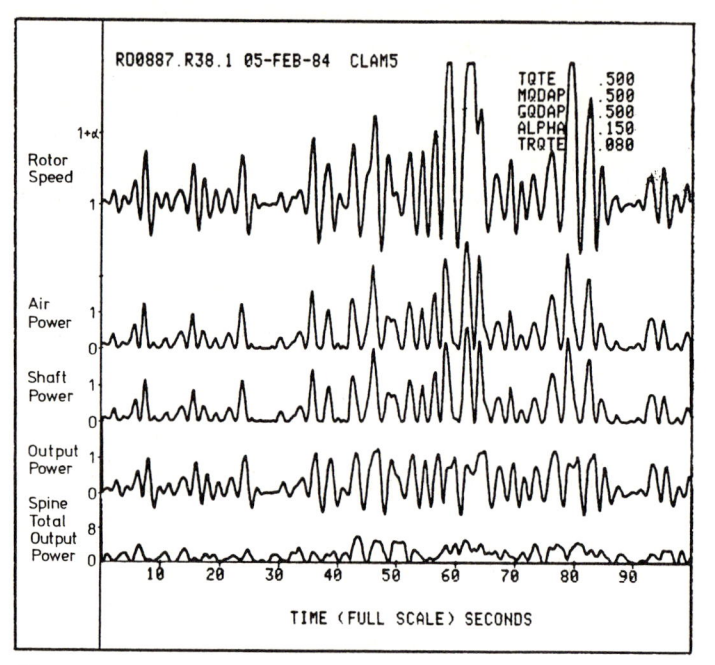

Figure 9.6

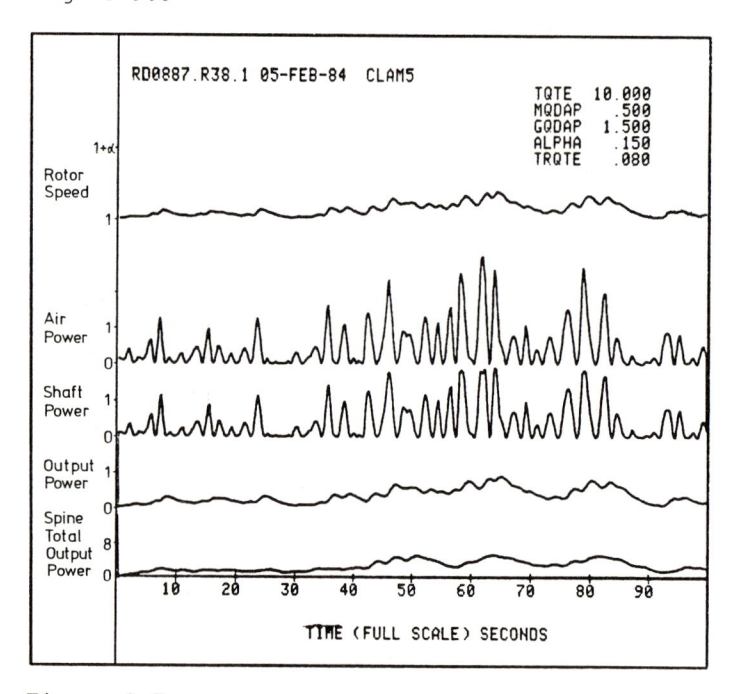

Figure 9.7

Clam Wave Energy Device

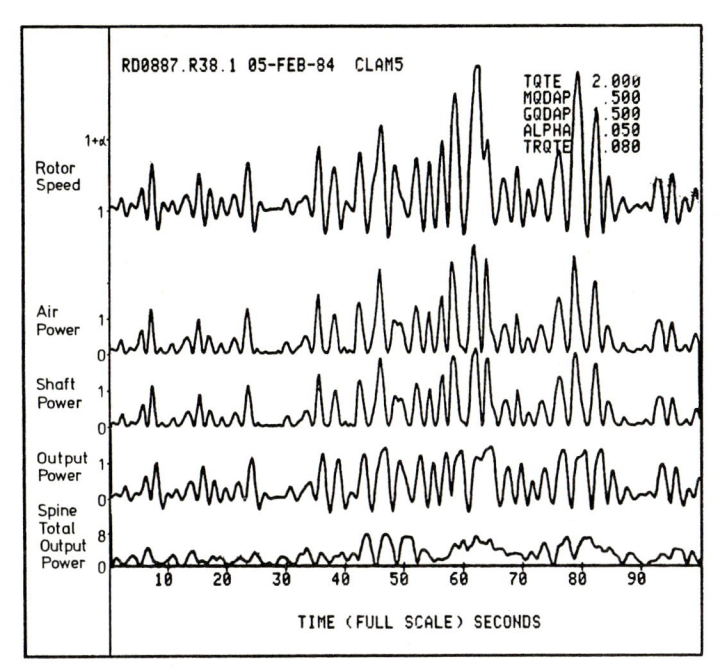

Figure 9.8

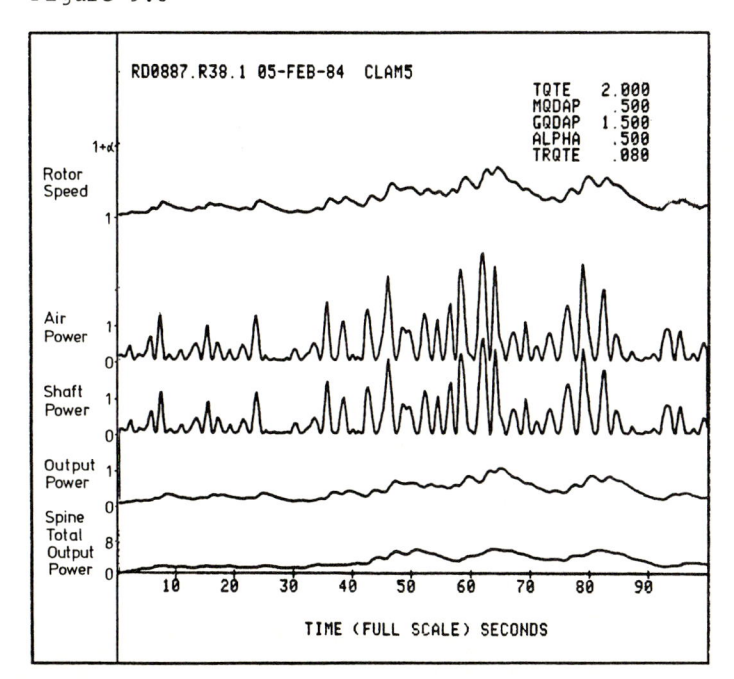

Figure 9.9

and Raabe, 1982), in which output increases linearly with speed from zero to maximum as normalized speed rises from 1 to 1 + α, staying constant thereafter. Low stiffness thus gives a meandering speed trace, roughly following the envelope of the airpower signal, with oscillations at double the wave frequency superimposed. Output is smoother but net efficiency is lower.

9.3.4 An Example

To put these dimensionless results in perspective, consider a device with say, eights bags which captures an annual mean of 40kw of airpower per bag. For typical resource scatter diagrams this implies an optimum design airpower DAP for the turbine of around 140kw. The design damping rate ε_d is the optimum rate determinable from model tests and is not very sensititve to power level. The design speed ω_d is the synchronous speed of a suitable induction generator (or equivalent) rated at around 85kw, for example 1500rpm for a 4-pole machine, with a magnetization time constant of about 0.5 seconds. Values for DAP, ε_d and ω_d then effectively fix a best design of Wells turbine, here a single stage 6-bladed machine of diameter of 1.5m and hub to tip ratio 0.6 (White, 1981). The working parts of such a turbogenerator would have comparatively low moment of inertia, perhaps 20kgm², so to make TQTE = 10, the highest inertia example shown in Figure 9.8, would require J = 340kgm² for airpower energy periods of about 6 seconds, which could be achieved with a shroud ring around the rotor blade tips with a mass of about 600kg.

Acknowledgements

The authors are grateful to the Department of Energy and SEA Energy Associates Ltd., for their support during the course of this work and to the Department for permission to publish the results presented in this paper.

References

Lockett, F. P. and Raabe, G-M., 1982
Operation of the Power Take-off System of the SEA-Lanchester Clam Wave Energy Device. International Conference on Systems Engineering, Coventry (Lanchester) Polytechnic.

Peatfield, A. M., Duckers, L. J., Lockett, F. P., Loughridge, B. W., West, M. J. and White, P. R. S., 1983
Tailoring the SEA-Clam wave energy device to meet community needs. 3rd International Conference on Energy for Rural and Island Communities, Inverness.

White, P. R. S., 1981
The Development and Testing of a 1/10th scale self-rectifying air turbine power conversion system. WESC (Commercial in confidence).

10. NUMERICAL MODELLING OF ILFRACOMBE SEAWALL

P. Hewson,
Plymouth Polytechnic, Plymouth, Devon.

P. Blackmore,
Building Research Establishment, Watford.

10.1 Introduction

The study of seawalls has been a subject for full scale (prototype) and laboratory testing dating back to before the start of this century.

Thomas Stevenson made the first recorded attempt at measuring wave pressures on sea walls using spring dynamometers in 1843. Since then work has proceeded on laboratory model work with occasional measurements in the field.

The only previous investigation known to measure structural response of a coastal structure subjected to wave loading was that of Kuribayashi et al (1959) who confined the investigation to measuring the period of vibration of the breakwater.

Seawall design is generally treated as a static problem with real seawaves beings replaced by a maximum or design wave for a given return period. This is probably sufficient for most components of wave loading which have a fairly low frequency and so generate small inertial forces in the structure which can be neglected. In a real sea there are wave loadings which can induce a dynamic response in the structure.

In this investigation an accelerometer was mounted on top of the seawall and the response recorded for comparison with theoretical results.

10.2 Finite Element Modelling of Ilfracombe Seawall

The particular finite element package used for this analysis was developed at Nottingham University and is known as PAFEC 75. The PAFEC package, whilst being able to calculate the usual stresses, deflections, mode shapes and natural frequencies, can also be used for non-linear problems such as creep, large displacements and plasticity and for dynamic problems with either sinusoidal or transient forces.

P. Hewson and P. Blackmore

10.3 The Need for a Finite Element Model Approach

The majority of proposed equations for estimating the maximum (design) wave pressure (including those developed from this investigation, and those actually used by many seawall designers are based upon the infrequently occurring impact pressure. Yet some authors (Ross, 1954, Hayashi et al, 1958), have suggested these impact pressures have no structural significance and hence should not be the basis for the design pressure.

It has been shown (Blackmore, 1982) that impact pressures occasionally act over large areas and have a significant duration, but the only evidence so far that impact pressures have any effect other than localised is found when wave pressure histories on the seawalls are compared with corresponding wall accelerations. Figure 10.1 is an example, here it can be seen that when an impact pressure spike occurs superimposed upon the hydrostatic pressure (impacts 1 and 4), the resulting wall accelerations, and hence displacements, are generally larger than for the hydrostatic pressures alone. This indicates that impact pressures are having a gross effect on wall response not just a localised effect.

The finite element models were developed for the Ilfracombe seawall using the main frame computer at Plymouth Polytechnic (a Prime dual 550) at level 3.4 in the PAFEC suite.

10.3.1 Modelling Techniques

The accuracy of the solutions obtained by finite element modelling is dependent upon the number of degrees of freedom allowed, which is in turn dependent upon the number of finite elements used. The more complicated the structure the more degrees of freedom are required to achieve a desired accuracy, but in most cases 60 degrees of freedom will result in accuracies of the order of 95% (this refers to the fundamental natural frequency), when using the PAFEC suite of programs.

The Ilfracombe seawall was modelled using eight noded quadrilateral isoparametric and six noded triangular isoparametric plate elements. Each plate element was allowed two degrees of freedom at each node in the $x(U_x)$ and $y(Y_y)$ directions, unless otherwise restrained. The elements (of unit width) could only be subjected to inplane forces, resulting in inplane stresses and deflections, all longitudinal bending and twisting effects were ignored.

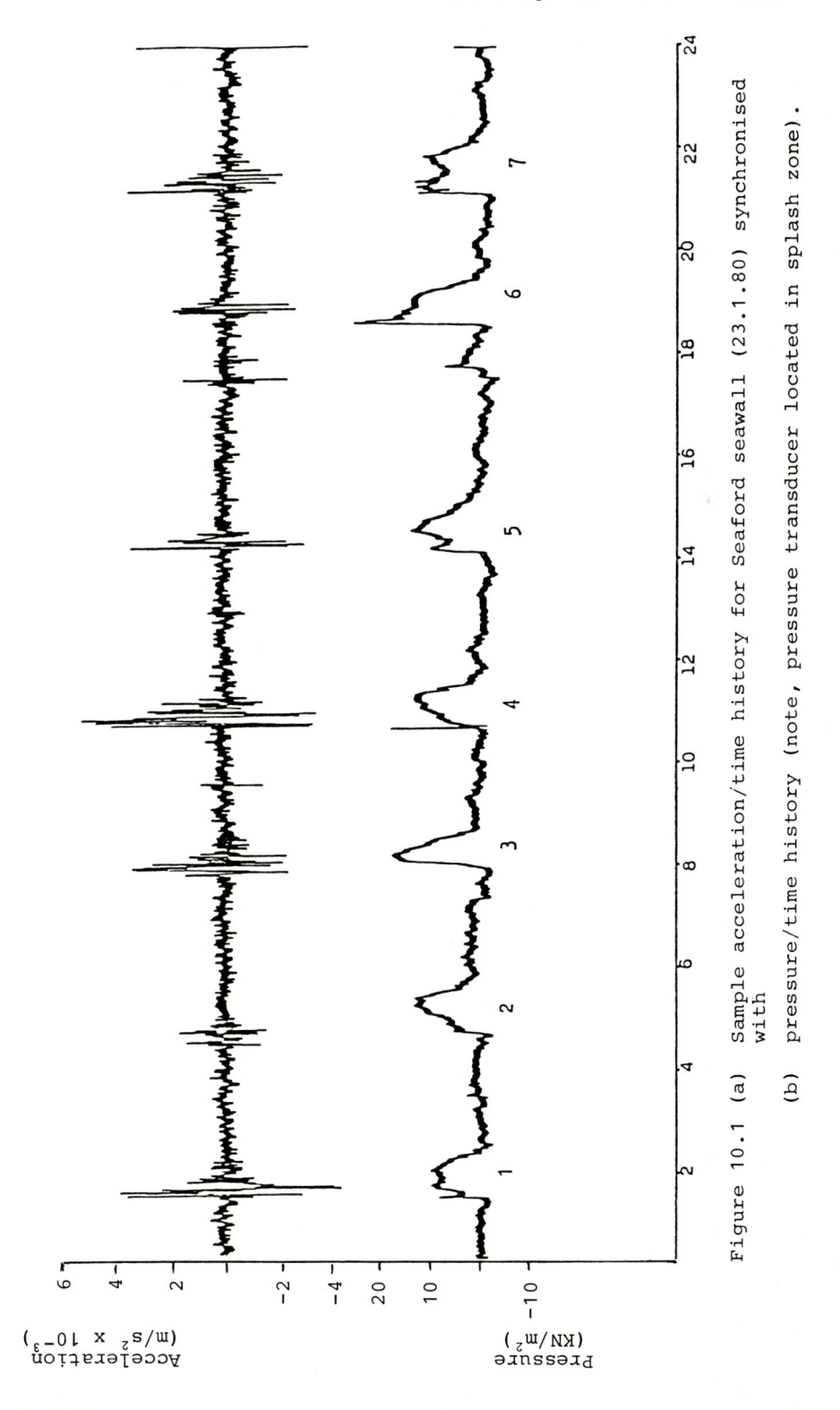

Modelling Ilfracombe Seawall

Figure 10.1 (a) Sample acceleration/time history for Seaford seawall (23.1.80) synchronised with
(b) pressure/time history (note, pressure transducer located in splash zone).

Since the Ilfracombe seawall was constructed from precast concrete blocks slotted onto dowel bars, these dowel bars were continuous over the height of the wall, so were included in the finite element model.

10.3.2 Model Constraints

The original seawall model was accurately constructed to the same physical shape and density as the real seawall, but this meant that natural frequencies and deflections of the model could only be corrected by modifying the properties or physical shape of the model, both of which were unacceptable. Therefore an improved model was developed to incorporate strips of foundation and backfill around the wall, the properties of which could be changed until the desired frequencies, deflections, etc were achieved, this enabled the seawall properties and shape to remain true. The width of this strip of foundation and backfill had to be a compromise between including sufficient area to allow local distortions around the structure to dissipate and keeping the size of the model within the limits of the computer core space available. The width of the strip of foundation and backfill was limited to about 1m by the amount of computer core available. This was insufficient to allow local distortions and stress concentrations to dissipate. Also the boundary conditions imposed on this strip were not representative of the actual boundary conditions existing at the real structures. These problems were partially solved by including very rigid, dense elements in the foundations and backfill at strategic points around the structure. This was not a mathematically rigorous solution but it did produce satisfactory results (Figure 10.2).

10.3.3 Modelling Seawall Response

The seawalls can be assumed to be subjected to both dynamic and static forces, where the dynamic component is due to breaking waves, and the static component is due to the head of water at the wall, including the contribution from the wave height. The (hydro)static pressure at the wall is a combination of the height of SWL above the toe and the wave height H. The pressure exerted by the height of the wave is actually of a dynamic nature but of such low frequency (of the order of 0.2Hz) that it may be considered as static, provided impact pressures do not occur.

Modelling Ilfracombe Seawall

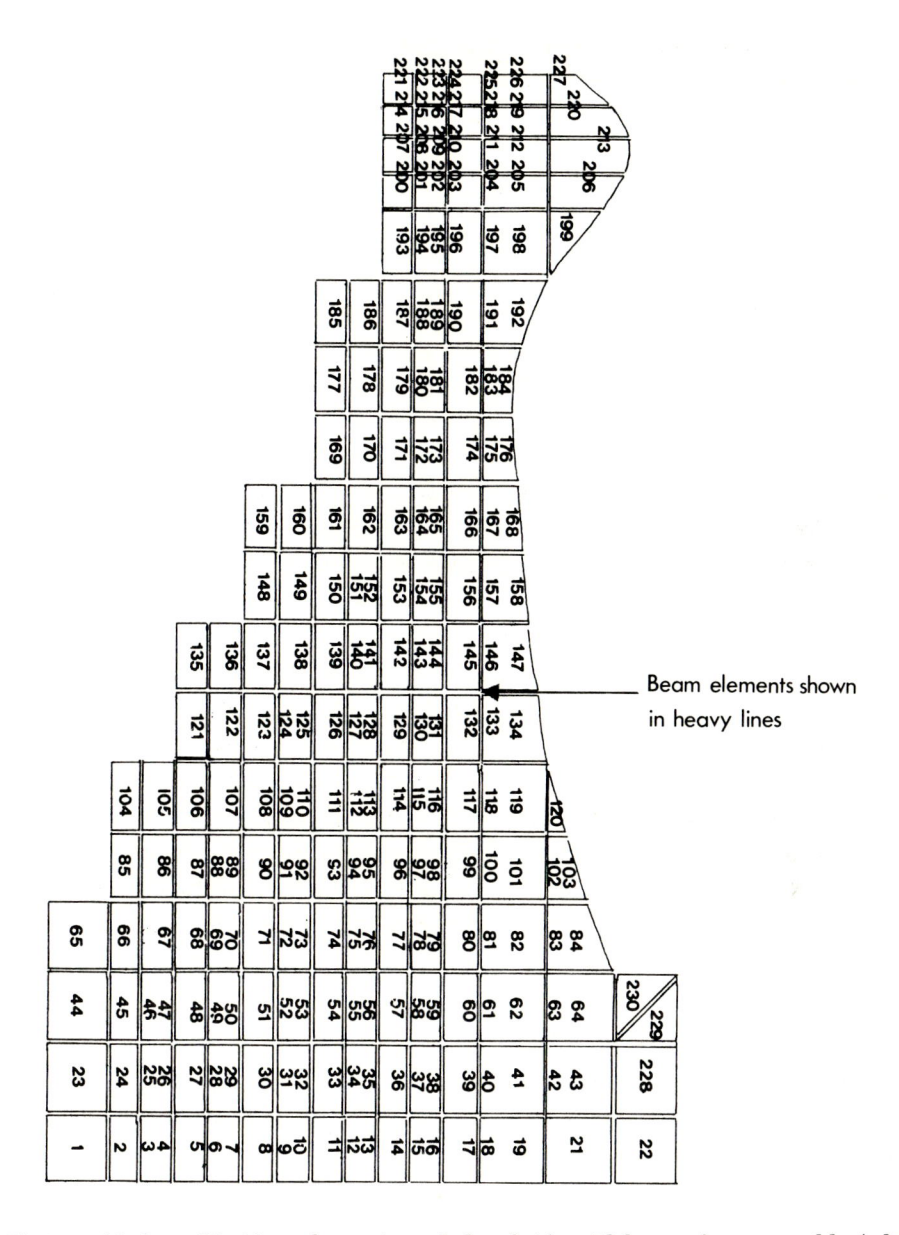

Figure 10.2   Finite element model of the Ilfracombe seawall (element numbers shown)

P. Hewson and P. Blackmore

10.3.4   Static Modelling Considerations

A static analysis of the Ilfracombe seawall was carried out to help provide a complete picture of the structural behaviour of the seawall, and as a check on the continuity between finite elements provided by the stress contours.

The modelling for a static analysis is fairly simple provided the exact physical shape and properties of the structures are known. In the case of the Ilfracombe seawall good estimations of the modulus of elasticity E and density ρ of the walls, foundations and backfill were obtained from the construction documents and drawings, as were the dimensions of the seawall.

In a static analysis the only structural parameters normally of interest are the stress and deflection, the accuracy of model deflections could be assessed against deflections measured on the real seawall. No stress measurements were made on the real seawall, therefore the accuracy of the stress contours could only be assessed by engineering judgement, that is checking that stress concentrations were occurring near the point of application of the loads, etc. The stress contours provided a useful check on the accuracy of the model because the degree of accuracy is proportional to the amount of mis-match in the continuity of stresses at element boundaries.

Seawall deflections were measured using a linear drive servo accelerometer which could respond to low frequency vibrations, but not static loads. Therefore, the only wall movements measured were those caused by the fluctuations in the wave height at the wall. The contribution of the static head of water at the wall could not be measured directly. The accelerometer was suitable for either horizontal or vertical operation, by rotating the unit through 90° from the horizontal to the vertical an output of +5 volts was produced, at any other angle of rotation between 0° and 90° a voltage of $V = \sin\phi \times 5$ volts was produced (where ϕ = the angle through which the accelerometer is rotated). Therefore, an estimate of the wall deflection caused by static loading can be made at any state of the tide by measuring the d.c. offset in the accelerometer output. This voltage is proportional to the angle of wall rotation, from which an estimate of horizontal wall deflection can be calculated. (The accelerometer must be zeroed with no water at the wall for the estimate to be valid, amplifier drift must also be avoided).

Modelling Ilfracombe Seawall

The only wall deflection which can be equated to the forces producing it with any degree of certainty is that caused by the hydrostatic wave pressure. Therefore only the hydrostatic pressure exerted by the wave height is applied to the finite element models, this pressure is calculated by assuming the waves to be fully reflected from the seawall (that is standing waves). The theory of Miche was used to calculate the standing wave pressure because Wiegel (1964) considers this theory to be better than that of Sainflou (1928). The contribution of the head of water at the wall was not included.

For example, from Miche, the pressure at the base of the wall due to the wave height is:-

$$P_b = (d + \frac{\bar{H}}{\cosh 2\pi d/L}) \rho g$$

and

$$\delta h = \frac{\pi H^2}{L} (1 + \frac{3}{4\sinh^2(2\pi d/L)} - \frac{1}{4\cosh^2(2\pi d/L)}) \coth 2\pi d/L$$

where $\bar{H}$ is the wave height that would exist at the wall were the wall not there, and δh is the mean level of the standing wave about SWL.

The characteristics of the waves used are given below. These waves are used because corresponding accelerometer measurements were available thus allowing the wall deflections to be calculated.

	$\bar{H}$(m)	L(m)	d(m)	h(m)	$\frac{\bar{H}}{\cosh 2\pi d/L}$	wall deflection from accelerometer
Ilfracombe	0.85	12.6	1.5	0.56	0.658m	0.0308mm

These pressures were applied to the Ilfracombe model at a SWL of 2.5m (AOD) because wall deflections were available at these states of the tide. The accelerometer on top of the Ilfracombe seawall (corresponding to node 193 in the model) indicated a mean wall deflection of 0.031mm for the above wave action. The properties of the foundations and backfill in the models were then modified so that models gave the same static deflections as measured on the real seawall. Deflections were only measured at one point on the real seawall and the model was then modified so that they gave a corresponding deflection at that point in the model, but there is no way of knowing if the overall deflection shapes are the same.

Stress contours were also plotted for the above loading conditions and checked for continuity especially at areas of high stress concentrations. If the mis-match was greater than 5% of the magnitude of the stress then the elements at that point were re-meshed. A sample stress contour plot is shown in Figure 10.3.

10.3.5 Dynamic Modelling Considerations

The relevant structural parameters necessary for a dynamic analysis are difficult to quantify because in many cases they cannot be measured. Two of the most important parameters affecting a dynamic analysis are the stiffness and the damping, neither of which could be measured directly during this investigation but had to be obtained by trial and error. Additional problems also occur due to the added mass of water constrained to move with the structure; this, and the structures natural frequencies change as the tidal level rises and falls. Then there is the soil/structure interaction and the hydrodynamic damping both of which are largely indeterminable. One further problem which might affect seawalls is the change in pore water pressure in the foundations due to the varying heads of water at the wall which can introduce uplift and/or back pressures.

When modelling the seawall, the natural frequency of the fundamental bending mode of the prototype wall can be modelled exactly, but all higher modes of vibration will be subject to slight errors because the model boundary conditions are not exactly representative of the real structure. This does not introduce large inaccuracies into the models because the higher modes of vibration are not greatly excited as the majority of the energy is contained in the fundamental mode of vibration.

The fundamental modes of vibration for the Ilfracombe seawall is 8.9Hz. This value was obtained from the spectral density plots of seawall accelerations. The properties of the foundation and backfill in the models were altered (by trial and error) until both static deflections and fundamental natural frequencies were the same as those of the real structures. When these conditions were achieved it was assumed that the model closely reproduced both the static and dynamic response of the real seawalls to the degree necessary for further analysis.

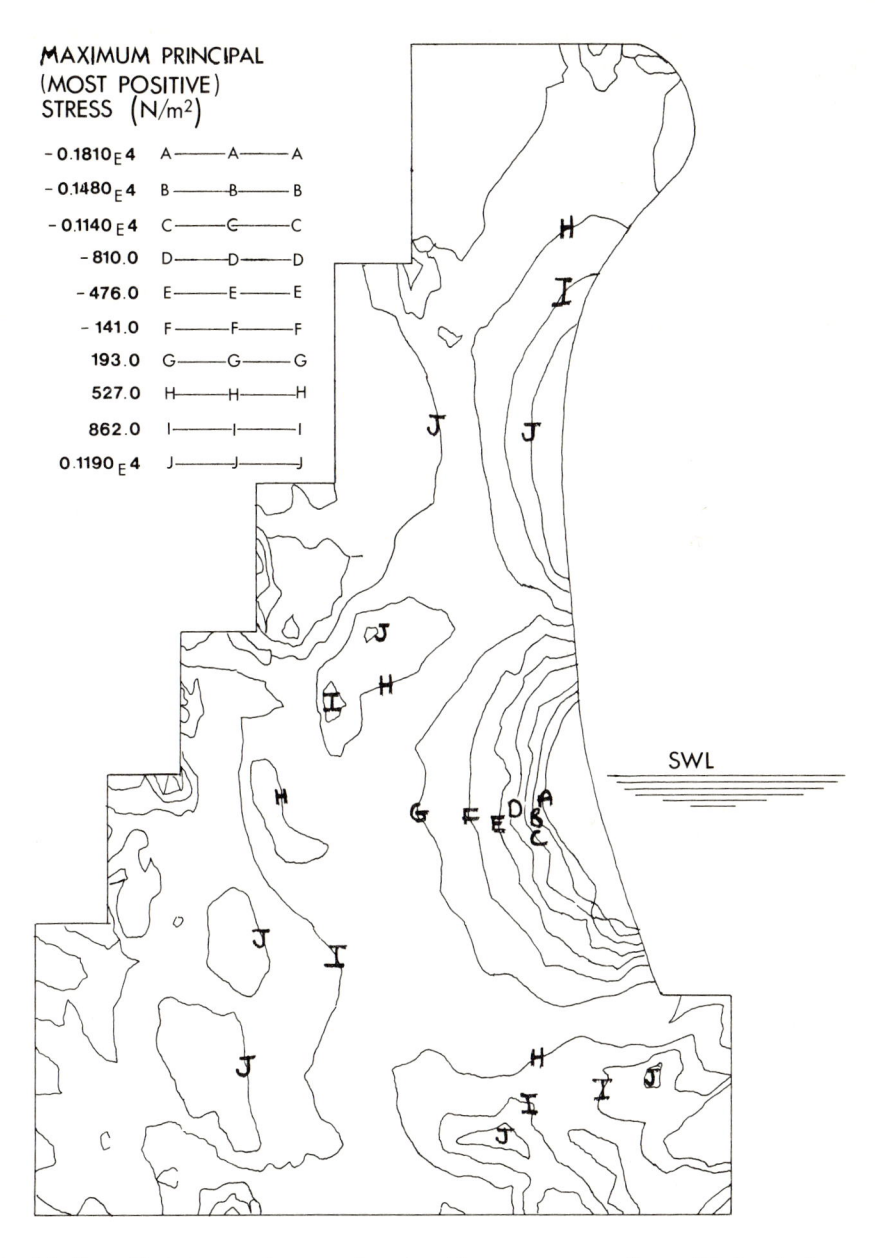

Figure 10.3 Stress contours due to the pressure from Miche being applied at a S.W.L. of 2.5m A.O.D. on the Ilfracombe seawall

10.3.6 Seawall Subjected to Transient Excitation

Many authors are of the opinion that transient impact pressures have no structural significance. For example, Ross (1954) states "the larger pressures are of too short duration for a structure of much weight to be moved appreciably". He concludes "usually the high pressures will not be important". Hayashi and Hattori (1958) conclude "when the duration (of the impact pressure) is very small, the wave pressure, even if its intensity is very large, may not have the effect of a force". These comments resulted from model experiments in wave tanks where the transients lasted in the order of one millisecond, whilst full scale transients generally have a duration of 0.1 to 0.2 seconds.

10.3.6.1 Transient Response Analysis

Because of the random nature of the transient wave forces, they cannot be represented by an excitation force derived by the superposition of a number of sinusoids. Hence a deterministic method was used, enabling the solution to proceed by a stepwise or numerical integration technique, (The forcing function must be known at all points in time before this technique can be used; statistical data alone is not sufficient).

10.3.6.2 Response to Wave Impact Pressure

Because the method of analysis is deterministic, the time history of the wave impact must be known for all time, so a vertical pressure distribution (containing both impact and hydrostatic components) as actually measured, was applied to the Ilfracombe model as shown in Figure 10.4(a,b).

Impact pressures were only measured at transducers number 1 and 3 (nodes 115 and 77 in the model) a distance of 1.6m apart. Pressures were assumed to act over the whole area between these two transducers with the pressures at the intermediate nodes (nodes 104 and 91) found by linear interpolation. The resulting structural response (deflection) is shown in Figure 10.5. This figure shows a large dynamic deflection due to the impact pressure followed by a lower, almost static, deflection caused by the second (hydrostatic) peak in the pressure history. The magnitude of this dynamic deflection is about four times the static deflection, although the dynamic (impact) pressure is only two to three times larger than the static pressure. Thus it seems likely that impact pressures occurring in real seas, with rise times of the order of 0.1 seconds, cause significant structural response.

Modelling Ilfracombe Seawall

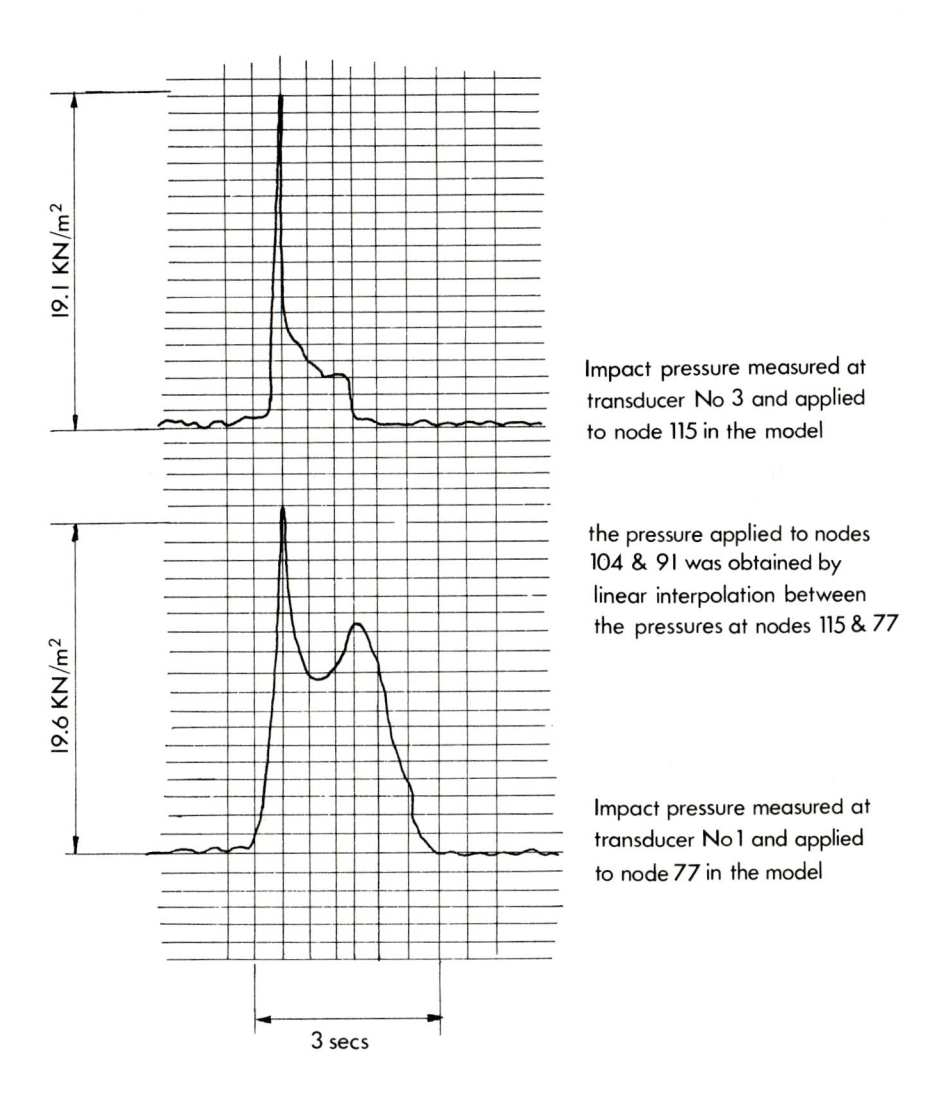

[Figure: two impact pressure traces on grid paper]

- Upper trace (scale 19.1 KN/m²): Impact pressure measured at transducer No 3 and applied to node 115 in the model
- Middle annotation: the pressure applied to nodes 104 & 91 was obtained by linear interpolation between the pressures at nodes 115 & 77
- Lower trace (scale 19.6 KN/m²): Impact pressure measured at transducer No 1 and applied to node 77 in the model
- Time scale: 3 secs

Figure 10.4(a) The impact pressure history as actually measured on the Ilfracombe seawall and as applied to the finite element model of this seawall.

P. Hewson and P. Blackmore

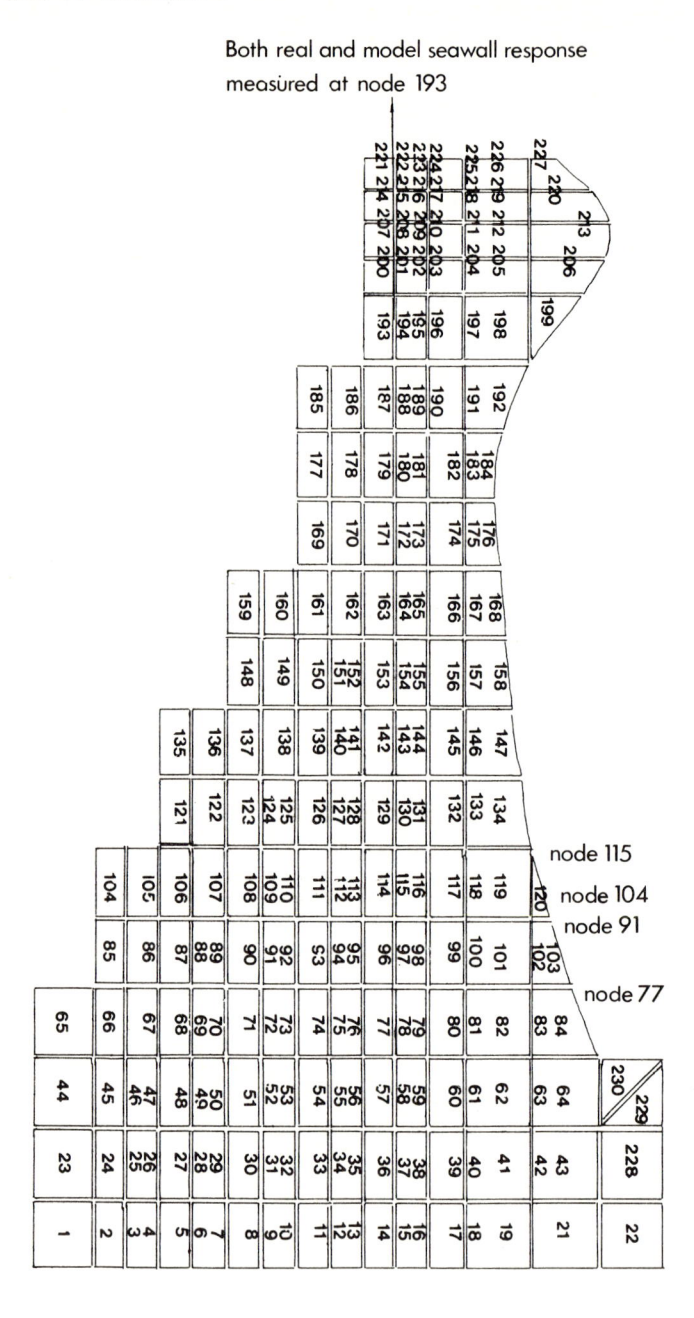

Figure 10.4(b)   The impact pressure history as actually measured on the Ilfracombe seawall and as applied to the finite model of this seawall.

Modelling Ilfracombe Seawall

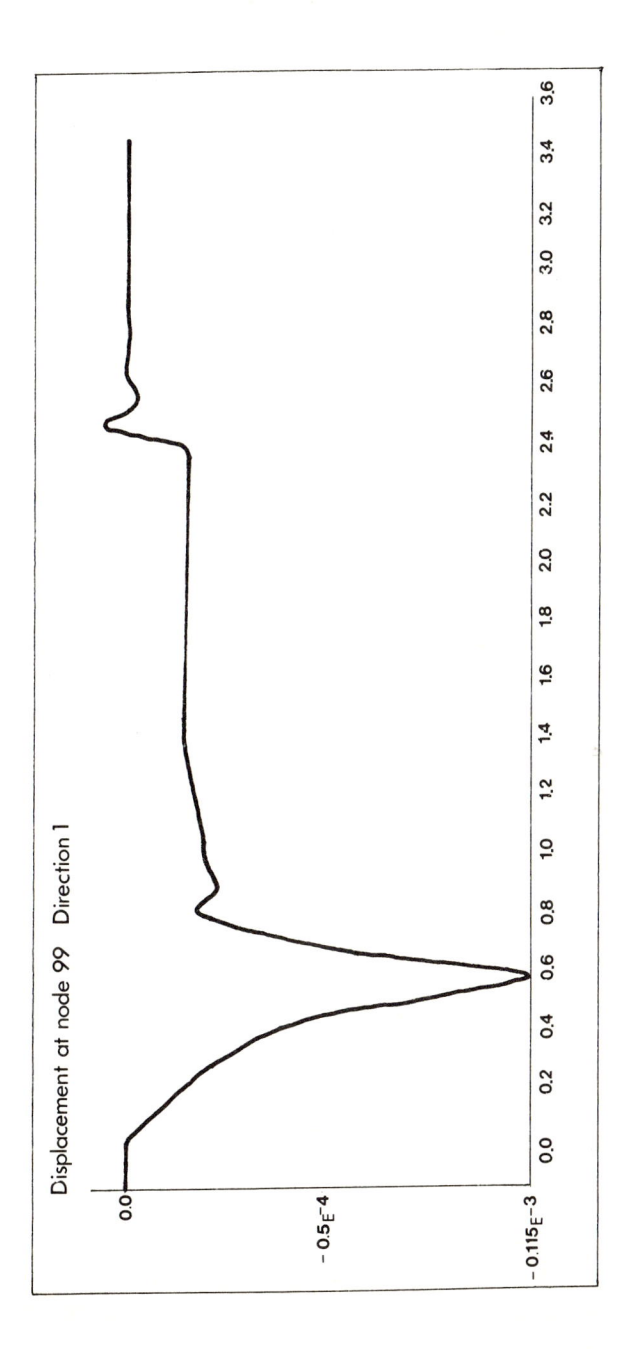

Figure 10.5 Ilfracombe (model) seawall displacement at node 193 for the impact pressure history shown in figure 10.4.

P. Hewson and P. Blackmore

The magnitude of this impact generated response is 0.115m (at node 193) and occurs after 0.50 seconds. The actual seawall deflection was not measured for this individual impact, but the average deflection for this period of impacts was about 0.035mm; so it seems that the predicted maximum deflection of 0.115m is of the correct order of magnitude. This also indicates that the percentage of critical damping used for this analysis (0.02%) is of the right order.

10.3.7 The Effects of Rise Time on Structural Response

Impact pressures have been measured in the laboratory with minimum rise times of the order of 0.5 milliseconds (Bagnold, 1939). It was on the basis of this sort of data that authors suggested that impact pressures have no structural significance. Yet full scale impacts in a real sea can have rise times of the order of 100 to 200 milliseconds, as measured during this investigation. Impact pressures with these longer durations have been shown in section 10.3.6.2 to cause appreciable deflections in a real seawall.

So with this in mind, an arbitrary impact pressure was applied to the Ilfracombe seawall model (ζ = 0.02) at nodes 115, 104, 91 and 77. The total impulse of this arbitrary impact was fixed at 1.22kN.s/m² (this was the average impact impulse measured on the Ilfracombe seawall), and the rise time was varied between 1 and 200 milliseconds to cover the range of both model and full scale wave impacts.

The six cases of loading chosen are shown below:-

Load Case	Rise Time (t)	Impact Pressure (P)	Impulse = Pt/2
1	1 mS	2440 kN/m²	
2	10 mS	244 kN/m²	
3	50 mS	48.8 kN/m²	
4	100 mS	24.4 kN/m²	
5	150 mS	16.3 kN/m²	The impulse is con-
6	200 mS	12.2 kN/m²	stant at 1.22kN.s/m²

In this table load cases 4 to 6 correspond to the type of rise times and impact pressures measured during this investigation, and load cases 1 and 2 show the very high pressures possible (for the same impulse)

Modelling Ilfracombe Seawall

but for very short rise times. The resulting dynamic deflections of the Ilfracombe seawall model for the above six loading cases are compared in Figure 10.6 with the deflections produced by the static application of the same loads.

From this figure it is seen that the ratio dynamic deflection/static deflection reaches a maximum value of 12.2% and remains constant at this value for all rise times greater than 40mS, but for rise times less than 40mS this percentage falls very rapidly to a value of 0.02% at a rise time of 1mS. This indicates that the seawall cannot respond fully to forces of less than about 40mS duration.

The dynamic deflection is likely to be reduced for larger values of the damping ratio.

For a wave impact lasting 7 milliseconds the Ilfracombe seawall model has a dynamic deflection which is only 1% of that produced by the static application of the same load, so it seems unlikely that impacts of this duration, or less, will have any significant gross effect on the deflection of the real seawall, although the high pressures associated with these impacts might have a localised effect. As most model investigations measured rise times less than 7 milliseconds then it would appear that the model test data is not suitable for the basis of full scale seawall design and if used, would produce highly conservative structures.

10.4   Conclusions

(1)   The finite element modelling suggests that the impact pressures measured during this investigation are of sufficient duration to generate a significant dynamic response in the seawalls, and as these impacts occasionally act simultaneously over large areas then the total dynamic deflection can be substantial and of the same order as, or greater than, the static deflection due to 'hydrostatic' waves.

(2)   The very high pressure, short duration wave impacts measured in model scale laboratory studies will have a negligible effect on the response of full size seawalls.

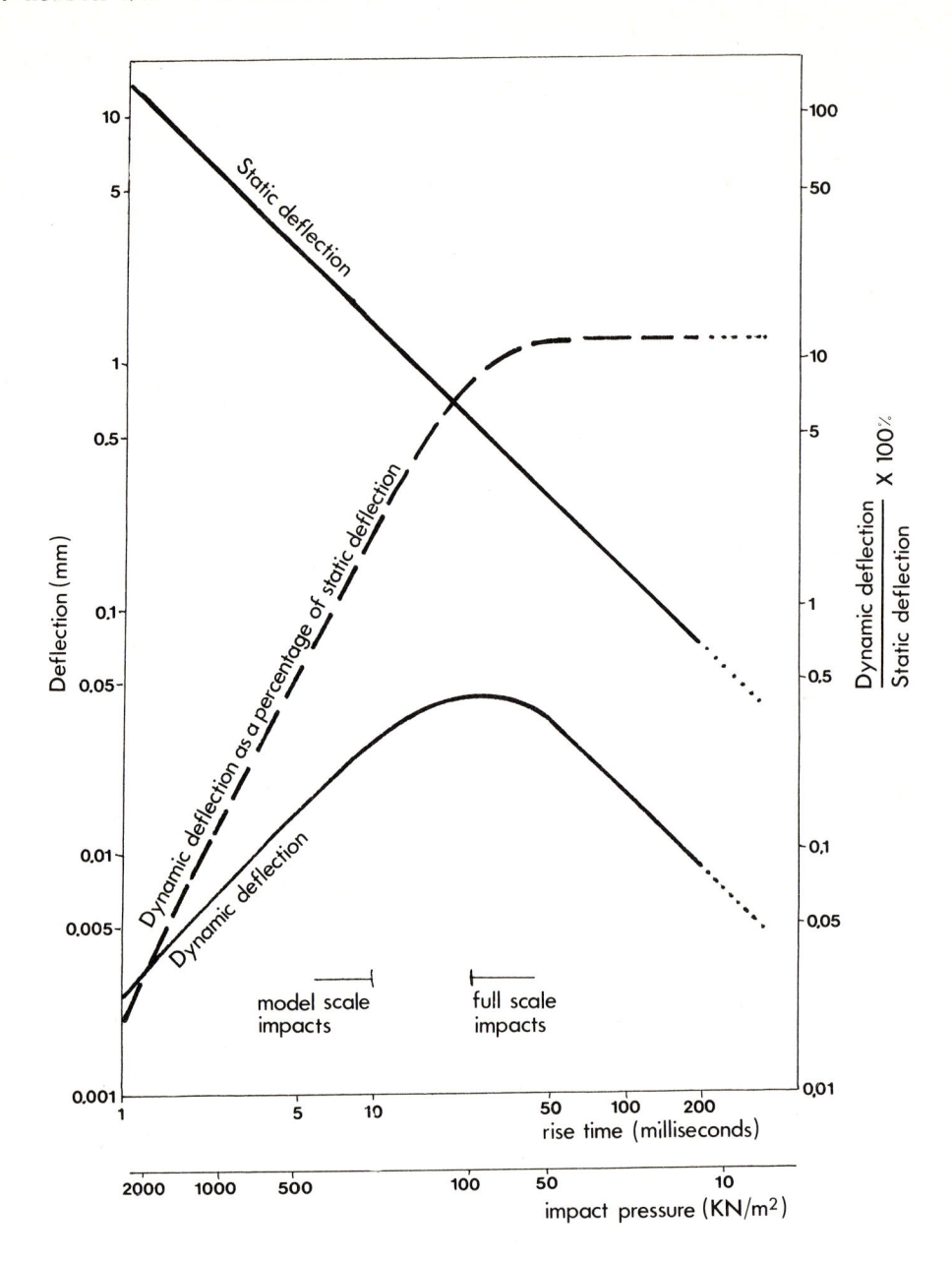

Figure 10.6 The effect of rise time and impact pressure on seawall response (deflection)

(3) Using pressures produced by wave impact as a static loading on a seawall will produce a conservative design.

References

Bagnold, R. A., 1939
Interim report on wave pressure research. Journal Institution, Civil Engineers, pp 202-226.

Blackmore, P., 1982
The evaluation of wave forces on seawalls, PhD Thesis, Plymouth Polytechnic.

Hayashi, T. & Hattori, M., 1958
Pressure of the breaker against a vertical wall. Coastal Eng. in Japan, Volume 1, pp 25-37.

Kuribayashi, T., Muraki, Y. & Udai, G., 1959
Field investigation of waves forces on breakwaters. Coastal Eng. in Japan, Volume 2, pp 17-27.

Ross, C. W., 1954
Shock pressures of breaking waves. Proc. 4th Conference Coastal Eng., pp 323-332.

Sainflou, E. J., 1928
Treatise on vertical breakwaters. Annales des Ponts et Chassees, No. 4.

Wiegel, R. L., 1964
Oceanographical Engineering. Prentice-Hall Int.

11. MODELLING THE PLAN SHAPE OF SHINGLE BEACHES

A.H. Brampton,
J.M. Motyka,
Hydraulics Research Station Limited, Wallingford,
Oxfordshire, OX10 8BA.

11.1 Introduction

The prediction of beach changes due to alongshore wave induced sediment transport has traditionally been made using physical modelling techniques. However, apart from the problems of scaling, physical models are generally too small to reproduce sufficiently long stretches of coast. With the recent development of mathematical models most problems dealing with beach plan shape changes are now dealt with numerically. Models dealing with sand beaches already exist and are used extensively at the Hydraulics Research Station to predict changes in coastline evaluation brought about by construction of harbour arms, artificial headlands, etc. Our view is that a relatively simple model is also needed which will allow the coastal engineer to predict the long-term changes of a shingle beach. Such a model clearly will not be able to incorporate short term fluctuations due to variations in the wave climate or due to extreme events. Nevertheless it should predict the mean annual rates of alongshore drift and of beach recession and advance.

At Hydraulics Research (HR) mathematical modelling of beach changes is made using a so-called one-line model (Price et al, 1972). In this type of model the coastline is represented by one characteristic beach contour, such as the crest of the beach berm, the swash line, etc., but more often than not, the high water line. Movement of material is assumed to take place parallel to the contours of the beach and these contours are assumed to move together in response to changes in the rate of littoral drift.

A number of "multi-line" beach models have previously been proposed by various authors. Indeed when the HR model began to be developed it was of two-line form proposed by Bakker (1968). Lack of knowledge about the response of the lower flat part of the beach profile by comparison with the well documented and more reliable upper beach contours (such as high water lines shown on Ordnance Survey maps going back over a period of more than a century) soon made it clear that a one-line model was simpler, produced more reliable results and was far cheaper to operate in terms of computer cost. Multi-line and certainly the three-dimensional models at present being developed cannot yet be relied upon, or justified for engineering use, given the scarcity of knowledge on the interaction between incident wave conditions and beach movement.

For shingle beaches however, it is likely that in the near future a much better understanding of profile response to waves and water levels will be obtained, both from hydraulic model studies and from prototype measurements. This will allow the introduction of a three-dimensional model perhaps within the next decade. Even so, this improvement is only likely to be of real value when major changes in waves/material take place, that is when a coarser material is artificially introduced as part of a beach nourishment scheme, or where major structures such as offshore breakwaters which affect the incident waves are built.

The one-line sand beach model developed at HR has been used extensively to predict shoreline changes brought about by the modification of littoral supply by jetties, sea walls, etc. It is hoped that this paper will show that with some modification and with some field verification it can also be used to predict changes in the configuration of shingle beaches. There are many beaches on the south-east and the north-west coasts of the United Kingdom where such a model could be used to predict beach changes; for example, in areas where, due to extensive sea defences at one part of the coastline, the rate of littoral supply to adjacent areas has been reduced drastically leading to coastal erosion problems. The model should be designed to predict future rates of shoreline erosion and to predict rates of beach renourishment necessary to halt or reduce this erosion. In the present paper we have decribed the first steps made by HR in developing such a model. It should become clear that due to our lack of knowledge about shingle movement only a very simple type of mathematical model can be used presently. Even then it will require verification in the form of historical beach changes before it can be used with any degree of confidence.

In section 11.2 below, a number of problems which affect the accuracy of beach mathematical models, for either sand or shingle beaches are described. These include the difficulty of specifying particular wave

Modelling the Plan Shape of Shingle Beaches

conditions to represent the long term wave climate.

Following this general discussion, section 11.3 traces the derivation of a formula to represent the alongshore transport of shingle, starting from the well known CERC sand transport formula (Shore Protection Manual,1977) which ignores the effect of grain size on the alongshore drift rate. The formula derived in this paper for shingle includes, additionally, a criterion for the threshold of motion and this term and its effects are described in Section 11.4.

However, the introduction of such a term itself causes problems, especially in the calibration of the model. These difficulties are described, although not resolved, in Section 11.5 which also recommends further research.

11.2 General Considerations When Modelling Beach Changes

11.2.1 Interpretation of Wave Data

When running a beach mathematical model for long time spans the data needs to be "compressed" to give a finite, manageable number of wave input values. The normal littoral transport formula is of the form $Q \propto H^2 n C \sin 2\alpha$ and since it can be simplified to $Q \propto H^{5/2} \sin 2\alpha$ in shallow water then the input values of H and α should be weighted accordingly to give the correct rate of sediment transport. Mehaute et al (1983) have shown that the errors due to averaging wave direction α can be kept small if angles of wave approach are grouped into sectors of $10°$ or less and if H is weighted so as to average wave energy. Instead of averaging energy it seems to us more appropriate to average $H^{5/2}$. This applies of course to the simplified form of the transport formula given above. In the proposed shingle transport formula developed later it is difficult to average wave conditions in this way and it is necessary therefore to group these into narrow bands. This of course leads to an increased amount of wave input but, providing the timesteps necessary to maintain stability are no greater than before, computing costs are unlikely to be increased significantly.

There is also the problem of deciding in what sequence to run the various wave events in the model. Because of the unpredictability of waves in

A. H. Brampton and J. M. Motyka

nature it is more realistic to input wave conditions randomly rather than in a pre-determined fashion. At HR the Monte Carlo technique is used and the predicted shoreline changes at any point are derived statistically. Usually the mean shoreline position, averaged over a period of say a year, is output, as well as the shoreline position at the end of that year. As one might expect during the first few years the difference between these two positions can be quite marked with increased duration the differences are smaller and usually become insignificant after about 40-50 years. With this form of output one can estimate the probability of the beach line being within certain specified limits and check for example that the shore has not receded to the point at which sea defences are endangered. This form of wave input also relieves the engineer from the responsibility of trying to choose a representative sequence of storms.

The accuracy of a model will also depend on good representation of nearshore wave transformation processes. At low water levels bed friction losses will reduce wave heights and slow down the rate of alongshore sediment transport. Changing beach contours will also affect the wave refraction pattern. It is necessary therefore to include a refraction subroutine within the model to assess the effect of changes in the beach contours on incident wave angle.

11.2.2   Beach Profile Assumptions

Necessarily, the distribution of beach accretion or erosion is made in a rather simplistic way. The active beach profile is assumed to move as a wedge over a horizontal plane. Such an assumption leads to some errors in the representation of shoreline movement and also affects the wave refraction pattern. This problem of the "closure depth" of the beach profile is clearly more serious for sand beaches than it is for steep shingle beaches.(The shingle toe is usually clearly defined and also the foreshore seaward of the toe is relatively flat). Where it is important to assess seasonal changes in beach profile as well as the mean changes in beach plan shape this is carried out empirically using the HR beach profile database. At present we are making a systematic collection of profiles for various types of beach, wave exposure, etc. From these we analyse the statistical variations of certain important parameters such as foreshore width, beach level variation, etc. In this

Modelling the Plan Shape of Shingle Beaches

way it is possible to complement the results of the beach mathematical model and find out, for example, how much the seasonal shore perpendicular variations in profile are likely to change the shoreline position.

Using the HR database we are also beginning to build up an understanding of the response of shingle beaches to both waves and tides. It is likely that this information will allow a simple three-dimensional shingle model to be developed in the next few years.

11.2.3 Calibration

Calibration of a one-line transport model is usually made by first reproducing historical changes in the position of high water or the crest of the beach. It is never satisfactory to calibrate against lower foreshore contours as these rarely show any meaningful trends. For example, in a recent study we examined changes in the position of high water along part of the Cumbrian coastline and from this areas of erosion and accretion could be clearly identified. No such clear pattern would have been deduced from changes in the low water line.

Where there is no input of material from alongshore or from offshore then even when the wave climate is not known precisely calibration can be made by altering the value of K_1 in the transport formula, see section 11.3, equation 11.6. If there is an onshore supply (or offshore loss) calibration can take this into account by introducing a renourishment volume into the continuity formula (Hydraulics Research Limited 1982). The way in which this volume is distributed along the coastline, however, is usually a matter of engineering judgement.

Calibration is thus very similar to the methods used in a physical model with the advantage that it is generally much easier to adapt a numerical than a physical model.

11.2.4 Boundary Conditions

These can consist of one or more of the following:

1. Free boundary - often, beyond the length of coast being examined there is a long stretch where alongshore sediment transport is

"unhindered". At such a boundary sediment must be allowed to cross in either direction and the beach to adjust as it wishes.

2. Fixed boundary - the presence of a rock outcrop, promenade, etc., may result in the beach line being constrained. A fixed boundary can be applied either within the model or at its boundary.

3. Specified transport boundary - transport across a boundary can be adjusted to reproduce for example zero movement past a long breakwater, or reduced movement past a groyne, offshore breakwater, etc. Now there is normally a very sharp beach curvature in the lee of a breakwater, due partly to diffraction effects. While this can be modelled for sand beaches the amount of computer time can be out of proportion to that necessary to get the general shape of the coastline predicted correctly. A "black box" approach can be used to determine the effect of say one or more breakwaters on the adjacent coastline. The basic principle is to assume the alongshore drift is a proportion of the "potential" drift past these structures. As the shoreline advances the trapping capacity of the breakwater is increased and the drift is reduced, and vice versa. The same type of approach is also useful for short groynes, and is believed to give reasonable results except close to these structures.

This type of condition is equally useful at the boundaries or inside the model.

4. Inerodible boundary - this condition can be used to describe beach movement in front of a seawall, allowing accretion to take place but preventing shoreline retreat (Ozasa and Brampton, 1980). When erosion reaches the wall and beach levels are lowered the rate of transport is reduced and then one reverts to the specified transport boundary condition described above. Again this type of boundary can be used at any point along the coastline being modelled.

Modelling the Plan Shape of Shingle Beaches

11.3 Derivation of an Alongshore Transport Formula

The beach mathematical model is a finite difference solution of an equation which expresses continuity of sediment volume moving along the shoreline:

$$\frac{\partial Q}{\partial x} + \frac{\partial A}{\partial t} = 0 \tag{11.1}$$

where Q is the volume rate of alongshore transport of beach material
     x is the distance along the shore
     A is the beach cross-sectional area and,
     t is time.

By denoting the co-ordinate perpendicular to the base-line by y, the area A can be expressed by the product of a depth D and the distance between the base line and the shore, y. Providing that D does not change with time equation 11.1 can be written as:

$$\frac{\partial Q}{\partial x} + \frac{D \partial Y}{\partial t} = 0 \tag{11.2}$$

Here the quantity D, the "closure depth", is normally measured from the toe of the beach to its crest. Any gain or loss of material at a point along the beach results in a seaward or landward movement of the profile parallel to its original position, down to this depth. The errors making this simplification are not significant particularly in the case of a shingle beach overlying a flat sand foreshore.

The usual process which induces the alongshore transport of material, Q, is due to waves breaking at an oblique angle to the coast. Such waves provide excess momentum, part of which drives a current parallel to the shoreline. Longuet-Higgins (1970) using his "radiation stress" theory equated the longshore component of wave momentum flux against the frictional resistance developed at the sea bed by this current. He arrived at a formula for longshore thrust H exerted by waves:

$$H = E \sin 2\alpha \tag{11.3}$$

A. H. Brampton and J. M. Motyka

where E is the energy density in deep water and α is the angle of the wave crests to the beach contours, measured in deep water.

It is generally accepted that the transport of sediment is stongly dependent on the alongshore flux of energy, or momentum. While Longuet-Higgins assumes that no wave energy loss occurs from deep to shallow water we prefer to use a relationship where the thrust, H, is proportional to the term E sin 2α at wave breaking. Both field and laboratory studies support this assumption and thus the CERC formula (Shore Protection Manual, 1977) gives the following relation between the longshore transport Q and the longshore component of wave energy flux, P, measured at the breaker line:

$$Q = 5.7 \times 10^3 \times P \qquad (11.4)$$

Q = longshore transport rate in m^3 p.a.

P = E (nC) sinα cosα

where E is the energy density = ρ g $H^2/8$ at breaking
    ρ is the water density
    g is the acceleration due to gravity
    nC is the group velocity of the waves at breaking
    α is the angle between the wave crests and the local depth contours at breaking.

The HR formula for sand beaches has the same structure except that a dimensionless constant is introduced and sediment transport due to alongshore changes in wave height (Ozasa and Brampton, 1980) is included:

$$Q = \frac{K_1}{\gamma} E \ (nC) \ (\sin 2\alpha - K_2 \frac{\partial H}{\partial x} \cot \beta \ \cos \alpha) \qquad (11.5)$$

where K_1 and K_2 are non-dimensional coefficients
    γ is the submerged weight of beach material in place
    x is the distance in an alongshore direction
    β is the mean slope of the beach face.

For shingle beaches $\partial H/\partial x$ can be ignored unless the wave height gradient is very large and the HR equation then reduces to the CERC type, (Shore Protection Manual 1977), that is

Modelling the Plan Shape of Shingle Beaches

$$Q = K_1 (\gamma)^{-1} E (nC) \sin 2\alpha \qquad (11.6)$$

For sand beaches the constant K_1 is usually quoted as 0.38.

Bruno et al (1981), however, argue that since this type of equation does not take into account sediment size, beach slope, porosity, etc., the value of K_1 is very unlikely to be a constant. However, they consider that the available data is too scanty to formulate the effects of these variables. In practise we have found it necessary to evaluate K_1 from site to site by calibrating the model against recorded beach changes (e.g. for the study described in Hydralics Research Ltd. (1982), $K_1 = 0.28$). Similarly, Pattearson and Patterson (1983) calibrated K_1 against beach changes along the Gold Coast of Australia and found it to be a factor of 3.4 less than the "recommended" CERC value.

For shingle beaches particle size obviously plays an important role in determining the incipience of motion as well as affecting the subsequent rate of movement. Studies carried out at the Delft Hydraulics Laboratory with regular and random waves (Van Hijum and Pilarczyk, 1982) shingle beaches have been used to derive what one might term an incipient motion factor.

For random waves the Delft longshore transport equation is:

$$\frac{Q}{g(D)^2 T} = 7.12 \times 10^{-4} \frac{H_s (\cos\alpha)^{1/2}}{D} \left[\frac{H_s (\cos\alpha)^{1/2}}{D} - 8.3 \right] \frac{\sin\alpha}{\tanh(Kh)} \qquad (11.7)$$

where Q is the volumetric rate of alongshore transport
     g is the acceleration due to gravity
     D is the ninety percentile particle diameter
     T is the significant wave period
     H_s is the significant wave height in deep water
     α is the angle of the wave crests to the beach contours at the toe of the beach
     K is $2\pi/L$ where L is the significant wave length at the toe of the beach
     h is the water depth at the toe of the beach.

The bracketed term indicates that incipient motion begins when $H_s (\cos\alpha)^{1/2} > 8.3D$.

A. H. Brampton and J. M. Motyka

A similar threshold conditon is given by Delft for uniform waves except that H_0 is used and 8.3 is replaced by 8.1. The threshold criterion for both irregular and regular waves thus appears to be very similar, but on an energy basis one might perhaps have expected an rms value of wave height ($0.71H_s$) to be used instead of the deep water wave height H_0. Despite these discrepancies the bracketed term in equation 11.7 can be considered a useful pointer to incipient sediment motion. Since the angle of the wave crests to the beach contours is generally small $(\cos\alpha)^{1/2}$ tends to be close to unity and the bracketed term in equation 11.7 can be simplified to:

$$H_s > 8.3 \, D \qquad (11.8)$$

11.4 Incipient Motion of Shingle

The effect of the Delft type "incipient motion factor" on the rate of shingle movement is shown in figure 11.1. One can see that the transport rate is strongly dependent on the threshold term until the ratio H_s/D is in excess of about 100.

Thus, for a typical D value of 50mm for UK south coast beaches shingle transport would seem to be dependent on particle size until the significant nearshore wave height is 5 metres or greater.

If one considers the threshold of movement in uni-directional flow one can see that the Shields criterion for a fixed particle and fluid density reduces to:

$$\frac{V^2}{D'} = \text{constant},$$

where V is the shear velocity and D' is the sediment diameter. The Delft threshold criterion is of the form H/D = constant and since in shallow water the wave celerity $C \propto H^{1/2}$ then $\frac{C^2}{D}$ is also approximately constant. Thus, the term H in the Delft formula is a surrogate for V^2 in the Shields equation.

It is difficult to judge whether the Delft formula still applies when the sediment is in motion under prototype size waves.

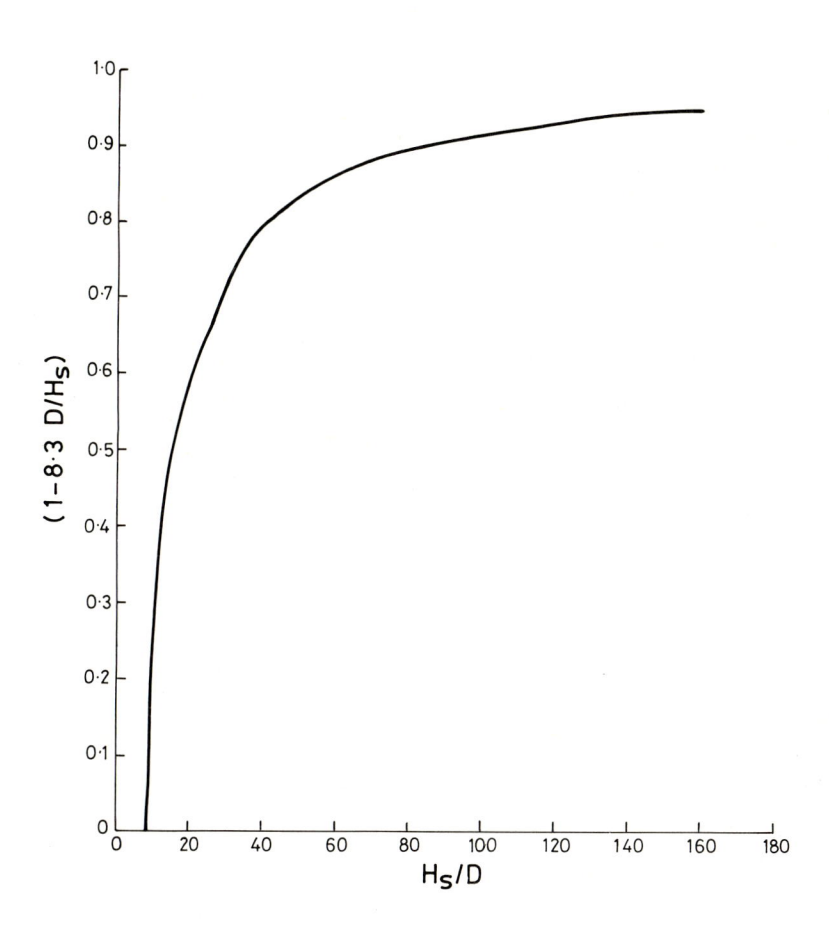

Figure 11.1   Incipient Motion Factor

A. H. Brampton and J. M. Motyka

The Delft random wave tests were limited in number, carried out for small waves and apparently only one particle size was tested. While the form of the threshold factor seems correct it may need to be raised to some higher power to reproduce movement at high energy levels. In uni-directional flow, for example, sediment transport is generally of the form $Q \propto (\tau-\tau_c)^{1.5}$ where τ is a dimensionless shear stress and τ_c is the stress at the incipience of motion. The Delft term suggests that $Q \propto (\tau-\tau_c)^\beta$ and for small waves $\beta \approx 1$. It seems likely that if the experiment had been carried out over a wider range of sediment sizes and greater wave heights a value of $\beta > 1$ may have been more appropriate. To test the sensitivity of the rate of sediment transport to the value of β we have carried out a number of beach mathematical model runs with values of equal to 1 and greater.

For these sensitivity analysis runs, the formula for the alongshore transport rate used was:

$$Q = (K_1/\gamma) \, E \, (nC) \, \sin 2\alpha \, (1-8.3D/H)^\beta \qquad (11.9)$$

where the wave conditions were taken at the point of breaking and K_1 equivalent to that quoted in equation 11.7 above. Using waves that are typical of conditions on the south coast of England, it was found that with $\beta = 1$ the predicted rates of transport were very high. From our experience, expected rates of shingle transport under the conditions used would only be about a tenth of that for sand. Indeed, studies carried out by Southampton University on two stretches of the south coast indicated that the rate of shingle drift may be 18 times less than for sand (Webber, 1980). From this evidence it appears that equation 11.9 may only be valid for small waves. It is interesting to note that the large number of laboratory experiments carried out by the Delft Hydraulics Laboratory with regular waves produced an equation with a different dependence on D than for the irregular wave experiments, namely:

$$Q \propto (D)^{-1} (1-8.1D/H_0) \qquad (11.10)$$

The dependence on the inverse of the sediment diameter recalls the work of Swart (1976).

Modelling the Plan Shape of Shingle Beaches

It would appear therefore that there are good grounds for introducing a dimensionless particle size into the transport equation. This coefficient we suggest should be of the form:

$$\left(\frac{L}{D}\right)^\varepsilon$$

where L is the wave length at breaking. The transport formula can then be modified to:

$$Q = K_1 (\gamma)^{-1} E (nC) \sin 2\alpha \left(\frac{L}{D}\right)^\varepsilon \left[1 - \frac{8.1D}{H}\right]^\beta \qquad (11.11)$$

It should be noted that in this proposed formula there are three, possibly four, quantities which may need to be evaluated from further prototype measurements.

11.5 Discussion

It has been shown that considerable complications arise in trying to develop a mathematical model to predict the evolution of the plan shape of shingle beaches. Laboratory experiments have been of great value in identifying the conditions under which shingle starts to be moved alongshore under oblique wave action. However, the formulae developed for the volume rate of alongshore transport on the basis of those experiments do not appear satisfactory for prototype conditions, and it has been suggested here that modification of the equations may therefore be necessary.

The suggested equation 11.11, presented above, has been formulated on the basis of both the Delft experiments, and on other work including methods for estimating the transport of coarse material is unidirectional flow.

There will be some considerable difficulty, however, in evaluating the various unknown quantities in that equation so that it can be successfully used in prototype situations. The threshold term, as shown in figure 11.1, has a considerable effect under wave heights which are much too large to be generated in suitable laboratory experiments. Any calibration will therefore necessarily have to depend on field measurements.

A. H. Brampton and J. M. Motyka

Unfortunately, suitable sites for prototype study are distinctly difficult to find. What is ideally required is a history of a shingle beach which has been allowed to accrete against a total barrier to alongshore transport such as a harbour breakwater. However, in almost all cases the transport of material in such situations has been hindered by the erection of groynes of variable and unknown efficiency! This is in addition, of course, to the usual difficulty of estimating the incident wave climate, and reducing that climate to a manageable number of wave conditions for use into the computational model.

In order to make progress it will be necessary to instigate a programme of field measurements. At present a number of beach nourishment schemes are being proposed or carried out using shingle. By carrying out regular surveys and wave measurements it may be possible to carry out a calibration. Of course, given the likely dependence on particle size distribution it may be that several such calibration exercises may be necessary before an entirely satisfactory model can be achieved.

11.6   Conclusions

1. A model is proposed for predicting the alongshore transport on shingle beaches. The transport formula includes a term for incipience of motion based on tests carried out by the Delft Hydraulics Laboratory.

2. Future work will concentrate on refining this formula and possibly introducing a dimensionless particle size diameter. Such a term would appear to be justified on the basis of Delft's tests with regular waves. (By comparison with Delft's irregular wave experiments their regular wave tests were carried out for a wide range of wave conditions and beach particle sizes.)

3. It is intended to modify the shingle beach model to allow a much larger range of wave conditions to be input. Because the wave height H appears in the "threshold term" it is not possible to compress the data and produce a weighted value of H as was previously the case.

Acknowledgements

We would like to thank Miss. C.E. Jelliman for modifying the existing beach plan shape model and for carrying out the proving tests for this study. Also we acknowledge support from Department of Environment.

References

Bakker, W.T., 1968.
The dynamics of a coast with a groyne system. Proc. 11th Coastal Eng. Conf., ASCE, pp 492-517.

Bruno, R.O., Dean, R.G., Gable, C.G. and Walton, T.L. Jr, 1981, April.
Longshore sand transport study at Channel Islands Harbour, California. U.S. Army Corps. of Engineers, Coastal Engineering Center. Technical Paper No. 81-2.

Hydraulics Research Limited, 1982.
Coastline changes near Pisa, Italy. A beach mathematical model study. Report No. EX1084.

Le Mehaute, B., Wang, J.D. and Chia-Chi, Lu., 1983.
Wave data discretisation for shoreline processes. Jnl. Waterway, Port Coastal and Ocean Eng. ASCE, Vol. 109 (1).

Longuet-Higgins, M.S., 1970.
Longshore currents generated by obliquely incident sea waves. Jnl. of Geophysical Research, Vol. 75 No. 33, pp 6778-6801.

Ozasa, H. and Brampton, A.H., 1980.
Mathematical modelling of beaches backed by sea walls. Coastal Eng. 4(1), pp 47-64.

Pattearson, C.C. and Patterson, D.C., 1983.
Gold Coast longshore transport. Proc. 6th Australian Conf. on Coastal and Ocean Eng.

Price, W.A., Tomlinson, K.W. and Willis, D.H., 1972.
Predicting changes in the plan shape of beaches. Proc. 13th Coastal Eng. Conf. ASCE, pp1321-1329.

Shore Protection Manual, 1977.
U.S. Army Corps. of Engineers, CERC, Fort Belvoir, Virginia.

Swart, D.H., 1976.
Predictive Equations regarding coastal transport. Proc. 15th Coastal Eng. Conf., ASCE, pp 1113-1132.

Van Hijum, E. and Pilarczyk, K.W., 1982.
Equilibrium profile and longshore transport of coarse material under regular and irregular wave attack. Delft Hydraulics Laboratory, publ. 274.

Webber, N.B., 1980.
Poole/Chrischurch Bay Research Project. Research on Beach Processes. Dept. of Cov. Eng., Univ. of Southampton.

12. MATHEMATICAL MODELLING APPLICATIONS FOR OFFSHORE STRUCTURES

P.J. Cookson,
Wimpey Offshore Engineers and Constructors Limited,
Middlesex. TW8 9AR.

12.1 Introduction

Of the order of 10,000 offshore structures are deployed in the world today. A wide range of disciplines have been involved in this achievement. The technologies required for the design of an offshore platform are given in table 12.1.

TECHNOLOGY	SUBJECT
Oceanography	Collection and interpolation of environmental data; wind, waves, current, tide, ice.
Foundation Engineering	Determination of soil characteristics. Modelling of soil/structure interaction, scour.
Structural Engineering	Materials and corrosion. Welding. Structural analysis. Design for fabrication and installation.
Marine Civil Engineering	Navigation and positioning. Installation methods. Installation equipment.
Naval Architecture	Flotation and buoyancy. Towing. Launching. Controlled flooding.

Table 12.1   Technologies involved in Platform Design

Each of these technologies requires at some time mathematical models in the execution of its work. New applications arise and improvements are continually being sought for mathematical representations of practical problems. It is necessarily important for the analyst to be aware of the practical aspects of any problems being addressed to ensure that the model adequately represents 'real life' behaviour.

P. J. Cookson

In this paper the practical constraints influencing the conception, design construction and operation of an offshore structure are reviewed. Some applications of mathematical modelling are highlighted and a summary of typical analytical tools for platform design is given.

12.2 Operational, Environmental and Foundation Condition

The spatial requirements and geometric arrangement of the topsides of any installation are dictated by the functional operations of the platform e.g. whether drilling, production and accommodation facilities are required. The gross weight of these facilities affects the form and size of the supporting structures.

For deep water developments (say over 100 metres), the cost of the supporting structure is generally so great that multiple platform installations at one location cannot be considered for economic reasons. In shallow water the economic case for single or multiple platforms is not usually clear-cut and the decision can be based on technical and other considerations, such as safety preferences and programme flexibility.

Ideally accommodation facilities should be located on a separate platform as this is clearly the safest arrangement. In addition locating the drilling and production facilities on separate platforms is sometimes advocated for safety reasons. There can be a programme advantage to this arrangement; it allows drilling to start before the production facilities are installed, bringing forward the 'first oil' date.

There may also be structural reasons for locating the facilities on multiple platforms. Multiple platforms enable a higher proportion of the topsides plan area to be cantilevered outside the area of the top of the jacket. This may result in a lighter total jacket weight for multiple jackets than for a single jacket.

The environmental factors; water depth, tides, currents, wave heights, wind speeds, temperatures and ice conditions, must be assessed as accurately as possible before the loading on the structure can be calculated. Environmental loads are predicted in probabilistic terms. Extrapolations from basic data, using mathematical models, must be undertaken with care if realistic design conditions are to be defined.

Modelling Applications for Offshore Structures

A comprehensive investigation of the sea bed is carried out to determine the soil characteristics at the proposed site of the installation. The soil conditions significantly affect the form of structure to be used, e.g. whether piled platforms or structures relying on self weight for stability are viable. Foundation movements during the life of an installation must not affect its structural integrity or functional operation and a platform's site and design must be determined accordingly.

12.3   Structural Concepts

The early steel jackets had four legs which served as templates for piles which were driven through the legs and welded at deck level, the annular space being grouted. With advance to deeper water the leg diameter and spacing were increased and the braces had to withstand heavier lateral and longitudinal loads. The large diameter legs required extensive stiffening, and with the large volume of welding plus a 100 percent radiography requirement the cost escalated. Because barges of adequate capacity did not exist those structures were built in specially constructed dry docks on flotation tanks, with radio/servo controlled valves for sequentially flooding the compartments of the tanks. The controlled flooding of the tanks enabled the structure to pass through the point of instability to a position of vertical equilibrium, whereupon the tanks were detached and returned to the dock for use on the next structure to be built.

The next structures were 'self-floaters' (figure 12.1). Two of the legs were of sufficient diameter to provide buoyancy for transportation to site. The large diameter legs had to be designed for a combination of longitudinal compression, punching shear, external pressure, and locally applied forces from the pile clusters. With the advent of large purpose-built launching barges it is now convenient to build structures with eight or more legs, at ground level. The legs can then be of smaller diameter, with less stiffening yet with moderate wall thickness. The braces are shorter and of smaller diameter, and the wave load attracted by the structure is reduced. Fabrication, construction and inspection are simplified. The foundation loads are better distributed and attachment of the jacket to the piles is easier. A barge launched jacket is shown in figure 12.2.

For suitable ground conditions concrete and steel gravity platforms can be used with the option of oil storage. The buoyancy/weight distribution

P. J. Cookson

Figure 12.1   Self Floating Platform

Modelling Applications for Offshore Structures

Figure 12.2   Barge Launched Platform

P. J. Cookson

of gravity platforms is such that stability can be maintained during submergence with a fully equipped deck so inshore deck installation, with shallow draught tow-out, becomes possible.

However, the weather window (the period when environmental conditions are likely to be sufficiently benign to permit offshore operations with minimal risk) can severely limit the programming of gravity platform installation. Analytical studies of towing stability to demonstrate that these structures can be installed in rougher sea states could significantly affect their competitiveness compared to piled structures.

For steel jackets the usual practice has been to install the platform structure and subsequently to place the deck equipment piecemeal. This necessitates modular assembly of equipment, the size of modules being dictated by the capacity and availability of crane barges. Each lift is subject to suitable weather. With suitably designed structures it is possible to install the deck and its equipment inshore, either in integrated or modular form. Deep water is then required both for installation and transportation, and during transportation and installation the whole platform is at risk. It is also possible to install a fully equipped deck offshore after the structure has been installed and secured; only a single short period of good weather is then required for the deck installation.

With the trends towards deeper waters floating and compliant structures are receiving increasing attention. Cable tethered boyant structures have all components of motion. Vertical cables limit vertical motion but allow considerable horizontal motion. Inclined cables reduce the horizontal motion but are more exposed to snagging. Rigid leg hinged structures are vulnerable to any malfunction of the hinges. The rigid structure can be operationally better and provides support for conductor tubes and risers. Design and analysis are simpler and displacements can be predicted with good accuracy. The economic depth threshold for rigid structures seems to be increasing, partly through confidence brought by experience, partly through improved design, and partly through advances in construction techology. The Cognac (313m water depth) and Hondo (260m) structures are examples of advancing construction technology, underwater pile driving making a key contribution to the Cognac structure.

Modelling Applications for Offshore Structures

For the exploratory phase of field development the requirement is for a platform which moves as little as possible, which can stay in position with a satisfactory probability whilst drilling a hole, which can move from station to station, and which has a good tolerance of the marine environment, both during drilling and in taking up position. The acceptable heave depends on the capacity of the heave compensator, whilst the acceptable translational and rotational movements depend on the flexibility of the drill string and depth of water. A drill string can accept a translation of the order of 1/20 of the water depth, with rotation of several degrees. The vertical force varies along its length. At the same time it is subject to currents (which can cause vibration) and to wave induced flows. An analytical basis for the deformation and stresses might provide useful comparative data for the drilling practitioner, if only to demonstrate the relative sensitivity to the various actions.

The drill ship with moon pool has the advantage of rapid transit between stations and if it is dynamically positioned it can take up and leave station instantly. However, the ship motions, and thruster (or anchor) capacity place limits on the weather conditions through which it can operate. The semisubmersible provides a much more stable platform, be it dynamically positioned or anchored, but it is slow and more costly.

The jack-up barge offers an almost stationery platform for water depths up to about 100 metres, but movement between stations is slow and jacking up and down must be performed in good weather. The behaviour of a jack-up during weight transfer, leg pre-loading and in service provides an interesting study in soil-structure and wave-structure interaction. A complete analytical prediction for these structures in real-life operation is complex but comparative studies coupled with service experience might lead to some economies in design.

The free motion of semi-submersibles is within the scope of existing computational techniques and programs exists to incorporate the restraint afforded by mooring cables, but the cable response is assumed to be linear.

P. J. Cookson

12.4 Fabrication

The deepwater jackets for the hostile environment of the North Sea demand high standards of material and welding specification and inspection and close construction tolerances. In general 100% inspection and recording is required.

There are three main types of steel used currently in steel platforms:-

a. Special structural steel is B.S 4360 50D (modified) Type II normalised. This grade is extensively used for node barrel fabrication where major forces are applied through the thickness of the material, and where high stress concentrations are caused by welding of complex stub branch configurations.

b. Primary structural steel is B.S 4360 50D normalised Type I and differs from modified 50D mainly in its permitted higher level of inclusions such as sulphides and silicates, with a lowering of the through thickness properties, toughness and ductility. The major part of a platform is fabricated using tubulars rolled from this steel.

c. Secondary structural steel is B.S 4360 43C normalised and is used on walkways, stairways, helidecks, plate stiffeners and other secondary parts of the structure.

Weldments must be of sufficient toughness to resist brittle fracture through the welds. The selection of consumables and processes to be used in fabrication depends on the size, shape and dimension of the sub-assembly. Assemblies are manipulated and positioned wherever possible for fully automatic welding. This process is standard practice for tubular manufacture in the pipe rolling mill where circumferential and longitudinal seams are relatively easy to automate.

The procedure adopted for node fabrication is as follows:-

1. Cut barrel plates to size and ultrasonically check edges for laminations.

2. Roll and weld tubes; after 48 hours check seams by X-ray or ultrasonics.

Modelling Applications for Offshore Structures

3. Cut and bevel tubes as required for perpendicular and angled intersections.

4. Place heating pads around weld area and heat to the specified temperature.

5. Fit and weld ring stiffeners in specified sequence and check by NDT.

6. Fit and tack-weld stubs and carry out dimensional check before fully welding.

7. Conduct NDT on finished node and check dimensionally.

8. Support in furnace for stress relieving.

9. Repeat NDT check on all welds and repeat dimensional checks.

12.5 Construction

Preliminary planning at the estimating stage leads to a basic approach to erection of steel jackets. Planning is based around a small number of major lifts in the 200-1000 tonne range and a flexible programme of infill erection.

The major parameters that govern assembly and erection sequences are:-

- o weight and size of component,
- o position within the jacket,
- o position of risers etc. within the jacket structure,
- o the capability of the proposed component to withstand the induced stresses during lifting,
- o available cranage.

Two approaches to erection are:-

i) Toast Rack Erection

In this method the transverse panels are set across the lower legs. This intermediate stage prior to upper leg installation and brace tube infilling, suggests the description used. The method relies

P. J. Cookson

on the transverse panels being supported by another part of the jacket and minimises temporary supports such as guys or steelwork. It is usual to start from the centre of a jacket and to work simultaneously out to the ends thus increasing the number of work areas available at any one time.

ii) Panel Erection

In this method, panels are erected as units and transverse infill panels and braces are added subsequently. The main legs in this case form part of the longitudinal panels. These main panels are very heavy and are usually rolled or lifted up with crawler cranes (see figure 12.3) which, being several in number, rely on the drivers' skill to maintain vertical and equal lifting forces. The first panel erected must be supported by cranes or guy wires until the second panel is erected and sufficient infill has been fixed in position to make that part of the jacket self-supporting.

Before any component is erected. it is checked dimensionally to ensure that it will fit the space in the jacket and that the lines of contact have a weld gap that will be within tolerance. If any extra fabrication is required, it must be known and completed before the cranes do any lifting to avoid the need to lower the component back to the ground. Construction by trial and error must be avoided.

Dimensions of space and component have a tolerance of plus or minus 2mm, all measurements being corrected for temperature. Direct measurements are made by steel tapes and levels. Indirect measurements are made using a system of multiple station three dimensional triangulation which enables a number of redundant measurements to be cross checked by computer. Photogrammetry is used when large assemblies have to be matched to openings in the structure.

12.6   Load Out

Barge launched steel jackets are constructed on their sides on a skid rail system. Following completion of the structural framework the jacket is winched or jacked on to the transportation barge. There is a surprising lack of information on frictional resistance. The barge is either grounded for this operation end-on to the quay wall or alternatively deballasted to counteract the jacket reaction on the barge. The lack of

Modelling Applications for Offshore Structures

Figure 12.3   Panel Erection in Jacket Construction

P. J. Cookson

vertical alignment or varying support stiffness during this operation requires careful design consideration to avoid overstressing the jacket structure. When the structure is in position on the barge, extensive sea fastenings have to be completed to withstand the horizontal shears and uplift forces resulting from the assumed sea state.

Self floaters or structures built on buoyancy tanks in a dry dock must be restrained against wind forces and manoeuvred out of the dock after flotation. The transfer of control from the fixed moorings to the tug lines involves careful planning, and has been accomplished by cutting the moorings by explosives.

12.7   Tow Out

Platform structures, their transporation tanks or barges, and the connections between them, must be designed to resist floating body motions in seas of an assumed severity. The floating characteristics of gravity platforms differ widely between submerged and non-submerged conditions. Tower moments due to roll can be quite high. Model tests are performed to confirm or augment calculation.

Sufficient towing power must be provided to at least maintain position in the worst weather, and in coastal areas or in narrow channels flanked by water of insufficient depth navigation and positional control must be of a high order. Standard mariners' charts are totally inadequate bathymetrically and special hydrographic surveys are required to define the route. Standard charts are based on soundings at rather wide centres, without the benefit of side-scan equipment, and with normal ship draughts in mind. The width of channel to be surveyed and the required precision of navigation and control are of course linked. Shore based precise navigation systems are then set up to enable the navigators afloat to inform the tow master of any deviation from route.

Tidal streams and currents are important and these also are not adequately defined on mariners' charts. In particular, flow at depth can be totally different in magnitude and direction from surface flows. The tow out operation is highly weather dependent; in bad weather seamanship must match the engineering and navigational provisions which has been made.

Modelling Applications for Offshore Structures

12.8 Installation

Where the jacket is launched from a barge a dynamic analysis of the jacket motions and stresses as it leaves the barge and enters the sea has to be performed. A critical stage of the loading occurs when the jacket pivots on the rocker arm at the stern of the barge. The drag forces on members entering the sea have to be determined. The motion of the jacket during launch must be such that it does not hit the sea bed and also that it does not overturn and achieves a stable floating position with the jacket upright. A similar but slightly less dramatic procedure is required with jackets which are self floating or supported on buoyancy tanks.

Once the jacket is upright it has to be manoeuvred into the correct position and orientation in the required target zone. This may be achieved by partially lifting the jacket with a crane barge or by mooring lines to tugs. The jacket is then lowered to the sea bed by flooding part of the structure. Pin piling of the jacket is performed to maintain its position during summer storm conditions. This is followed by driving and grouting the main piles.

12.9 Mathematical Modelling in Platform Design

The sequence of events describe above must be considered in the design of a platform. The principal conditions are:-

- o   Fabrication
- o   Load out
- o   Transportation
- o   Installation
- o   Storm loadings
- o   Fatigue history

Analyses undertaken by Wimpey Offshore Engineerings and Constructors Limited centre around the program OFFPAF, an in-house development of the PAFEC analysis system. The principal features are summarised in Table 12.2. The OFFPAF system is fully compatible with other packages operating from the same database. In addition to the PAFEC system a substructuring program, interactive colour graphics and a topside weight control system are used.

Element types:	Simple tubular, composite tubular, tubulars with end offsets and eccentricities, variable section tubulars. (Plus all standard elements from the PAFEC system).
Joint types:	Rigid or flexible joints.
Wave theories:	Linear, 5th order Stokes, low and high order streamfunction theory.
Loading types:	Wave loadings (Morison's equation), wave slam, buoyancy, current, marine growth, gravity and thermal loads.
Analysis types:	Static deterministic, static spectral, dynamic spectral, earthquake.
Foundation modelling:	Fully non-linear pile-soil-structure interaction, accounting for non-linear p-y curves and non-linear beam-column behaviour. Single piles or pile groups may be modelled. This coding makes use of substructuring techniques for computational efficiency.
Launch and Float:	Three dimensional launch and flotation analysis, using time integration to determine jacket motions. A structural analysis may be performed at each time step, to check member and joint stabilities.
Member Code checks:	AISC, API and BS5400 (for composite members).
Joint Code checks:	API and EUG (for grouted joints).
Fatigue Analysis:	Deterministic or spectral. An SN curve is permissible, specified by the user. Fatigue damage is calculated at a minimum of 8 points around circumference of joint. Stress concentration factors are generated using parametric formulae (Kuang; Marshall and Kellogg; Smedly and Wordsworth).
Fatigue Fracture Mechanics	Using an influence coefficient approach, data file of member stresses is written to perform crack growth using the weight function program FPDSSM.

Table 12.2   Main Features of OFFPAF

Modelling Applications for Offshore Structures

Substructuring is the scheme whereby individual components of a complete structure may be generated and analysed individually. Substructuring capabilities are required in the following types of analyses:-

o   Conceptual designs. By using substructuring, only the portions of the structure which have been modified require re-analysis.

o   Structures in which there is repetition or symmetry. In particular, by using substructuring in this manner to achieve the principal of 'divide and conquer', very large analyses (e.g. gravity platforms) may be successfully accomplished.

o   Non-linear problems. The linear portions of the structure do not require to be re-analysed for each load increment when substructuring is used, and hence major savings in computer time may be affected.

o   Pile/structure interaction. If the foundations are contained in a separate substructure, solution of the non-linear foundation problem may be performed efficiently.

Because of the growing importance of weight engineering and control, it has become necessary to develop purpose written software as an extension of the structural analysis systems. A program has been written as a database system for storing equipment details, and calculating the weights and centres of gravity of platform components, including topside modules, deck support frames and appurtenances. The information for each platform is divided into several categories, e.g. equipment list, topside steel, miscellaneous, jacket steel and template steel. The program is fully interfaced with the front end of the OFFPAF system, providing an efficient link between the functions of weight control and structural analysis.

To expedite the determination of suitable structural configurations for offshore structures and components a program (OFFGEN) was specifically developed. Jacket structures are handled by an interactive generation of 'stick models'. Plots of code check values and fatigue lives on drawings of the jacket are produced for reporting purposes. Finite

P. J. Cookson

element meshes (shell and 3D) are generated for the analysis of simple and complex tubular joints.

Other more specialised analysis programs used deal with:-

- o   Fracture mechanics - for both linear and elasto-plastic problems
- o   Non-linear buckling - for plastic and large deformation analysis
- o   Floating stability
- o   Pile driving
- o   Earthquake analysis
- o   Non-linear pile analysis

12.10   Conclusions

Offshore engineering involves a broad spectrum of technical disciplines. Each of these technologies is frequently encountering new problems which are often best addressed through mathematical modelling. However, in using such techniques it is important that the analyst fully understands the practical constraints applying to his problem. Only if these conditions are adequately represented in the analysis will mathematical modelling fulfill its potential as a safe, accurate and economic problem-solving tool.

13. MATHEMATICAL MODEL OF A MARINE HOSE-STRING AT A BUOY: PART 1 - STATIC PROBLEM

M.J Brown,
Department of Applied Mathematical Studies,
The University of Leeds, Leeds,
West Yorkshire, LS2 9JT

13.1 Introduction

Single point mooring (SPM) installations enable very large oil tankers to be loaded or unloaded without the need of deep water facilities. The most common system is shown in Figure 13.1 and comprises of a floating buoy secured to the sea-bed by chains. The tanker's hawser and the surface hose are connected to a turntable on top of the buoy in response to the prevailing wind, wave and current forces. The oil is transferred from the buoy to the sea-bed manifold by an underbuoy submarine hose and then to the onshore facilities by a steel pipeline.

The surface hose-string consists of a large number of reinforced rubber hoses joined together by steel flanges at each end. Usually, all of the hoses apart from the first off-the-buoy have floatation covers along their entire length. This first hose, which is specially designed to withstand the severe stresses and strains imposed by the buoy's motion, is initially free of floatation covers so that it can act as a damper by absorbing small horizontal movements of the buoy. This unfloated section also has extra reinforcement which tapers gradually along its length so that the large stresses at the end of the first flange are transferred into the main body of the hose, as described in Dunlop (1971).

The performance of a SPM terminal is highly dependent upon that of the surface hose-string since hose failure whilst a tanker is discharging or loading will have both economic and ecological consequences, with the cost of de-moorage and replacement exceeding that of the original hose.

The likelihood of hose failure occuring is dependent upon the period for which it has been operating. The failure rate for a surface hose can be separated into three distinct phases, as shown in Figure 13.2. In the first phase, in which the rate of failure decreases, failure occurs due to defective or badly handled hoses. There is then a period of time with a constant rate of failure which corresponds to the useful working

M. J. Brown

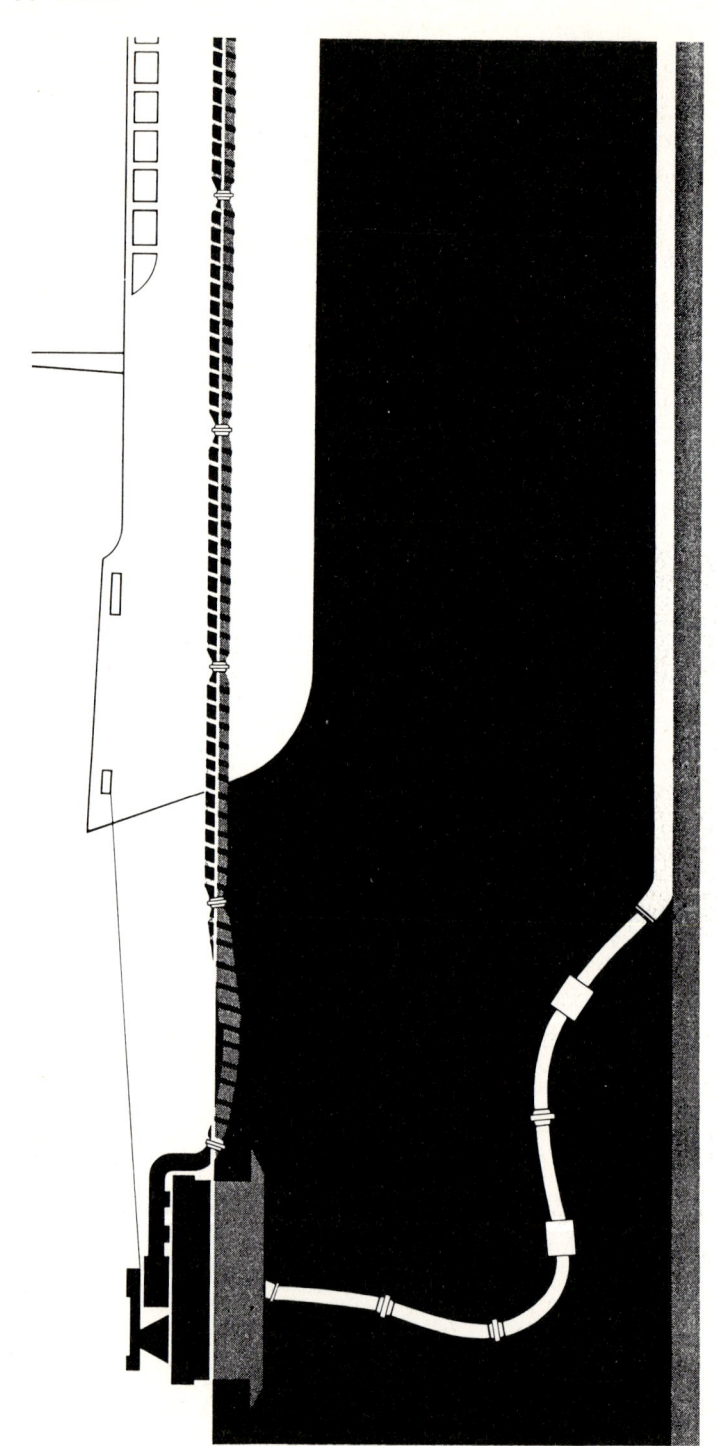

Figure 13.1 Single Point Mooring System

Mathematical Model of Hose String - Static

life of a hose. The third phase indicates when a hose has become worn out and is characterised by an increasing failure rate. Due to the stresses induced in the hose by the motion of the buoy, the life-expectancy of the first and second hoses off the buoy is shorter than for the rest of the hoses in the string.

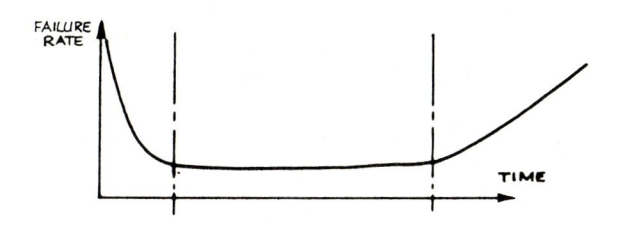

Figure 13.2   Hose Failure Rate

Since any stresses caused by wave-action and the movement of the buoy have to be combined with those inherent in the system, it is hoped that by obtaining a mathematical model of a static hose-string, these inherent stresses can be minimised with a resulting reduction in the hose failure rate.

In order to gain more information about hose failure, hose manufacturers have undertaken various studies to measure the loads that the hoses are subjected to and their performance under them. These include field measurements (Brady, 1974, Dunlop, 1973, Saito, 1980), test rig experiments (Young, 1980) and scaled model tests (Tscheope, 1981, Graham, 1982). Nevertheless, attempts at modifying the design of the hoses so as to increase their life-spans are still a matter of trial and error with the modified hose being studied on a dynamic test rig or observed in operation for a trial period. Due to the cost of manufacturing and testing these prototypes the manufacturers have developed various numerical models to provide an alternative for the assessment of design parameters. Dunlop (1976) and Bridgestone (1976) have both presented models for calculating the stresses in a static hose-string. However, whilst both took the same assumptions as those used in this paper, their approach in solving the resulting equations requires many simplifications to be made to the hose parameters and, hence, considerably limits their use in simulating the physical problem.

M. J. Brown

13.2 Assumptions

In order to model this complex physical problem it has been necessary to retain only its most significant features so as to reduce it to manageable proportions. Consequently, the following assumptions have been made so that the hose can be modelled as a beam subject to pure bending.

1. The effect of shear forces and horizontal forces upon the curvature can be neglected so that the curvature is only dependent upon the bending moment.

2. The curvature of the hose can be approximated by

$$\frac{1}{r} = \frac{d^2y}{dx^2} \qquad (13.1)$$

i.e. the term $(\frac{dy}{dx})^2$ is neglected in comparison with unity. This linearised model introduces a constraint on the maximum slope of the hose-string, yet in some of the calcuations this has been relaxed to allow the manifold slope to be set as high as $15°$ so as to match certain buoy design data. The error introduced into the bending moment by this value is under 10%, so it is felt that the linear model meets the objective of this paper, namely an understanding of the various hose parameters and the effect of the different expressions for the load, whilst reducing the computing time and satisfying all the available design data.

3. Since the gradient of the hose has been assumed small the hydrostatic forces are treated as those arising from the horizontal situation. (Such an approximation introduces and equivalent order of error to that of linearising the curvature).

4. Since the hose will contain either oil under pressure or sea-water, the hose is assumed to be solid with the bending stiffness being constant over any cross-section perpendicular to the axis of the hose. However, the bending stiffness can be made dependent upon the pumping pressure or the radius of curvature.

5. Longitudinal forces will be induced into the hose by the action of internal and external forces. This phenomenon has been studied extensively in relation to cables, piplines and risers,

Mathematical Model of Hose String - Static

e.g. Nordgren (1982), Felippa and Chung (1981) and Bernitsas and Kokkinas (1983), but as the depths involved in the present problem are small such effects can be neglected.

Under these assumptions the hose can have flanges and the length of each hose or flange can be varied. The radius of the hose can also be changed so that the hose includes both floated and unfloated sections.

13.3 Equations

The equilibrium equations for the above small strain model are as given by Timoshenko (1955), namely

$$\frac{d^2y}{dx^2} = -\frac{M}{EI_z} \qquad (13.2)$$

$$\frac{dM}{dx} = V \qquad (13.3)$$

$$\frac{dV}{dx} = -Q \qquad (13.4)$$

It is convenient to non-dimensionalise the physical quantities arising in the problem in the following way

$$x = \bar{x} L \qquad y = \bar{y} L \qquad Q = \bar{Q} q'$$
$$M = \bar{M} m' \qquad V = \bar{V} v' \qquad (13.5)$$

where L is the total length of the hose-string, and was chosen so as to try and avoid instabilities in the numerical method at large values of x (the displacement behaves like e^{mx} - see later).

The values of m' and v' are given by

$$m' = \frac{d\, EI_r}{L^2} \qquad v' = \frac{d\, EI_r}{L^3} \qquad (13.6)$$

where m' is a ratio of the bending stiffness to the radius of curvature, whilst v' is a ratio of the bending moment ot the length of the hose-string.

Substitution of equations (13.5) in equations (13.2), (13.3) and (13.4) gives

$$\frac{d^2\bar{y}}{dx^2} = -\bar{M} \qquad \frac{EI_r}{EI_z} = -\frac{\bar{M}}{F} \qquad (13.7)$$

$$\frac{d\bar{M}}{dx} = \bar{V} \qquad (13.8)$$

$$\frac{d\bar{V}}{dx} = -\bar{cQ} \qquad (13.9)$$

where $F = \frac{EI_z}{EI_r}$ is the bending stiffness ratio and $\bar{c} = \frac{L^4}{d} \frac{q'}{EI_r}$.

For the remainder of this paper the non-dimensional equations (13.7), (13.8) and (13.9) will be used and for convenience the bar will be omitted.

13.4 Boundary Conditions

The effect of wave-action upon the buoy results in the manifold being designed so that the initial hoses are at least partially submerged, so enabling the hose-string to absorb small horizontal movements of the buoy. For the static situation the initial displacement, Y_0, and slope θ_0, of the hose at the buoy are govened by the design criteria of the manifold and are treated as specified for this model.

Providing that a sufficient length of hosing is considered, the stresses on the hose at the buoy can be treated as independent of those at the tanker. For the static situation, the stresses on the hose at a large distance from either the buoy or the tanker will be very small, since this section of hosing will tend asymptotically to its equilibrium position i.e. lying flat at its equilibrium depth, Y_e. Since Y_0, θ_0 and Y_e are easily obtainable, the boundary conditions used throughout this paper are;

at the buoy, $x = 0$ $y = Y_0$ $\theta = \theta_0$
at the hose end, $x = 1$ $y = Y_e$ $\theta = 0$ (13.10)

Mathematical Model of Hose String - Static

13.5 Hose Radius

The first hose off-the-buoy is designed differently to the rest of the hose-string since it has to withstand the severe stresses imposed on it by the motion of the buoy. Consequently, the first hose has extra reinforcement and initially has no floatation covers. For simplicity the following has been taken to represent the hose radius.

$$r = r_u \qquad\qquad 0 \leq x < x_u$$

$$r = \frac{(r_f - r_u)}{(x_f - x_u)}(x - x_u) + r_u \qquad\qquad x_u \leq x \leq x_f$$

$$r = r_f \qquad\qquad x_f < x \leq 1 \qquad (13.11)$$

where the suffixes u and f indicate an unfloated or floated section of the hose.

13.6 The Load

If a hose is partially submerged a buoyancy force per unit length, B, is generated which is equal to the weight of the water displaced by the hose, and is consequently a function of the depth below sea-level of the axis of the hose. The buoyancy force is given by

$$B = \rho \left[r^2(\pi - \alpha) + y^2 \tan\alpha \right] \qquad (13.12)$$

where $\alpha = \cos^{-1}(y/r)$.

When the hose is completely submerged the buoyancy force becomes independent of depth and is a constant. This maximum value, B_{max}, is given by

$$B_{max} = r^2 \pi \rho \qquad (13.13)$$

No buoyancy force is generated when the hose is out of the water. In all three cases the load per unit length, Q, is given by the weight per unit length of the hose, W, minus the buoyancy force.

Therefore, the load for what is referred to as the buoyancy approximation throughtout this paper is given by

$$Q = W \qquad -r > y$$
$$Q = W-B \qquad -r \leq y \leq r$$
$$Q = W-B_{max} \qquad y > r \qquad (13.14)$$

Two alternative expressions were developed for the load. Firstly, there is the expression that is referred to as the elastic foundation approximation. Its name derives from the fact that, as can be seen in Figure 13.3, the variation of the buoyancy force with depth for $|y| \leq r$ is almost linear and so the hose can be treated as if it lies on an elastic foundation, whereby if the hose is partially submerged a restoring force is generated proportional to the distance from the equilibrium depth. As with the buoyancy approximation, if the hose is out of the water the load equals the weight of the hose, and if it is completely submerged it is given by its weight minus the maximum buoyancy force.

Therefore, the load for the elastic foundation approximation is given by

$$Q = W \qquad -r > y$$
$$Q = W-K(r+y) \qquad -r \leq y \leq r$$
$$Q = W-B_{max} \qquad y > r \qquad (13.15)$$

where B_{max} is given by equation (13.13) and $K = \dfrac{B_{max}}{2 \cdot r}$.

Both Dunlop (1976) and Bridgestone (1976) used this approximation for the load. Although some information is lost by using a linear polynomial to approximate the buoyancy force when the hose is partially submerged it has the advantage that for simplified problems, the system of equations can be solved analytically and hence it can be used to test the numerical integration.

Another expression for the load can be obtained by using a tanh function to approximate equation (13.14). This approximation was considered since it has continuous derivatives which it was thought might improve the rate of convergence, whereas equation (13.15) has a discontinuous first derivative at $y = \pm r$ and equation (13.14) has a discontinuous second derivative, also at $y = \pm r$.

Mathematical Model of Hose String - Static

The equation for the tanh approximation is given by

$$Q = W - \tfrac{1}{2} B_{max} (1 + \tanh \beta y) \qquad (13.16)$$

where β is a constant chosen such that the load obtained has the smallest difference (least squares) from the buoyancy approximation and is dependent upon the radius and weight of the hose and of the density of the sea-water. For the values used in this paper (as given in Appendix I) a value of $\beta = 2.4d$, where d is the vertical scaling factor, was taken.

Figure 13.3 compares the load obtained by the three approximations for the typical values as given in Appendix I. It should be noted that each approximation has a different equilibrium depth i.e. the depth where the buoyancy force equals the weight of the hose. The equilibrium depth for the tanh and the elastic foundation approximations are respectively given by

$$Y_e = \frac{1}{2\beta} \log_e \left(\frac{W}{B_{max}-W}\right) , \quad Y_e = \frac{W}{K} - r . \qquad (13.17)$$

No analytic expression could be derived for the equilibrium depth of the buoyancy approximation and so a computer subprogram was written to calculate this.

By using these three approximations for the load it is possible to detect any errors caused by taking the more approximate forms for the load, as was the case in Dunlop (1976) and Bridgestone (1976).

13.7 Method of Solution

By putting $\theta = \frac{dy}{dx}$, equation (13.7) can be separated into two first order differential equations. Hence, equations (13.7), (13.8) and (13.9) can be expressed as four first order differential equations in terms of the vertical displacement, the slope, the bending moment and the shear. These four equations together with the four boundary conditions (13.10) form a two-point boundary value problem which is solved numerically. The unknown boundary values for the moment and the shear have to be estimated and then the equations are integrated from the boundary values at each end into a specified matching point by the Runge-Kutta Merson method. For the true solution, the difference

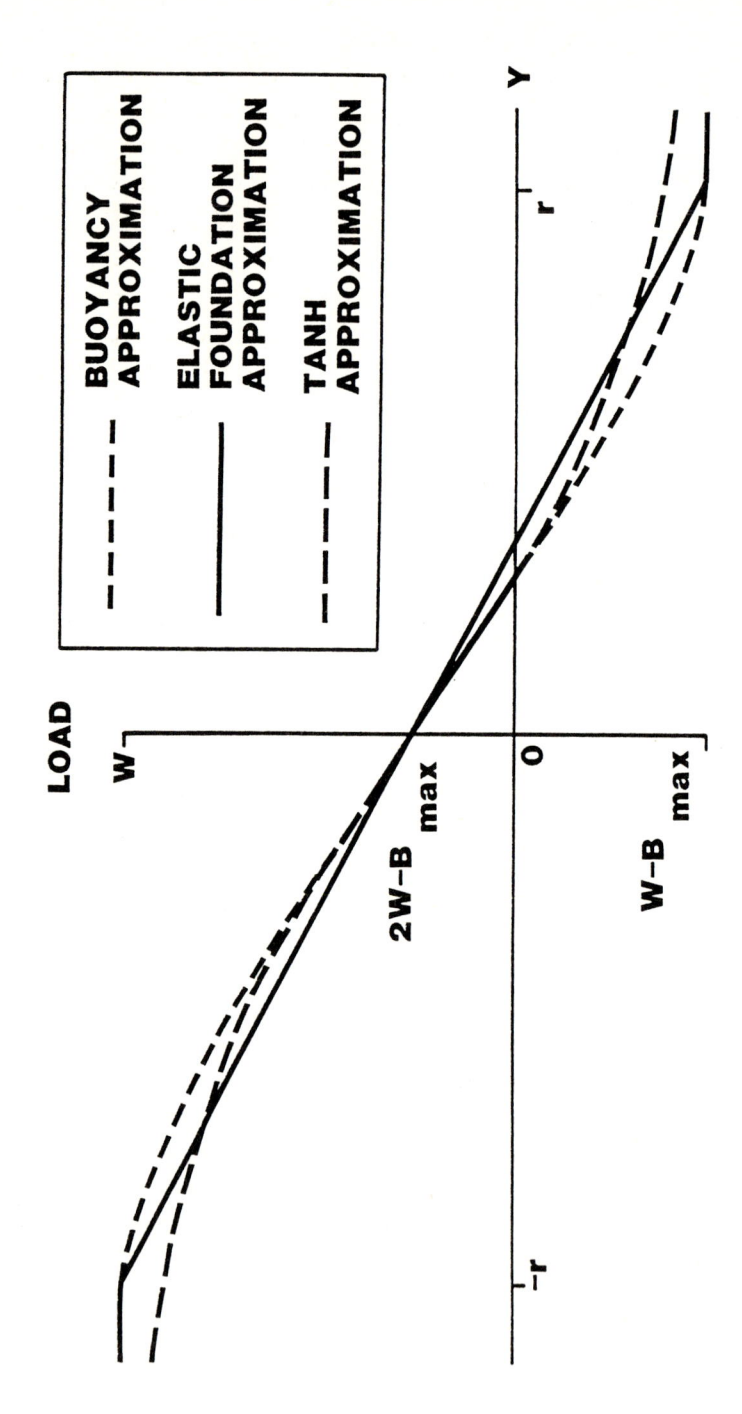

Figure 13.3 The Three Load Approximations

Mathematical Model of Hose String - Static

between the values obtained by the forward and the backward integrations, known as the residual, should be zero for each equation at this point. If the residual is larger than the required accuracy a generalised Newton method is used to reduce it by calculating corrections to the estimated boundary values. This is repeated until convergence to the required accuracy is obtained.

13.8 Analytical Solutions for Simplified Models

As shown in Dunlop (1976) and Timeshenko (1955), analytic solutions exist for certain simplified problems when the load is assumed constant or proportional to the displacement. By using a numerical approach it was felt that a more sophisticated model of the physical problem could be obtained. Nevertheless, analytic solutions were derived for specific cases during the development of this paper so that a means of testing the accuracy of the numerical integration was available.

For a hose with continuous bending stiffness on an elastic foundation the differential equations become

$$\frac{d^4 y}{dx^4} = - c K y , \qquad (13.18)$$

the general solution of which is

$$y = e^{mx}(A \cos mx + B \sin mx) + e^{-mx}(C \cos mx + D \sin mx) \qquad (13.19)$$

where $m = (\frac{Kc}{4})^{1/4}$.

The constants of integration are specified by the boundary conditions, equations (13.10). Table 1 compares the analytical and numerical solutions for equation (13.19). From this, and other, analytic solutions it was felt that the correlation between the two sets of results justified the use of this method when flanges, floatation covers and the alternative approximations of the load are all present, and the problem becomes such that no analytic solution is possible.

x	analytical soln y	numerical soln y
0.00	1.250000e-2	1.250000e-2
0.10	2.125219e-2	2.125219e-2
0.20	1.332792e-2	1.332792e-2
0.30	4.975570e-3	4.975572e-3
0.40	4.975745e-4	4.975759e-4
0.50	-8.681748e-4	-8.680890e-4
0.60	-7.982872e-4	-7.984232e-4
0.70	-4.032291e-4	-4.033546e-4
0.80	-1.220547e-4	-1.221161e-4
0.90	-1.338718e-5	-1.016440e-5
1.00	0.000000e-0	0.000000e-0

Table 1   Loaded Hose without Flanges

13.9 Results

The case of the unloaded hose was initially investigated so that, independent of the load, the general effect of the parameters could be obtained. For this problem, the profile of the hose is mainly determined by the boundary conditions since there is no load to affect the results. The general behavour of the hose was as expected, e.g. if its stiffness was increased then its bending moment would also increase, and since this case exhibited the same general features as the loaded case no results for it will be presented.

For the unloaded hose approximately 4 iterations to the estimated boundary values, taking in all approximately 0.25 seconds of computing time on an Amdahl V7, were normally required to achieve convergence with a maximum residue of $O(10^{-5})$.

The loaded hose was considered after an understanding of the parameters had been obtained. The profile of the hose was then determined by the load, with the boundary conditions having an increasing effect as either the load or the length of hosing was decreased. For the values given in Appendix I it was found that four hoses were sufficient to ensure that the hose-string had reached its equilibrium state.

Numerous runs were carried out, with each of the parameters being varied for all three approximations. The effect of varying any of the parameters was the same for all three approximations, with there being a slight

Mathematical Model of Hose String - Static

difference in the results due to their different load distributions. This can be seen in Figure 13.4 which compares the three approximations for the same hose-string, namely four flanged hoses with an unfloated section and the parameters as given in Appendix I. The following examples also use this hose-string, differing only when a parameter is being varied.

For the static problem, for which wave action has been neglected, the optimum manifold design would be one which was flat at the equilibrium depth, since this would produce no curvature and hence no stresses. However, as explained earlier, the manifold has to be designed such that the initial hose is at least partially submerged so that it can absorb small horizontal movements of the buoy. Nevertheless, Figure 13.5 indicates that careful consideration has to be given to the advantages gained by using the initial hose to damp out movement against the disadvantages due to the inherent stresses caused by having a deep manifold. A similar result is obtained if the angle of inclination of the manifold is varied. The larger the angle, the greater is the curvature of the hose and consequently the greater are the stresses upon it.

The load on a hose is a function of its weight, external radius and sea-density. The effect of varying the external radius is shown in Figure 13.6 (a similar effect can be obtained by varying either the sea-density of the weight of the hose). When the radius increases a larger upward buoyancy force causes the hose to reach equilibrium earlier, increasing its curvature and the stresses upon it.

Another factor which affects the stresses is the stiffness of the hose. This is dependent upon the bending stiffness of the rubber sections of hosing, of the flanges (which in reality can be considered rigid) or the percentage of flange per hose. In Figure 13.7 more stiffness is built into the hose by stiffening the flanges. Although the stiffer hose takes longer to reach equilibrium its stiffness causes it to generate the larger stresses. However, in practice the stiffer hose may be more able to sustain these increased stresses as is the case with the first hose off the buoy.

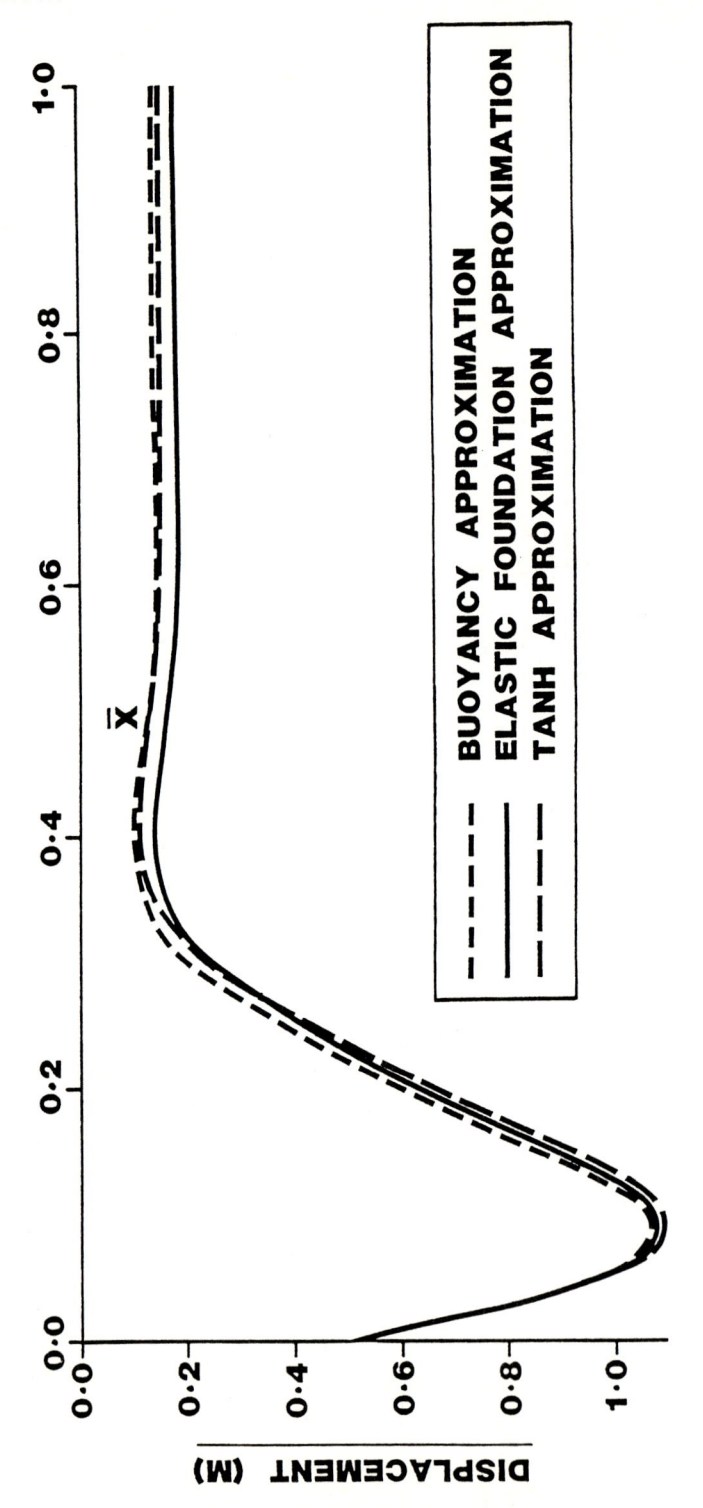

Figure 13.4 Comparing the three load approximations

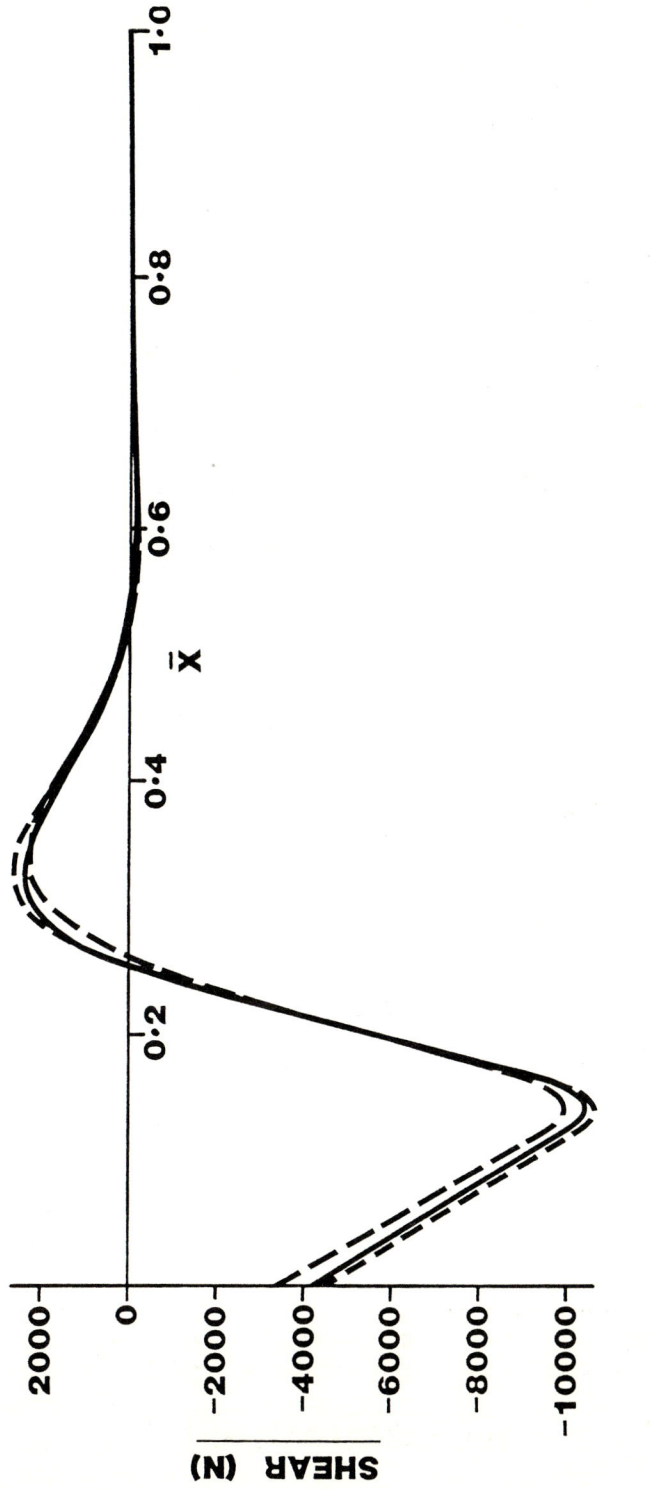

Mathematical Model of Hose String - Static

Figure 13.5 Shear stresses produced by the three load approximations

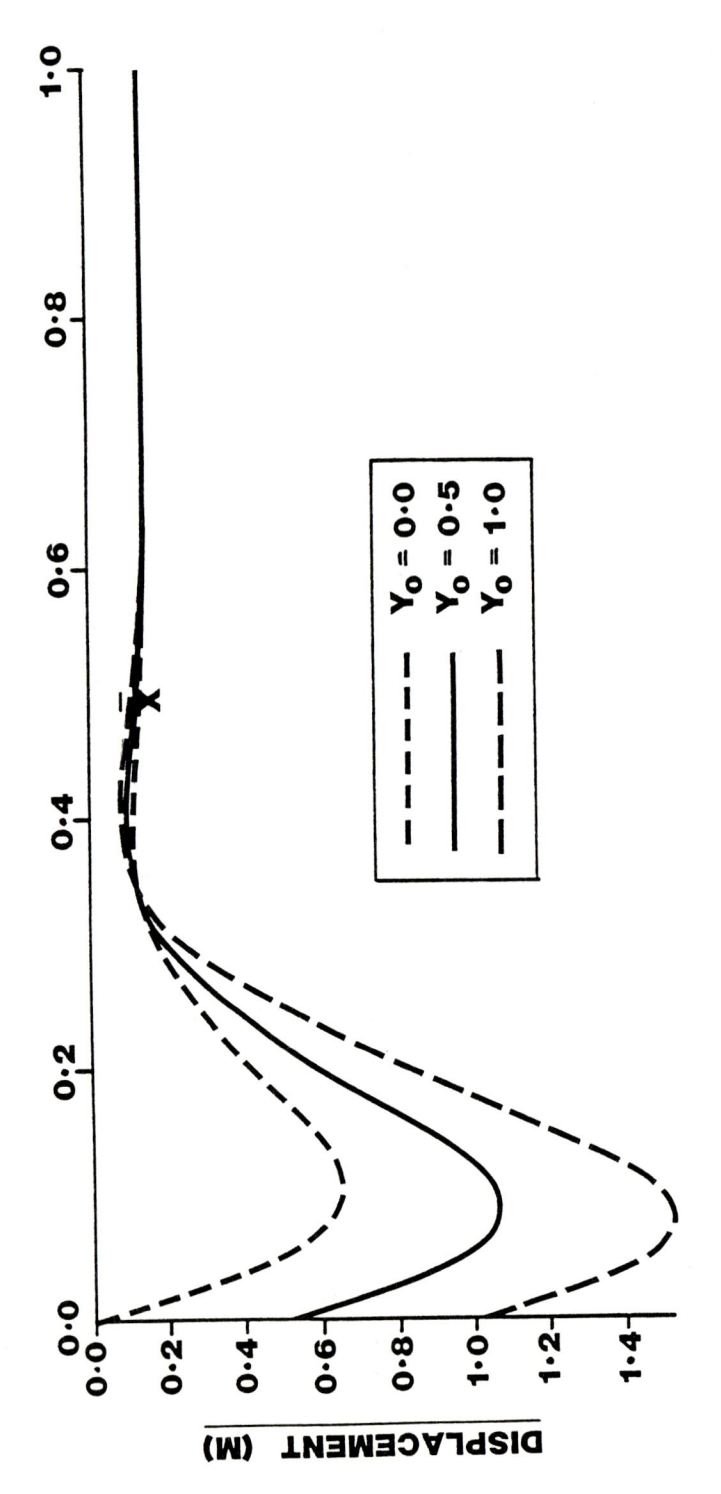

Figure 13.6   Varying the depth of the manifold

Mathematical Model of Hose String - Static

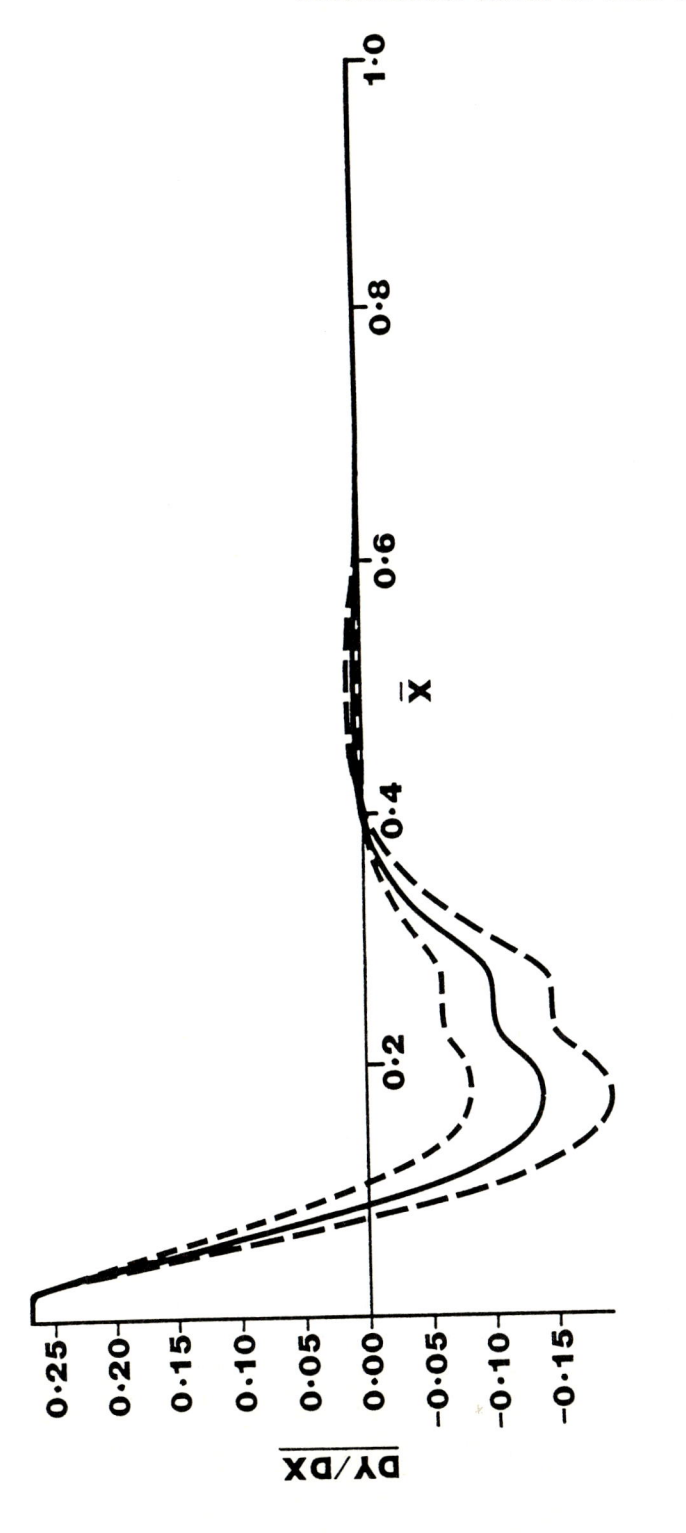

Figure 13.7 Effect of manifold depth upon the slope of the hose-string

M. J. Brown

The examples shown in Figures 13.5 to 13.11 all use the buoyancy approximation, although the tanh or elastic foundation approximations could equally have been used. The effects of two of the parameters are clearly illustrated in these examples. The application of the floatation covers causes the shear force to change direction, as shown in Figures 13.8 and 13.9, whilst the flanges cause the hose slope to remain nearly constant due to their curvature being negligible; this latter effect arising from the large increase in bending stiffness and the necessity for the bending moment to remain continuous at the flange.

The buoyancy approximation always produces the largest bending moments and shear forces whilst the tanh approximation produces the smallest, suggesting that the elastic foundation approximation is closer to the true buoyancy than is the tanh approximation. It was felt that this was due to most of the stresses being generated in the initial, submerged hose. For a submerged hose, as can be seen in Figure 13.3, the same value for the load is given by both the elastic foundation and the buoyancy approximation, whilst the tanh approximation gives a smaller value. There was no noticable difference in the number of iterations of the boundary values required for convergence between the three expressions for the load, whilst the buoyancy approximation was found to use about 10% more computing time than the other two methods. By using 'informed' estimates of the boundary values and the optimum matching point (found by trial and error) the tanh and elastic approximations needed seven iterations on average for a maximum residual of $O(10^{-5})$, each iteration taking approximately 1.1 seconds of computing time.

13.10 Application

In 1980/81, a new SPM terminal was installed off the coast of Angola. The buoy operator requested that the first two surface hoses off the buoy should be of the under-buoy variety (i.e. hoses without any floatation cover) as well as outlining the general configuration that the hose-string was to take up; this configuration to be obtained by placing floatation collars at suitable intervals along these first two hoses. The model outlined in this paper was modified so as to be able to accurately represent the variation in the hose parameters, as given in Dunlop (1971), of those hoses that were used. The resulting program was run with the floatation collar distribution that was finally applied in obtaining the outlined configuration.

Mathematical Model of Hose String - Static

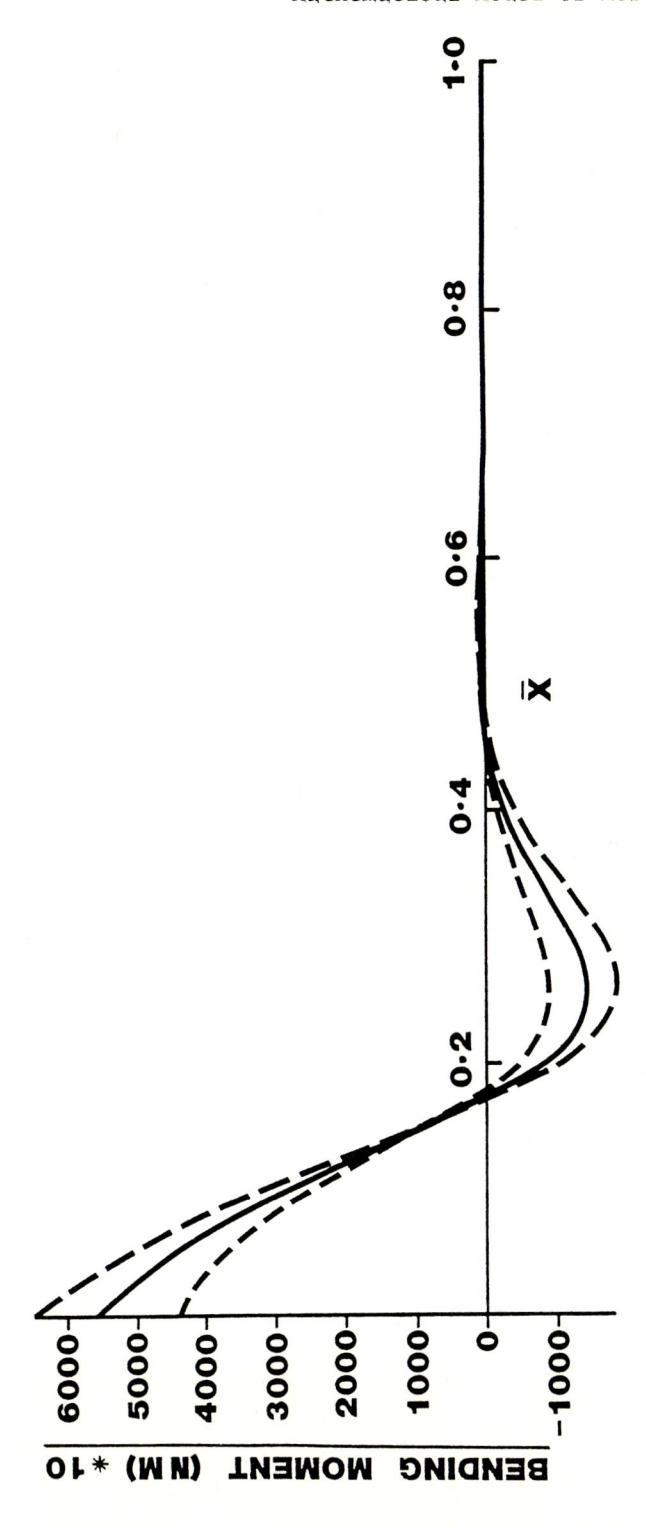

Figure 13.8 Effect of manifold depth upon the bending moment of the hose-string

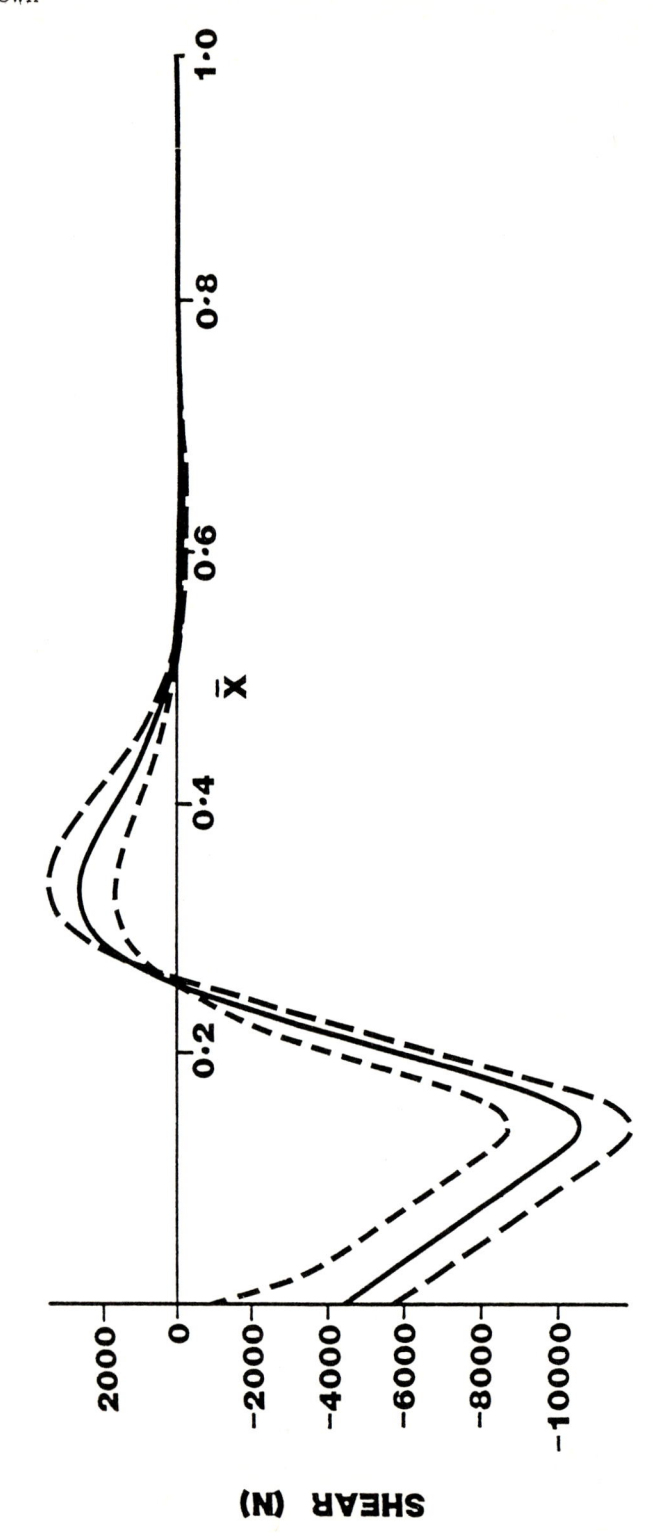

M. J. Brown

Figure 13.9 Effect of manifold depth upon the shear stresses in the hose-string

Mathematical Model of Hose String - Static

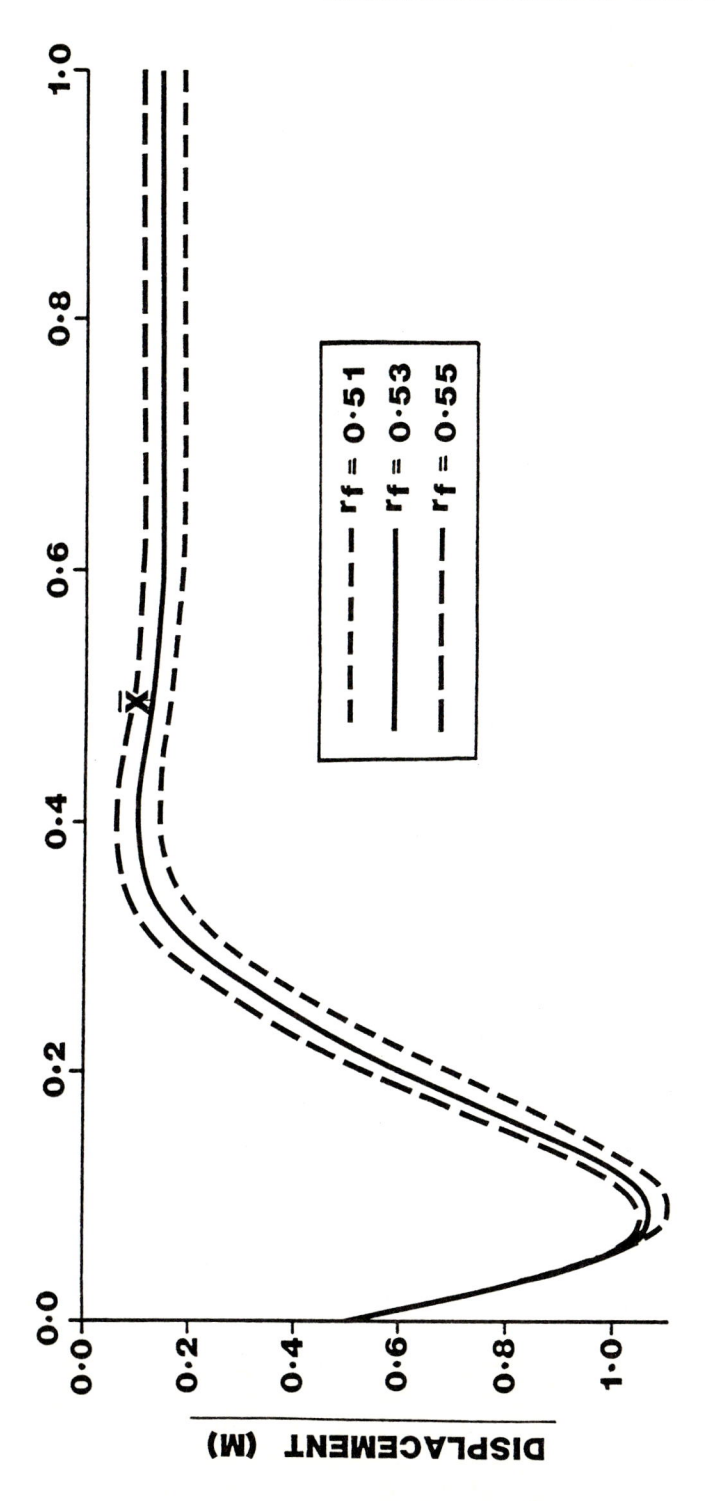

Figure 13.10 Varying the radius of the floatation covers

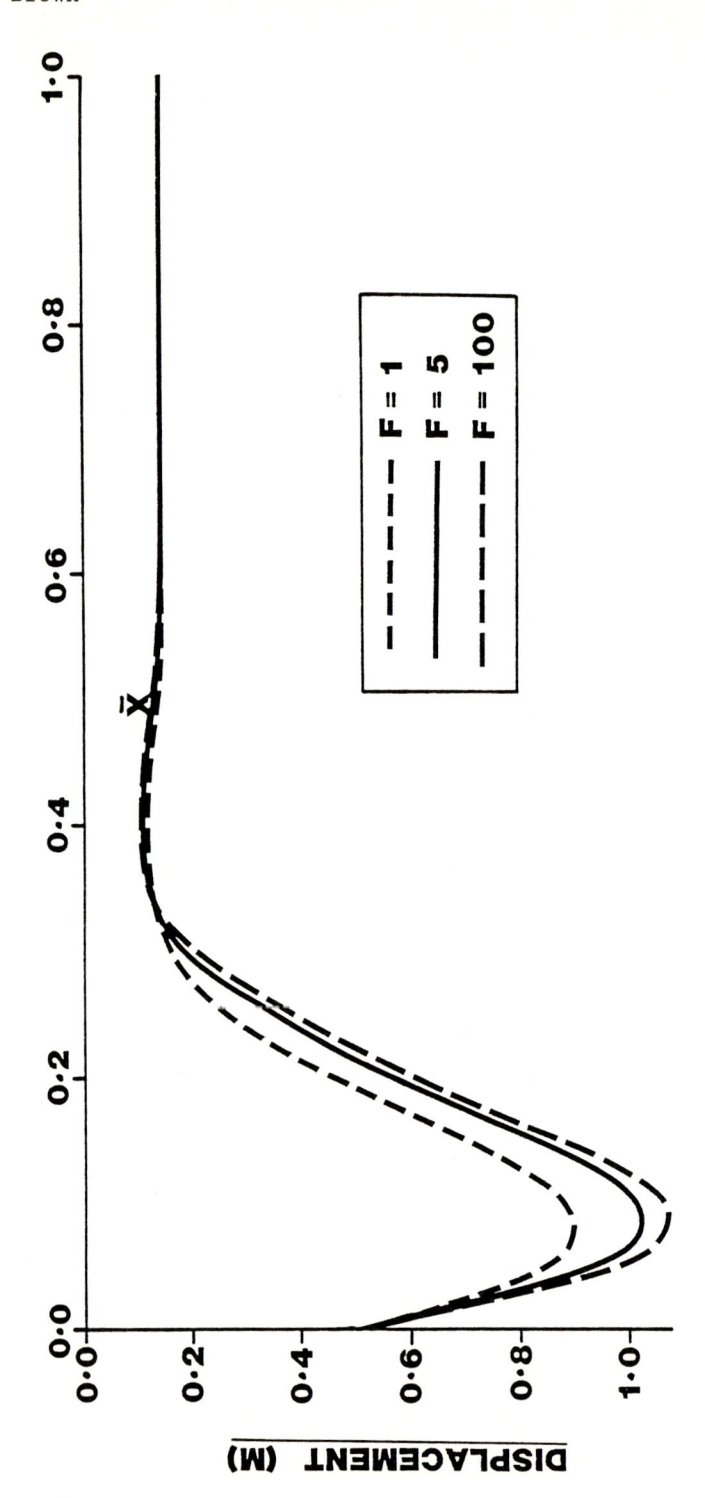

Figure 13.11 Varying the stiffness of the flanges

Mathematical Model of Hose String - Static

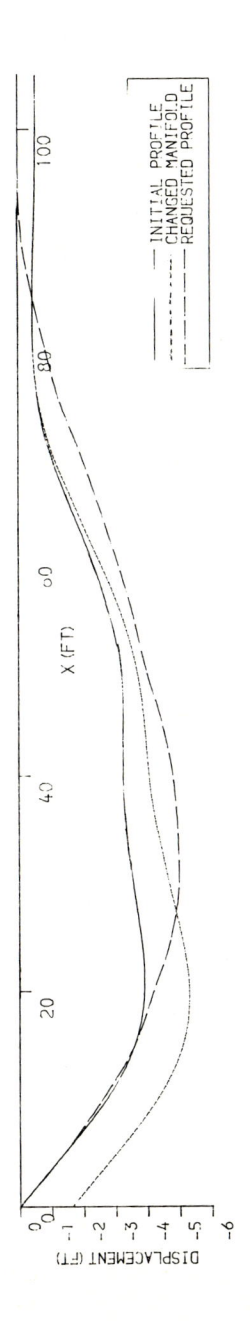

Figure 13.12   Modelling physical problem

M. J. Brown

As can be seen from Figure 13.12, the profile predicted by this model was much shallower than that requested. However, it was discovered that the buoy had been incorrectly set up in that, whilst its manifold bisected the waterline where there were no hoses attached to it, the addition of the hose-string resulted in the manifold lying approximately 0.5m beneath the surface due to the weight of the hose-string. The profile for this modified value of Y, which is also shown in Figure 13.12, is similar to that requested.

13.11   Conclusions

A model has been developed to calculate the profile and stresses on a static hose-string attached to a monobuoy. It has been shown that these stresses increase as both the bending stiffness and the load (which is itself dependent upon the weight and radius of the hose) increase. It is also clear that the stresses are dependent upon the design of the buoy's manifold.

Between the three approximations for the load there was no noticable difference in their ability to converge to a solution, with the buoyancy approximation taking slightly longer than the other two approximations. However, they do produce differing results due to their load profiles so that there is an automatic error built into any model using the tanh or elastic foundation approximations.

By using a numerical method it has been possible to compare the effect of the different parameters in the design of the hose. The understanding gained through this model will be of assistance in the development of the dynamic hose system, where movement will arise from the combined effects of the buoy motion and waves.

Acknowledgements

I would like to express my appreciation and gratitude to Dr. M.I.G. Bloor, Dr. D.B. Ingham and Dr. L. Elliott for their supervision and guidance during this work.

I would also like to gratefully acknowledge the financial support provided during this work by Dunlop Ltd. (Oil and Marine Division) and the Science and Engineering Research Council through a CASE studentship.

References

Bernitsas, M.M. and Kokkinis, T., 1983.
Buckling of risers in tension due to internal pressure: nonmovable boundaries. Transactions of the ASME, Journal of Energy Resources Technology, Vol 105, pp 277-281.

Brady, I. et al, 1974.
A study of the forces acting on a monobuoy due to environmental conditions. Proc. 6th Annual Offshore Technology Conference, Vol 2, Paper OTC 2136, pp 1051-1060.

Bridgestone, J., 1976.
Study of causes of kinking in floating hoses at Petrobras/Tefran terminal. Bridgestone Tyre Company of Japan. Report 6YMT-0011.

Dunlop Oil and Marine Division, England, 1971.
Offshore hose manual.

Dunlop Oil and Marine Division, 1973.
A study of forces acting on a monobuoy due to environmental conditions.

Dunlop, 1976.
SBM floating hose configuration. Central Research and Development Division, Birmingham, England. Report no. PR3408.

Felippa, C.A. and Chung, J.S., 1981.
Non-linear static analysis of deep ocean mining pipe. Transactions of the ASME, Journal of Energy Resources Technology, Vol 103, pp 11-25.

Graham, H., 1982.
Newcastle model hose tests. Report for Dunlop Oil and Marine Division.

Nordgren, R.P., 1982.
Dynamic analysis of marine risers with vortex excitation. Transactions of the ASME, Journal of Energy Resources Technology, Vol 104, pp 14-19.

Saito, H. et al, 1980.
Actual measurements of external forces on marine hoses for SPM. Proc. 12th Annual Offshore Technology Conference, Vol 3, Paper OTC 3803, pp 89-97.

Timoshenko, S., 1955.
Strength of materials, Part 1, 3rd Edition, pp 170-175.

Tschoepe, E.C. and Wolfe, G.K., 1981.
SPM hose test program. Proc. 13th Annual Offshore Technology Conference, Vol 2, Paper OTC 4015, pp 71-80.

Young, R.A. et al, 1980.
Behaviour of loading hose models in laboratory waves and currents. Proc. 12th Annual Offshore Technology Conference, Vol 3, Paper OTC 3842, pp 419-428.

Mathematical Model of Hose String - Static

Appendix I

Except for the application of this model to the physical problem, the values for the parameters used in this paper are those given in Dunlop (1976). Unless a parameter is being varied all of the results are based upon these values. However, the program is such that any other value could easily be used.

The length of each hose is taken as 10m, with each hose having a flange of length 1m at either end. The manifold is assumed to lie 0.5m below sea-level, at a slope of $15°$. The bending stiffness of a rubber section of hose, EI_r, is taken as $400kNm^{-2}$, and that for a flange, EI_f as $40000kNm^{-2}$, which gives F=100. The weight in air of a hose, W, is taken as $6000Nm^{-1}$ and the density of the sea-water as $1010Nm^{-3}$. The first 5m of the initial hose has no floatation covers and have an external radius of 0.39m. Floatation covers are fully applied after 7m of hosing and their radius is 0.53m.

These values produce a maximum buoyancy force of $8913.56Nm^{-1}$, with equilibrium depths of 0.146m for the buoyancy approximation, 0.150m for the tanh approximation and 0.183m for the elastic foundation approximation.

14. MATHEMATICAL MODEL OF A MARINE HOSE-STRING AT A BUOY
PART 2 - DYNAMIC PROBLEM

M.J. Brown,
Department of Applied Mathematical Studies,
The University of Leeds, Leeds,
West Yorkshire, LS2 9JT

14.1 Introduction

In Chapter 13, a mathematical model was presented for predicting the stresses in a static hose-string. The model presented there and in Brown (1982) was extended by Brown (1983) to obtain a mathematical model of a hose-string attached to a dynamic buoy, such that the stresses due to the hose parameters could be investigated independently from any stresses caused by wave-motion. The model is further extended in this paper to show that the effects arising from wave-motion can be studied.

As far as the author knows, no other dynamic model exists of this problem. Ghosh (1981) presented the governing equations for an oil carrying hose subject to time-dependent loads, together with a suggested iterative procedure for their integration. Unfortunately, no work has been conducted on this complex set of equations to substantiate whether they can be solved numerically as they stand. To obtain the results presented in this, and the previous paper, certain assumptions have had to be made to this highly non-linear problem. Nevertheless, all the important physical features of the hoses have been retained so that the stresses produced by particular hose designs or hose-string configurations can be compared and, hence, this model can be used to assist in the designing of hoses with longer life-spans.

In this paper the mathematical model of the hose-string is presented together with the numerical method of solution. Simplified problems, for which analytic solutions have been derived, are then considered so that the accuracy of the numerical integration can be shown by comparing its results with the corresponding analytic solution. The effect of the stiffness of the first hose, the buoy's motion and the steepness of the waves are then considered.

14.2 Equation of Motion

If the motion of the hose is restricted to being in the direction perpendicular to the mean water level then, if the hose-string is

M. J. Brown

modelled by the same method as was used in part 1, the equation of motion is that given by Bishop and Johnson (1960), namely

$$EI_z \frac{\partial^4 y}{\partial x^4} + m \frac{\partial^2 y}{\partial t^2} = Q \qquad (14.1)$$

where the load, Q, is dependent upon the weight and radius of the hose and upon its depth below the surface.

In attempting to minimise the severe stresses that the first hose off the buoy can be subjected to, it is designed such that its initial section is unfloated and contains extra reinforcement. For simplicity the following expression has been taken for the radius.

$$r = \begin{cases} r_u & 0 \leqslant x < x_u \\ r_u + \frac{(r_f - r_u)}{(x_f - x_u)}(x - x_u) & x_u \leqslant x \leqslant x_f \\ r_f & x_f < x \leqslant L \end{cases} \qquad (14.2)$$

In Chapter 13, three expressions were used to approximate the load with their respective results being compared for several problems. For all of the problems that were considered there was only a slight difference in the results obtained by the three approximations. Therefore, the load has been represented by the elastic foundation approximation, modified to include the wave-motion, since it simplifies the numerical integration and hence saves on computing time.

$$Q = \begin{cases} W & y - \eta < -r \\ W - K(r + y - \eta) & -r \leqslant y - \eta \leqslant -r \\ W - B_{max} & -r < y - \eta \end{cases} \qquad (14.3)$$

where $B_{max} = \pi r^2 \rho g$, $K = \frac{B_{max}}{2r}$ and η is the sea-surface. A third order Stokes wave has been used in this paper so that η is as given in Kinsman (1965), namely

$$\eta = a \cos(\kappa x - \omega t) - \tfrac{1}{2} a^2 \kappa \cos 2(\kappa x - \omega t) + \tfrac{9}{8} a^3 \kappa^2 \cos 3(\kappa x - \omega t) \qquad (14.4)$$

where $\omega^2 = \kappa g (1 + a^2 \kappa^2)$.

Mathematical Model of Hose String - Dynamic

The higher order corrections in equation (14.4) to the sinusoidal wave result in a more realistic representation of the sea-surface by producing steeper crests and shallow troughs.

The problem was non-dimesionalised by introducing the following variables

$$x = L\bar{x} \qquad y = d\bar{y} \qquad t = s\bar{t}$$
$$EI_z = EI_r \bar{F} \qquad Q = dK_f \bar{Q} \tag{14.5}$$

where EI_r is the bending stiffness of the reinforced rubber sections of the second and later hoses, EI_z is the bending stiffness of a general section of hosing and K_f is the elastic stiffness of the load for a fully floated section of hosing. Substituting equation (14.5) into equation (14.1) gives

$$\bar{F}\frac{\partial^4 \bar{y}}{\partial \bar{x}^4} + \bar{b}\frac{\partial^2 \bar{y}}{\partial \bar{t}^2} = \bar{c}\,\bar{Q} \tag{14.6}$$

where $\bar{b} = \dfrac{L^4 m}{s^2 EI_r}$ and $\bar{c} = \dfrac{L^4 K_f}{EI_r}$.

For convenience the bar has been omitted from above the non-dimensional variables for the rest of this paper.

14.3 Boundary Conditions

The boundary condition at the buoy are dependent upon the design of the manifold and upon the buoy's motion. Operating experience has resulted in most buoys being designed such that their manifold is bisected by the mean water level and slopes into the water at an angle of 15°. This design ensures that the first hose is not subjected to any unusual bending stresses when the system is in its neutral position and so prevents premature hose failure, as described in Dunlop (1971). The buoy's motion will vary depending upon the steepness of the waves. Graham (1982) observed that in steep seas the buoy undergoes pitching with no noticable heave and that, as the lengths of the waves increased, the buoy could follow the waves with greater ease (such that the buoy underwent mainly heave) whilst its motion tended to lag behind that of the waves. Consequently, the heave and pitch of the buoy, as well as the phase difference between the buoy and the wave, ε, are dependent upon the length and amplitude of the wave. Therefore, if the depth and

M. J. Brown

inclination of the manifold are respectively Y_0 and θ_0 then the boundary conditions at the buoy are:

$$y = Y_0 + \eta \ (x=\varepsilon_h/\kappa, \ a=a_h) \qquad \theta = \theta_0 + \frac{\partial \eta}{\partial x} \ (x=\varepsilon_p/\kappa, \ a=a_p) \qquad (14.7a)$$

where the subscripts h and p indicate the buoy's heaving and pitching motion.

In a calm sea the free end of the hose-string will lie flat at its equilibrium depth, Y_e. For a hose subject to wave motion the free end will try to follow the wave profile but at its equilibrium depth below the surface. Its ability to do this will depend upon the wave's steepness and upon the stiffness of the hose. Therefore, the conditions applied to the free end of the hose are:

$$y = Y_e + \eta \ (x=1, a=a_1) \qquad \theta = \frac{\partial \eta}{\partial x} \ (x=1, a=a_1) \qquad (14.7b)$$

where a_1 is the amplitude of the free end.

The two initial conditions that have been taken are that the hose starts from rest at its equilibrium position. If $f_0(x)$ is the solution of the static equation, as given in part 1, then the initial conditions are:

$$y = f_0(x) \qquad \frac{\partial y}{\partial t} = 0 \qquad (14.7c)$$

14.4 Method of Solution

In Chapter 13, the Runge-Kutta method was used to integrate the equations for a static hose. However, the space integration used a large amount of computing time which it was felt restricted its suitability for the dynamic case in which results would be required over several time periods. Consequently, a fixed mesh finite-difference scheme was used to solve the equation of motion.

The usual finite-difference notation has been used such that if there are N internal space intervals, the space and time step lengths are given by

$$\delta x = h = \frac{1}{(N+1)} \qquad \delta t = k \qquad (14.8)$$

and the value of y at the point (x,t) is denoted by

Mathematical Model of Hose String - Dynamic

$$y(x,t) = y(ih,jk) = y_{i,j} \qquad (14.9)$$

Initially, a central difference approximation was taken for the second order time derivative so that the equation (14.6) became

$$F\left(\frac{\partial^4 y}{\partial x^4}\right)_{i,j} + b\left(\frac{\partial^2 y}{\partial t^2}\right)_{i,j} = c\, Q_{i,j} \qquad (14.10)$$

where $\left(\frac{\partial^2 y}{\partial t^2}\right)_{i,j} = \frac{y_{i,j+1} - 2y_{i,j} + y_{i,j-1}}{k^2} + O(k^2)$. $\qquad (14.11)$

However, it was found that this scheme was unstable and so a backward difference approximation was used. Hence, equation (14.6) became

$$F\left(\frac{\partial^4 y}{\partial x^4}\right)_{i,j+1} + b\left(\frac{\partial^2 y}{\partial t^2}\right)_{i,j+1} = c\, Q_{i,j+1} \qquad (14.12)$$

where $\left(\frac{\partial^2 y}{\partial t^2}\right)_{i,j+1} = \frac{y_{i,j+1} - 2y_{i,j} + y_{i,j-1}}{k^2} + O(k)$. $\qquad (14.13)$

This expression was found to be stable. Nevertheless, the backward difference approximation gives exactly the same finite-difference relationship as does the central difference approximation, but is an order of accuracy lower than it. This suggested that the accuracy of the results could be improved if some account could be taken of the values obtained at the jth time-level. This was achieved by introducing a weighting constant, α, in order to weight the time derivative towards the central difference approximation. Therefore, the finite-difference scheme taken was

$$\alpha F\left(\frac{\partial^4 y}{\partial x^4}\right)_{i,j+1} + (1-\alpha)F\left(\frac{\partial^4 y}{\partial x^4}\right)_{i,j} + b\left(\frac{\partial^2 y}{\partial t^2}\right)_{i,j+1} = c\, Q_{i,j+1} \quad (14.14)$$

The fourth order space derivative was approximated by

$$\left(\frac{\partial^4 y}{\partial x^4}\right)_{i,j} = \frac{y_{i+2,j} - 4y_{i+1,j} + 6y_{i,j} - 4y_{i-1,j} + y_{i-2,j}}{h^4} + O(h^2) \qquad (14.15)$$

As shown in Brown (1982), after substituting equations (14.13) and (14.15) into equation (14.14), and applying the boundary conditions, equations (14.7), the resulting system of finite-difference equations for N internal mesh points reduces to the form

$$\underline{A}\, \underline{Y}_{j+1} = \underline{C}_j \; , \qquad (14.16)$$

where $\underline{y}_{j+1}^T = (y_{1,j+1}, y_{2,j+1}, \ldots, y_{N-1,j+1}, y_{N,j+1})$,

$\underline{\underline{A}}$ is a NxN matrix and $\underline{C}_j$ a N-dimensional vector, both of which contain known values at each time-step.

For a hose with continuous bending stiffness, matrix $\underline{\underline{A}}$ is a banded matrix with a half-band width of three.

The solution $\underline{y}_{j+1}$, at each time-step is obtained by inverting equation (14.16). On an Amdahl V7 the finite-difference method required approximately 0.40 seconds of computing time per time-step for N=2000 whilst, for an equivalent step-length, the Runge-Kutta method required approximately 8 seconds to solve the static ordinary differential equation.

14.5 Flanges

In obtaining the Taylor series used to derive equations (14.15), and hence (14.16), the space-derivatives have been assumed continuous. However, for a hose with flanges the bending stiffness is discontinuous at an interface between a steel flange and a reinforced rubber section of hosing. This results in discontinuous derivatives of the displacement of second order and higher.

As the displacement, the slope, the bending moment and the shear force are all continuous variables, then by using the equations linking these variables, as given in Chapter 13, the following relationships can be derived at a discontinuity of the bending stiffness.

$$\frac{\partial y_r}{\partial x} = \frac{\partial y_s}{\partial x} , \quad \frac{\delta^i y_r}{\delta x^i} = F \frac{\delta^i y_s}{\delta x^i} \quad \text{and} \quad \frac{\delta^6 y_r}{\delta x^6} = F \frac{\delta^6 y_s}{\delta x^6} \qquad (14.17)$$

where $F = \dfrac{EI_s}{EI_r}$,

i = 2,3,4,5 and the subscripts r and s indicate that the point is on the rubber or steel flange side of the interface respectively. Therefore, if an interface is at $x-X_0$, then by using equations (14.17) the Taylor series on either side of the discontinuity are given by

Mathematical Model of Hose String - Dynamic

$$y_r = y_0 + x\frac{\partial y}{\partial x}(X_0) + \frac{Fx^2}{2!}\frac{\partial^2 y}{\partial x^2}(X_0) + \frac{Fx^3}{3!}\frac{\partial^3 y}{\partial x^3}(X_0) + \ldots + \frac{F^2 x^6}{6!}\frac{\partial^6 y}{\partial x^6}(X_0) + \ldots$$
(14.18a)

$$y_s = y_0 + x\frac{\partial y}{\partial x}(X_0) + \frac{x^2}{2!}\frac{\partial^2 y}{\partial x^2}(X_0) + \frac{x^3}{3!}\frac{\partial^3 y}{\partial x^3}(X_0) + \ldots + \frac{x^6}{6!}\frac{\partial^6 y}{\partial x^6}(X_0) + \ldots$$
(14.18b)

To derive the approximation for the fourth derivative across this discontinuity to the same accuracy as for the rest of the hose requires that six points are used, namely at X_i, where $i = -3,-2,-1,1,2,3$, with each being expanded as far as the sixth derivative. If

$$d_i = \frac{h^i}{i!}\frac{\partial^i y}{\partial x^i}(X) , \quad i = 1,2,\ldots,6 \quad (14.19)$$

then the Taylor series of these points about X_0 can be written as

$$\underline{y} = \underline{y}_0 + \underline{\underline{P}}\,\underline{d} \quad (14.20)$$

where

$$\underline{y} = \begin{bmatrix} y_{-3} \\ y_{-2} \\ y_{-1} \\ y_1 \\ y_2 \\ y_3 \end{bmatrix}, \quad \underline{d} = \begin{bmatrix} d_1 \\ d_2 \\ d_3 \\ d_4 \\ d_5 \\ d_6 \end{bmatrix}, \quad \underline{y}_0 = \begin{bmatrix} y_0 \\ y_0 \\ y_0 \\ y_0 \\ y_0 \\ y_0 \end{bmatrix}$$

and $\underline{\underline{P}}$ is a 6x6 matrix containing the known coefficients of the Taylor series.

To determine each d_i, $i = 1,2,\ldots,6$, explicitly in terms of the surrounding values requires that each expansion be assigned a weighting function. If $\underline{W}_i$ is defined by

$$\underline{W}_i = (w_{1i}, w_{2i}, w_{3i}, w_{4i}, w_{5i}, w_{6i}) \quad (14.21)$$

then multiplying equation (14.20) by $\underline{W}_i$ results in

M. J. Brown

$$\underline{W}_i \cdot \underline{y} = \underline{W}_i \cdot \underline{y}_0 + \underline{W}_i \cdot (\underline{\underline{P}} \; \underline{d}) \quad ,$$

$$= y_0 \sum_{k=1}^{6} w_{ki} + \sum_{k=1}^{6} w_{ki} \left(\sum_{j=1}^{6} p_{kj} d_j \right) \quad ,$$

$$= y_0 \sum_{k=1}^{6} w_{ki} + \sum_{j=1}^{6} \left(\sum_{k=1}^{6} w_{ki} p_{kj} \right) d_j \quad . \tag{14.22}$$

The w_{ki} are chosen so that the coefficient of every derivative, except for the ith, is zero; thus enabling the required derivative to be expressed explicitly in terms of $\underline{y}$. This condition requires that

$$\sum_{k=1}^{6} w_{ki} p_{kj} \begin{cases} = 0 & i \neq j \\ \neq 0 & i = j \end{cases} \tag{14.23a, 14.23b}$$

where equation (14.23a) gives 5 equations in terms of the 6 unknowns w_{ki}, $k=1, \ldots, 6$.

The solution of equation (14.23a) is not unique since if $\underline{W}_i$ is a solution then so too is $c\underline{W}_i$ for any non-zero constant c. Therefore, providing that the ℓth element of $\underline{W}_i$ is non-zero, a solution of equation (14.23a) can always be found such that $w_{\ell i}$ is any non-zero constant, i.e. take $w_{\ell i}=1$, where $1 \leq \ell \leq 6$. Hence, equation (14.23a) can be solved since there are now only five unknowns. However, if a non-zero value has been assigned to a zero element of $\underline{W}_i$ then there is no constant c ($\neq 0$) such that $c\underline{W}_i$ satisfies equation (14.23a). This is easily overcome by assigning a value to a different element of $\underline{W}_i$. Therefore, switching the known values in equation (14.23a) onto the right-hand side gives

$$(\underline{\underline{P}}^T)^{i\ell} \cdot \underline{W}_i^{\ell} = -w_{\ell i} \; \underline{p}_\ell^i \tag{14.24}$$

where $(\underline{\underline{P}}^T)^{i\ell}$ is the transpose of $\underline{\underline{P}}$, minus its ith row and ℓ^{th} column,

$\underline{p}_\ell^i$ is the ℓth column of $\underline{\underline{P}}$ minus the ith row,

and $\underline{W}_i^{\ell}$ is $\underline{W}_i$, minus the ℓth element.

From equation (14.24) the unknown w_{ki}, $k=1, \ldots, 6$, $k \neq \ell$, can be found by matrix inversion. Substituting equation (14.23a) into equation (14.22) gives

Mathematical Model of Hose String - Dynamic

$$\underline{w}_i \cdot \underline{Y} = y_0 \sum_{k=1}^{6} w_{ki} + d_i \sum_{k=1}^{6} w_{ki} p_{ki} \quad ;$$

resulting in

$$d_i = \frac{h^i}{i!} \frac{\partial^i y}{\partial x^i}(X_0) = \frac{\underline{w}_i \cdot \underline{Y} - y_0 \sum_{k=1}^{6} w_{ki}}{\sum_{k=1}^{6} w_{ki} p_{ki}} \quad (14.25)$$

Equation (14.25) provides the derivatives across the discontinuity in terms of the displacement at the surrounding points.

Differentiating equations (14.18a) and (14.18b) produces

$$\frac{\partial^4 y_r}{\partial x^4} = F \frac{\partial^4 y}{\partial x^4}(X_0) + Fx \frac{\partial^5 y}{\partial x^5}(X_0) + \frac{F^2 x^2}{2} \frac{\partial^6 y}{\partial x^6}(X_0) + \ldots \quad (14.26)$$

$$\frac{\partial^4 y_s}{\partial x^4} = \frac{\partial^4 y}{\partial x^4}(X_0) + x \frac{\partial^5 y}{\partial x^5}(X_0) + \frac{x^2}{2} \frac{\partial^6 y}{\partial x^6}(X_0) + \ldots \quad (14.27)$$

Therefore, if the discontinuity in the bending stiffness is at X_0 then equation (14.15) can be used at all points except at X_{-1}, X_0 and X_1, whilst at these three points equations (14.26) and (14.27) must be used, with the differentials at X_0 being given by equation (14.25).

At the actual discontinuity either equation can be used since the shear force is continuous. Hence, if the point X_0 is in a rubber section of hosing then the shear force is given by

$$V = EI_r \frac{\partial^4 y_r}{\partial x^4} = EI_r F \frac{\partial^4 y}{\partial x^4}(X_0) \quad (14.28)$$

whilst if the point X_0 is in a section of flange then

$$V = EI_s \frac{\partial^4 y_s}{\partial x^4} = EI_s \frac{\partial^4 y}{\partial x^4}(X_0) \quad (14.29)$$

Both of these possibilities give the same results since $EI_s = F \, EI_r$.

By using six points to calculate the fourth order space derivative, the half band width of matrix A in equation (14.16) is increased to four.

M. J. Brown

14.6 Comparison of Analytical and Numerical Results

All of the important physical properties of a hose-string can be modelled by using a numerical, rather than an analytical, method for solving the equations. However, analytic solutions do exist for much simplified problems (Dunlop, 1973, Brown, 1982, 1983) and whilst they are of little practicable use for modelling the hose, they can be used to test the accuracy of the numerical integration.

In Brown (1983), analytic solutions were derived to test both the accuracy of the basic finite-difference equations used and the equations across a discontinuity in the bending stiffness. It was found that the accuracy of the numerical integration slightly increased as α decreased from 1.0 until, for values of α less than 0.5, the integration became unstable. Similarly, a simplified problem was derived in the present situation in order to test the accuracy and stability of the method when wave-action is incorporated into the model. If the hose is fully floated along its entire length, has no flanges and its load is taken as proportional to the displacement from the sea-surface, which is restricted to first order, then the hose's motion is governed by

$$\frac{\partial^4 y}{\partial x^4} + b \frac{\partial^2 y}{\partial t^2} = - c \{y - a \cos(\kappa x - \omega t)\} \qquad (14.30)$$

If the following boundary conditions are taken

$$\begin{aligned} y(x=0) &= A \cos(-\omega t) & y(x=1) &= A \cos(\kappa - \omega t) \\ \theta(x=0) &= -A\kappa \sin(-\omega t) & \theta(x=1) &= -A\kappa \sin(\kappa - \omega t) \end{aligned} \qquad (14.31)$$

where

$$A = \frac{ac}{c + \kappa^4 - b\omega^2}$$

then the analytic solution simplifies to

$$y = A \cos(\kappa x - \omega t) \qquad (14.32)$$

Since the numerical integration requires that the solution be known at the two previous time levels the analytic solutions at times $-\Delta t$ and $-2\Delta t$ were used in calculating the numerical solution at time t=0. After this time-level the numerical integration was completely independent of the analytic solution.

Mathematical Model of Hose String - Dynamic

The numerical and analytical solutions for the above problem at times t=0 and after ten periods (period=4.0, Δt=0.1) are given in Table 14.1 for α=1.0 and α=0.9. For a hose forced into motion by the movement of the buoy, the accuracy of the numerical integration has been found to be improved by weighting the derivatives towards the jth time level, (Brown,1982). However, as can be seen in Table 14.1, this does not seem to be the case when the behaviour of the hose-string is dependent upon wave-motion. Consequently, the results given in the next section have been obtained with the full backward finite-difference approximation of the time derivative i.e. α=1.

	analytic solution	numeric solutions			
		time 0.0		time=40.0	
x	t=0, t=40	α=1.0	α=0.9	α=1.0	α=0.9
0.0	0.00000	0.00000	0.00000	0.00000	0.00000
0.1	1.74813	1.76960	1.75904	1.74340	1.79823
0.2	1.11221	1.10712	1.10897	1.07977	1.07052
0.3	-1.08433	-1.08873	-1.11998	-1.11085	-1.33719
0.4	-1.80210	-1.79970	-1.84262	-1.86734	-2.08376
0.5	0.00880	0.01463	0.01440	0.04564	0.06212
0.6	1.80770	1.80900	1.82060	1.80900	1.89274
0.7	1.07008	1.06502	1.01632	1.03857	1.03280
0.8	-1.12688	-1.13119	-1.16363	-1.15450	-1.37941
0.9	-1.74263	-1.74069	-1.78013	-1.74155	-1.94747
1.0	0.01817	0.01817	0.01817	0.01817	0.01817

Table 14.1   Comparing analytic and numeric solutions for one unflanged hose for α=1.0 and α=0.9

14.7   Numerical Results

A model of a hose-string in a calm sea forced into motion by the movement of the buoy was given in Brown (1983). This enabled the relationship between the stresses in the hose-string and the hose parameters to be investigated independently from any stresses arising due to wave-motion. After including wave-motion into the model, it was found that the effects of the hose parameters upon the stresses on the hose were similar to those obtained in Brown (1983). Consequently, only those results which do not correspond to those given in Brown (1983) will be presented.

M. J. Brown

For reasons that have been described earlier, the first hose off the buoy is designed differently from the rest of the hoses in the string. One of these differences is the extra reinforcement which it possesses so as to transfer the large stresses generated at the buoy into the main body of the hose. If it is assumed that the stiffness of the hose has a negligible effect upon the buoy's motion then the effect which this extra reinforcement has upon the stresses occuring in the rubber sections of each of the hoses in the string is shown in Table 14.2. (The parameters used in obtaining these, and subsequent, results are given in Appendix I). The extra stiffness can be seen to have a significant effect upon the stresses applied to the first two hoses off the buoy. The extra reinforcement of the first hose should be able to handle the increased stresses to which it is subjected. However, the second hose has no extra reinforcement and will subsequently be more prone to failure than the rest of the hoses in the string.

Hose off the buoy	$F_1=1$ maximum moment kNm	$F_1=1$ shear kN	$F_1=2$ maximum moment kNm	$F_1=2$ shear kN	$F_1=5$ maximum moment kNm	$F_1=5$ shear kN	$F_1=10$ maximum moment kNm	$F_1=10$ shear kN
1	48.6	11.50	87.0	16.22	185.8	28.16	321.6	43.33
2	17.6	2.82	19.8	3.14	24.1	8.37	42.4	17.08
3	11.4	1.82	11.6	1.75	12.5	2.22	20.4	4.05
4	12.2	1.82	12.1	1.84	12.2	1.90	11.8	1.70
5	11.7	1.69	11.7	1.68	11.6	1.67	11.6	1.68
6	11.9	1.74	11.9	1.74	11.9	1.74	11.9	1.74
7	11.8	1.72	11.8	1.72	11.8	1.72	11.8	1.72
8	11.7	1.77	11.7	1.77	11.7	1.77	11.7	1.77
9	12.7	1.90	12.7	1.90	12.7	1.90	12.7	1.90
10	12.0	2.27	12.0	2.27	12.0	2.27	12.0	2.27

where $F = \dfrac{\text{bending stiffness of reinforced rubber of first hose}}{\text{bending stiffness of rubber for subsequent hoses}}$

Table 14.2   The effect upon the bending moment and shearing forces of reinforcing the rubber section of the first hose

Mathematical Model of Hose String - Dynamic

It has just been shown that the stresses in the hose-string are dependent upon the stiffness of the first hose off the buoy. These stresses are also going to be dependent upon the design of the manifold and the forces restoring the hose-string to the surface. The relationship between the manifold design and the extra reinforcement of the first hose can be seen in Tables 14.3 and 14.4. These tables correspond to Table 14.2 in that they show that the larger stresses are produced by the stiffest initial hose. It can also be seen that the stresses produced are dependent upon the displacement and the slope of the manifold, with the optimum manifold design being that which will impose the least excess curvature upon the hose. Therefore, for the most common manifold-hose combination (the manifold sloping into the water at $15°$ from the MWL and the first hose initially having no floatation covers) the benefits obtained by reducing the axial loads through damping out the effects of horizontal movements of the buoy have to be weighed against the increased bending moments and shear forces arising from the larger curvature.

θ_0	Hose off the buoy	$Y_0=-0.10$m maximum moment kNm	shear kN	$Y_0=0.00$m maximum moment kNm	shear kN	$Y_0=0.10$m maximum moment kNm	shear kN
$0.0°$	1	39.2	14.66	34.8	12.31	30.4	9.96
	2	13.9	2.79	12.9	2.52	13.2	2.30
	3	12.1	1.72	12.0	1.72	11.9	1.72
	4	12.1	1.81	12.1	1.81	12.1	1.81
	5	11.7	1.71	11.7	1.71	11.7	1.71
$7.5°$	1	45.2	7.61	49.6	8.58	53.6	9.60
	2	15.0	2.45	15.9	2.67	16.9	2.92
	3	11.9	1.74	11.8	1.74	11.7	1.75
	4	12.2	1.80	12.1	1.80	12.1	1.81
	5	11.7	1.71	11.7	1.71	11.7	1.71
$15.0°$	1	83.7	14.80	87.0	16.22	90.8	18.32
	2	19.0	3.11	19.8	3.14	20.4	3.15
	3	11.7	1.76	11.6	1.75	11.6	1.74
	4	12.2	1.84	12.1	1.84	12.1	1.85
	5	11.7	1.69	11.7	1.68	11.7	1.68

Table 14.3 $F_1=2$; The effect of the manifold design upon the stresses in the hose-string

θ_0	Hose off the buoy	$Y_0=-0.10$m maximum moment kNm	shear kN	$Y_0=0.00$m maximum moment kNm	shear kN	$Y_0=0.10$m maximum moment kNm	shear kN
0.0^0	1	81.9	19.04	74.2	16.05	66.6	13.05
	2	17.1	4.56	16.0	4.29	15.7	4.15
	3	12.3	1.75	12.3	1.74	12.2	1.72
	4	12.1	1.84	12.1	1.84	12.1	1.84
	5	11.7	1.71	11.7	1.71	11.7	1.71
7.5^0	1	103.5	14.01	111.2	15.52	118.0	17.22
	2	15.8	5.35	17.0	5.08	18.0	4.89
	3	12.5	1.83	12.4	1.81	12.3	1.79
	4	12.1	1.82	12.1	1.83	12.1	1.85
	5	11.7	1.71	11.7	1.69	11.7	1.69
15.0^0	1	179.3	26.42	185.8	28.16	192.3	31.16
	2	23.6	8.10	24.1	8.37	24.6	8.50
	3	12.7	2.15	12.5	2.22	12.4	2.32
	4	12.2	1.90	12.2	1.90	12.1	1.90
	5	11.7	1.67	11.7	1.67	11.6	1.67

Table 14.4 $F_1=5$; The effect of the manifold design upon the stresses in the hose-string.

In Tables 14.2, 14.3 and 14.4, the extra stresses generated by the motion of the buoy, the manifold construction and the design of the first hose can be seen to only have a significant effect upon the first two hoses in the string. The stresses on the rest of the hose-string arise mainly from it trying to follow the wave profile. The ability of the hose to do this will depend upon its stiffness and upon the steepness of the wave. A comparison of the stresses generated by different length waves is given in Table 14.5. For the shortest wave, $\lambda=20$m, the steepness of the crests was such that the hose-string could not follow the wave profile, resulting in the hose being completely submerged by a crest. The crest produced by the next shortest wave, $\lambda=30$m, was only just sufficient to completely immerse the hose whilst, for the other wavelengths, the hose-strings were always able to follow the wave. An example of the motion of a hose-string for one wave period ($\lambda=30$m, T=4.29 secs) and the stresses which it has to withstand, is given in Figures 14.1, 14.2 and 14.3 (only the first four hoses off the buoy have been displayed). The fact that the hose-string could not follow

Mathematical Model of Hose String - Dynamic

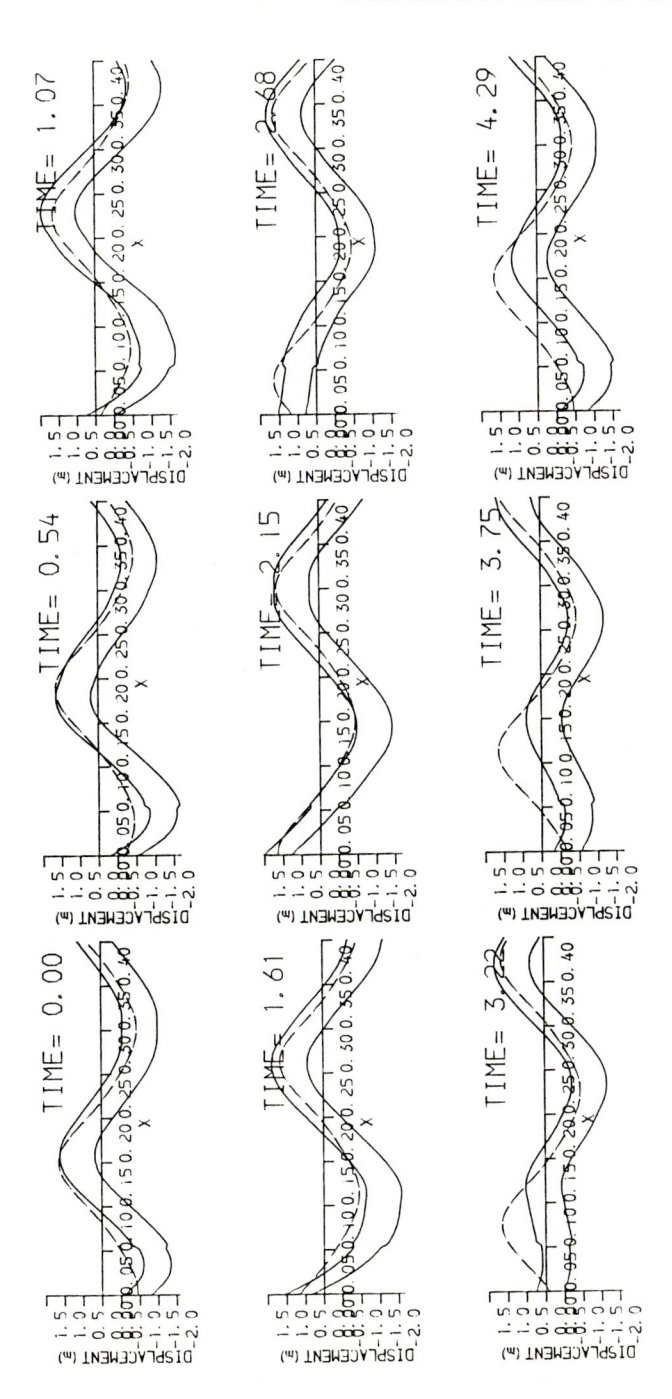

Figure 14.1 Hose motion over one period, $\lambda = 30\text{m}$

M. J. Brown

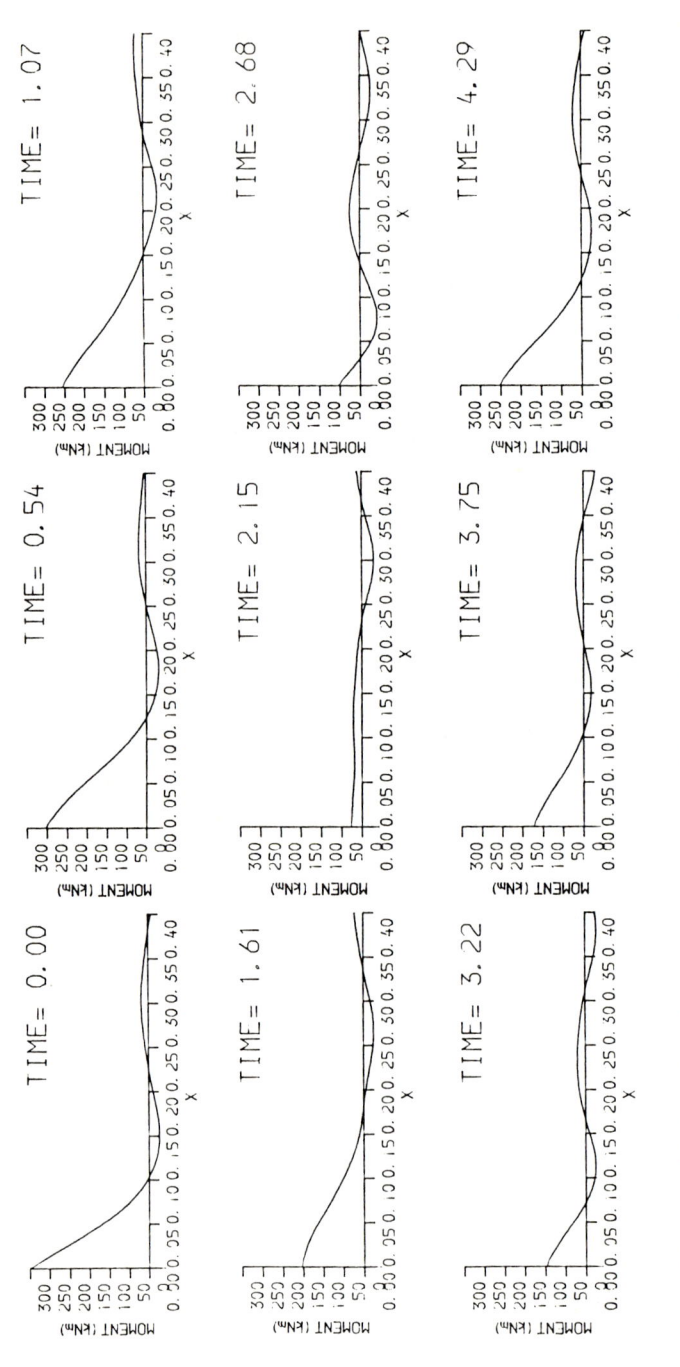

Figure 14.2 Banding moment generation with time along the hose-string

Mathematical Model of Hose String - Dynamic

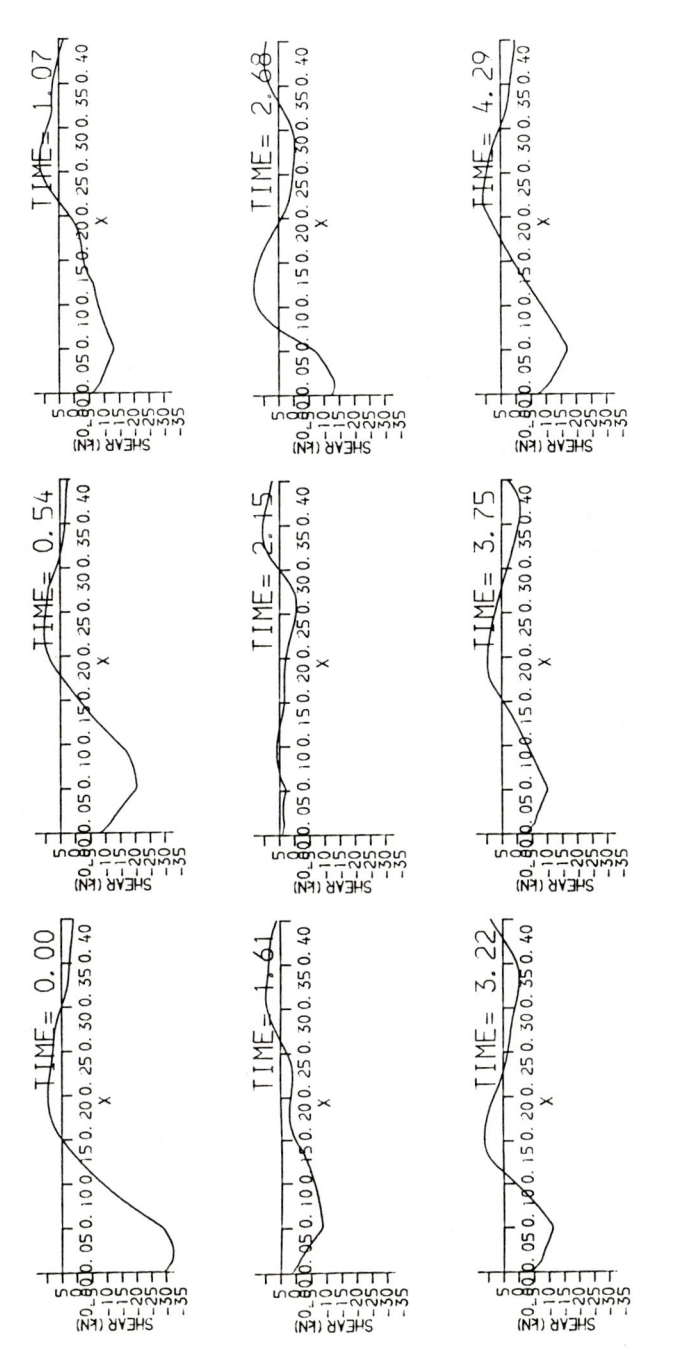

Figure 14.3   Shear stress generation with time along the hose-string

the shorter waves is highlighted by the increased stresses in the last hoses in the string. These stresses arise because the boundary condition applied at the end of the hose, namely that the hose follows the wave-profile at its equilibrium depth beneath the surface, is inappropriate for this type of wave.

Hose off the buoy	λ=20m maximum moment kNm	shear Kn	λ=30m maximum moment kNm	shear kN	λ=40m maximum moment kNm	shear kN	λ=50m maximum moment kNm	shear kN
1	306.9	75.82	250.7	43.23	208.2	32.46	173.2	26.90
2	57.8	22.69	33.2	14.60	30.0	9.85	19.8	7.44
3	34.8	12.22	35.7	7.57	20.1	3.73	9.2	1.44
4	29.1	10.76	29.0	6.32	18.3	3.27	8.3	1.16
5	31.9	10.81	28.3	6.66	18.2	3.21	8.9	1.04
6	31.0	10.78	31.6	6.62	18.1	3.16	8.1	0.98
7	31.8	10.81	28.3	6.12	20.0	3.28	8.1	0.99
8	32.9	10.90	28.4	6.90	18.7	3.38	8.0	0.90
9	42.9	13.33	32.1	7.11	18.9	3.25	8.3	1.39
10	88.3	22.84	35.6	7.99	18.6	4.70	8.3	1.72

Table 14.5 The effect upon the bending moment and shearing forces of varying the length of the waves

In Table 14.5, the steepness of the waves was changed by keeping their amplitude constant whilst varying their length. An alternative method of varying the steepness of the wave is to vary their amplitude whilst keeping their length constant - the results of doing this are given in Table 14.6. It can be seen from Table 14.6 that changing the wave steepness by varying their amplitude has a similar effect upon the stresses as that obtained by altering their length.

Mathematical Model of Hose String - Dynamic

Hose off the buoy	a=0.75m maximum moment kNm	a=0.75m maximum shear kN	a=1.00m maximum moment kNm	a=1.00m maximum shear kN	a=1.25m maximum moment kNm	a=1.25m maximum shear kN
1	174.7	27.22	185.8	28.16	197.1	30.33
2	19.7	7.66	24.1	8.34	27.8	8.83
3	9.4	1.61	12.5	2.22	15.9	3.11
4	8.7	1.22	12.2	1.90	17.0	2.61
5	8.1	1.13	11.7	1.67	16.0	2.63
6	8.2	1.13	11.9	1.74	16.1	2.53
7	8.2	1.12	11.8	1.72	16.0	2.50
8	8.1	1.18	11.7	1.78	15.8	2.43
9	9.2	1.22	12.7	1.90	17.5	3.06
10	8.3	1.59	12.0	2.27	16.3	3.16

Table 14.6   The effect upon the bending moment and shearing forces of varying the amplitude of the waves

In all of the problems considered so far, the buoy's motion has followed that of the waves. This will not always be the case since, as described earlier, the motion of the buoy is dependent upon type of waves present. The effect of phase difference between the motion of the buoy and the waves is shown in table 14.7. From this table, it can be seen that the stresses in the hose-string are a function of this phase difference. Since very large stresses can cause serious damage to these hoses, even if they are applied for only a short period of time, out of phase motion of the buoy relevant to the waves will have a large effect upon the life-spans of these hoses.

ε_0	Hose off the buoy	$\varepsilon_h=0^0$ maximum moment kNm / shear kN		$\varepsilon_h=90^0$ maximum moment kNm / shear kN		$\varepsilon_h=180^0$ maximum moment kNm / shear kN		$\varepsilon_h=270^0$ maximum moment kNm / shear kN	
0^0	1	185.8	28.16	167.7	21.99	144.8	32.86	157.8	22.12
	2	24.1	8.37	26.7	8.33	27.1	10.04	26.4	9.10
	3	12.5	2.22	12.6	2.17	15.7	2.61	12.4	2.58
	4	12.2	1.90	12.2	2.00	12.2	1.86	12.2	1.84
90^0	1	161.4	22.00	137.9	18.18	168.8	49.06	174.0	27.25
	2	18.8	8.80	19.2	8.00	32.1	10.02	27.8	10.62
	3	11.2	3.05	15.4	2.98	16.6	4.06	18.8	4.36
	4	12.2	1.85	12.1	1.96	12.3	1.78	12.3	1.85
180^0	1	201.0	28.16	200.1	26.73	160.3	32.86	183.7	31.13
	2	26.4	12.65	16.7	9.35	29.6	10.86	27.7	11.89
	3	19.9	4.17	14.4	2.37	17.3	4.02	23.6	4.71
	4	12.3	1.87	12.2	1.85	12.4	1.80	12.4	1.88
270^0	1	255.4	44.36	211.5	33.19	179.7	23.92	189.2	23.73
	2	24.5	12.42	22.0	9.36	28.3	10.99	27.2	12.03
	3	18.0	3.17	14.8	2.36	14.4	2.86	16.9	3.98
	4	12.0	1.74	12.2	1.85	12.2	1.79	12.3	1.81

Table 14.7   The effect of the phase difference between the motion of the buoy and that of the wave-train

14.8   Conclusions

A mathematical model of an oil-transferring floating hose-string subjected to wave-motion has been obtained. This model can be used in predicting the stresses and the motion of a hose-string attached to a buoy, which can undergo heaving and pitching motions, for varying sea states.

It has been shown that the crests of very steep waves will completely submerge the hose-string. As their steepness decreases, either through increasing their wavelength or by decreasing their amplitude, the hose-string becomes increasingly more able to follow the waves with a corresponding drop in the stresses in the hose-string.

Mathematical Model of Hose String - Dynamic

If extra reinforcement is built into the first hose off the buoy, this has the effect of increasing the stresses on the subsequent hoses. If the second hose off the buoy is of the same design as the rest of the hose-string (as normally is the case) such that it has no extra reinforcement, then its likelihood of failure will be increased by these extra stresses.

Phase difference between the motion of the buoy and that of the waves can lead to very high stresses being applied to the hoses nearest the buoy. This could have a serious affect upon the life expectancy.

It has also been shown that the benefits obtained by choosing a particular manifold design in order to reduce the axial loads on the hose-string, have to be balanced against the increased bending moments and shear forces.

Acknowledgements

I would like to express my appreciation and gratitude to Dr. M.I.G. Bloor, Dr. D.B. Ingham and Dr. L. Elliott for their supervision and guidance during this work.

I would also like to gratefully acknowledge the financial support provided during this work by Dunlop Limited (Oil and Marine Division) and the Science and Engineering Research Council through a CASE studentship.

References

Bishop, R.E.D. and Johnson, I., 1960.
The mechanics of vibration. Cambridge University Press, pp 282-284.

Brown, M.J., 1982.
Analysis of the stresses on a floating hose, Part 1 - Attached to a static buoy. University of Leeds, report for Dunlop Oil and Marine.

Brown, M.J., 1983.
Analysis of the stresses on a floating hose, Part 2 - Attached to a dynamic buoy. University of Leeds, report for Dunlop Oil and Marine.

Dunlop Oil and Marine Division, England, 1971.
Offhsore hose manual.

Dunlop Oil and Marine Division, 1973.
A study of forces acting on a monobuoy due to environmental conditions.

Ghose, P.K., 1981.
A numerical method for non-linear transient analysis of oil carrying offshore hose. University of Newcastle upon Tyne, report for the SRC Marine Technology Program.

Graham, H., 1982.
Newcastle model hose tests. Report for Dunlop Oil and Marine Division.

Kinsman, B., 1965.
Wind waves their generation and propogation on the ocean surface. Prentice-Hall, p 295.

Mathematical Model of Hose String - Dynamic

Appendix I

Unless it has been otherwise indicated, either in the text or in the tables, the following values have been used in obtaining the presented results.

Hose: Ten hoses in the string, each of length 10m.

Length of steel flange	= 1.0m
Mass, m	= 528.5 kgm^{-1}
Radius of unfloated hose, r_u	= 0.365 m
Radius of floated hose, r_f	= 0.469 m
Floatation covers first applied, x_u	= 4.750 m
Floatation covers fully applied, x_f	= 5.250 m
Bending stiffness of a steel flange	= 40000 kNm^2
Bending stiffness of reinforced first hose	= 2000 kNm^2
Bending stiffness of rubber of other hoses	= 400 kNm^2

Wave:

Length, λ	= 50.00 m
Period, T	= 5.61 secs
Amplitude, a	= 1.00 m

Boundary Conditions:

Manifold depth, Y_0	= 0.0 m
Manifold slope, θ_0	= 15.0°
Equilibrium depth, Y_e	= 0.236 m
Heave phase difference, ε_h	= 0.0°
Pitch phase difference, ε_p	= 0.0°
Amplitude of pitch, a_p, heave, a_h, and free end, a_l	= 1.0 m

Numerical computation: $\Delta t = 0.1$ $\Delta x = 0.001$

15. THE DESIGN OF CATENARY MOORING SYSTEMS
FOR OFFSHORE VESSELS

A.K. Brook,
The British Ship Research Association,
Wallsend, Tyne & Wear.

15.1 Introduction

Catenary mooring systems are used for a number of offshore applications which may either be of a permanent or temporary nature. Permanent moorings are required for certain semi-submersible rigs which are used for production or accommodation platforms and occasionally for monohulls which are used for the storage of oil. In this case the mooring system needs to be able to survive extreme conditions due to wind and wave action. Temporary moorings are required for rigs or monohulls involved in a number of offshore activities including drilling, diving support and the transfer of oil to a tanker from another storage tanker. These applications often require that the vessel motion and heading variation should be restricted and the moorings may be used together with a dynamically positioned system. Typical mooring arrangements for monohulls and semi-submersibles are shown in Figures 15.1 and 15.2. Figure 15.3 shows a temporary mooring system for a ship where mooring lines are attached to permanently moored buoys and the ship's anchors are also used.

It is desirable that the design of a catenary mooring system should include both static and dynamic analyses. A static analysis needs to determine the mooring loads which are required to achieve a static balance with assumed steady environmental conditions and the effect of varying the number, position, length and size of chains in the mooring system can be investigated.

A dynamic analysis needs to determine the motion of the vessel about the equilibrium position due to random excitations which arise from wind gusting and wave action. The peak tensions which occur together with vessel movements need to be calculated to determine if the mooring system is satisfactory.

This paper describes a mathematical model which can be used to determine the response of a moored vessel to wind, wave and current action. A

A. K. Brook

discussion is included on the nature of the environmental forces and moments which act on a vessel together with practical methods of calculating both steady and time varying forces and moments. Expressions from which the mooring forces and moments can be calculated are also discussed and consideration is given into the validity of linearising these restoring forces and moments. The equation of motion can be solved numerically and typical results have been presented for a semi-submersible.

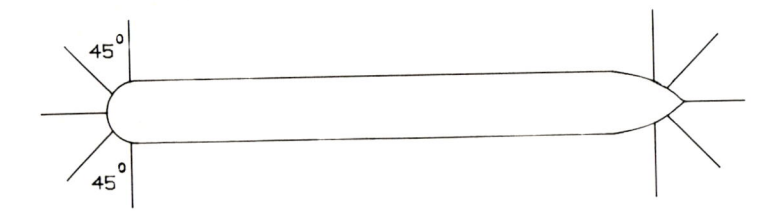

Figure 15.1   Mooring Arrangement of a Ship in the Open Sea

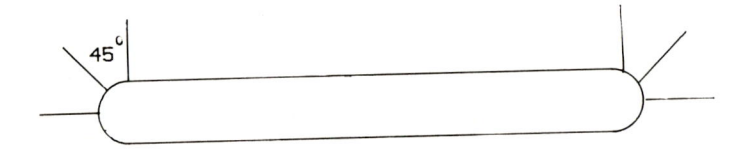

Figure 15.2   Mooring Arrangement of a Semi-submersible in the Open Sea

Catenary Mooring Systems

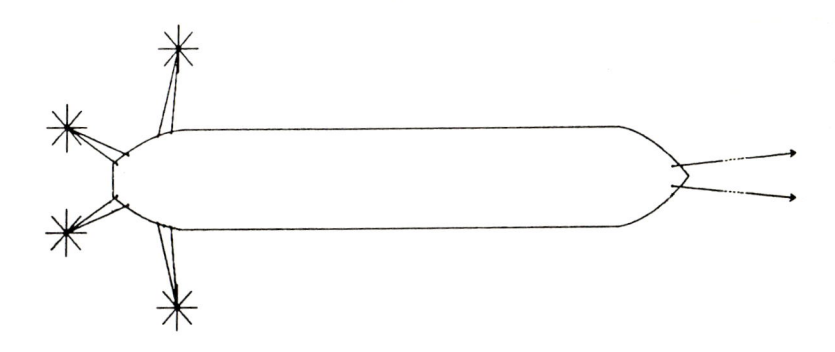

Figure 15.3   Mooring Arrangement of a Ship in the Open Sea Using Buoys and Anchors

15.2   Representation of the Environment

A moored vessel in the open sea is exposed to wind, waves and currents and methods are required from which these environmental forces and moments can be calculated for a ship or semi-submersible. Time varying forces arise due to wind gusting and wave action but current effects can be assumed steady since tidal variations in the open sea vary slowly.

Wave action on a free floating stationary vessel produces motions in all six degrees of freedom and in regular seas the vessel oscillates about the stationary position. In a random sea the mean position of the vessel calculated over a long period of time will coincide approximately with the equilibrium position. The mooring system of a vessel is not designed to prevent these motions due to first order wave action, although the vessel responses will be affected by the mooring system.

A second order wave effect is also produced by waves on an unrestrained stationary ship due to wave action which causes the vessel to drift in surge, sway and yaw and the mooring system needs to be capable of counteracting this so called "wave drift effect" which varies slowly in time.

A static design analysis of a mooring system needs to determine the mooring loads which are required to balance assumed steady environmental surge and sway forces and yaw moments due to wind, wave drift and current effects. A dynamic analysis needs to consider the behaviour of a moored

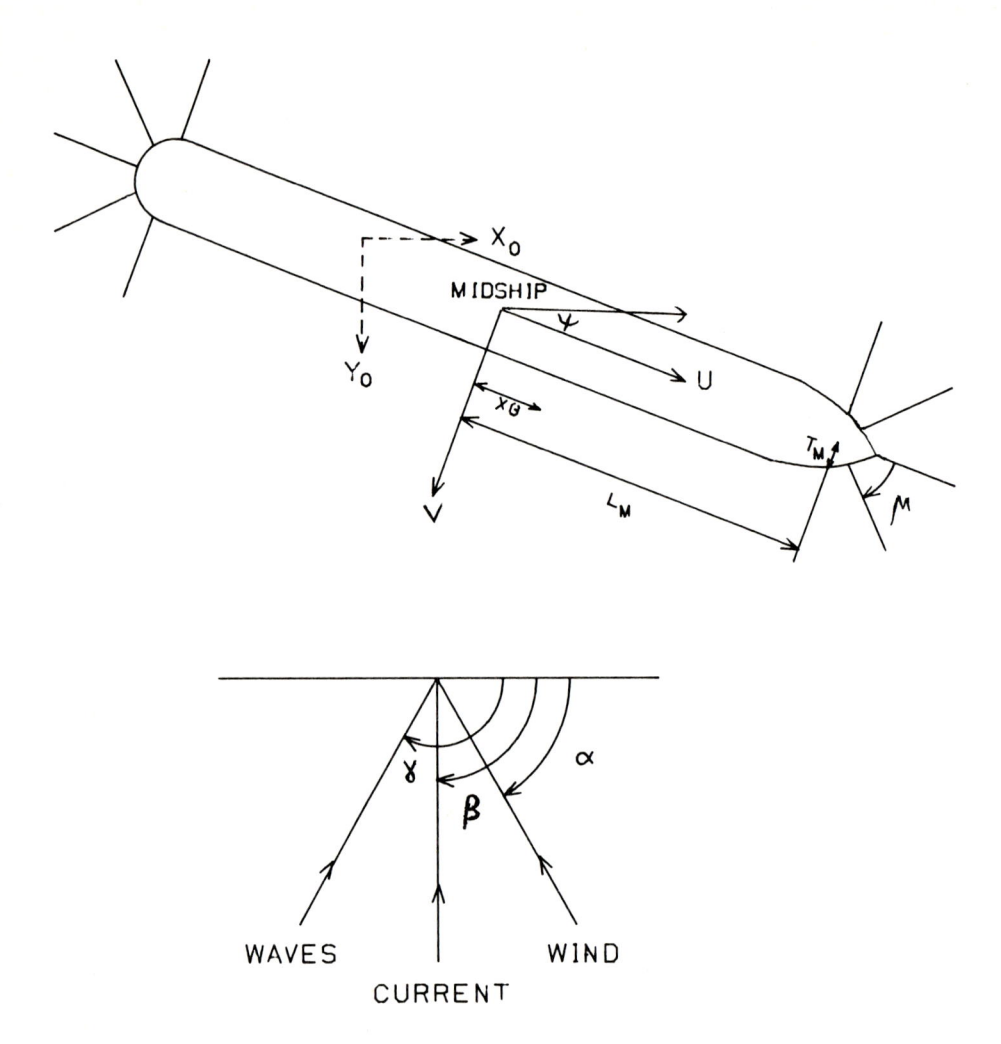

Figure 15.4 Axes and Co-ordinate System used in Mooring Simulation

Catenary Mooring Systems

vessel to the irregular or time varying forces and moments which arise due to wind gusting, and first and second order wave effects.

The inclusion of first order wave action, however, requires the calculation of vessel motions in six degrees of freedom and this has not been included in the present analysis. This paper focusses upon the surge, sway and yaw responses due to steady wind, wave drift and current effects and considers the effects of wind gusting and slow varying wave drift action.

15.3 Mathematical Model of Moored Vessel

The equations of motion which describe the behaviour of a moored vessel are based on commonly accepted manoeuvuring equations as described in Ratcliffe et al (1981) and Brooke et al (1983). The equations of motion which describe the surge, sway and yaw motions with reference to axes fixed in the centre of the vessel as shown in Figure 15.4 are given by:-

$$\Delta(\dot{u} - rv - x_G r^2) = X = X_{HI} + X_W + X_C + X_{\bar{\xi}} + X_M$$
$$\Delta(\dot{v} + ru + x_G \dot{r}) = Y = Y_{HI} + Y_W + Y_C + Y_{\bar{\xi}} + Y_M \quad (15.1)$$
$$I_z \dot{r} + x_G(\dot{v} + ru) = N = N_{HI} + N_W + N_C + N_{\bar{\xi}} + N_M$$

where x_G is the position of the LCG measured from midships. The mass moment of inertia I_z is approximately $(0.25\ L_{pp})^2 \Delta$ for most ships. X,Y,N include the added mass terms; wind, wave drift and current effects; and the forces and moments due to the mooring lines. The position and heading of the ship with reference to fixed earth axes are given by:-

$$\dot{x}_0 = u\cos\psi - v\sin\psi + V_c\cos\beta$$
$$\dot{y}_0 = u\sin\psi + v\cos\psi + V_c\sin\beta \quad (15.2)$$
$$\dot{\psi} = r$$

These equations can be solved numerically using an integration procedure which varies the time step as required. The added mass terms are assumed to have the same form as given in Ratcliffe et al (1981) where:-

A. K. Brook

$$X_{HI} = X_{\dot{u}}\dot{u} - Y_{\dot{v}}rv - Y_{\dot{r}}r^2$$

$$Y_{HI} = X_{\dot{v}}\dot{v} + Y_{\dot{r}}\dot{r} + X_{\dot{u}}ur \qquad (15.3)$$

$$N_{HI} = N_{\dot{r}}\dot{r} + (Y_{\dot{v}} - X_{\dot{u}})uv + Y_{\dot{r}}(\dot{v} + ur)$$

The added mass terms vary with water depth and for monohulls the expressions given in the discussion (Ratcliffe et al, 1981) have been used where:-

$$X_{\dot{u}} = -0.05\Delta$$

$$\frac{Y_{\dot{v}}}{\Delta} = \frac{3(1 + \frac{T}{B})}{2 + \frac{B}{T}} + \frac{(1 + C_B)}{3B} Lpp \left(\frac{T}{H}\right)^{3.5}$$

$$Y_{\dot{v}} = x_G Y_{\dot{v}}$$

$$\frac{N_{\dot{r}}}{\Delta Lpp^2} = \frac{1 + \frac{T}{B}}{6(2 + \frac{B}{T})} + \frac{(1 + C_B)}{100B} Lpp \left(\frac{T}{H}\right)^{3.5} \qquad (15.4)$$

If the vessel is excited by waves the added mass and linear damping terms vary with wave frequency but since the first order response of the ship to waves has not been included the above expressions can be used in the present study.

15.4 Calculation of Environmental Forces and Moments

(a) Steady Forces and Moments

Steady surge and sway forces and yaw moments due to wind, current and wave drift action can be calculated from available model test results. OCIMF (1977) contains wind and current coefficients for oil tankers where wind data is presented for load and ballast conditions and current data for a range of water depth to draft ratios. Wind and current data for tankers and general cargo vessels are given in BSRA (1973), Page (1971) and Gould (1967) contain wind data for a number of ship types including tankers, general cargo ships, ferries and trawlers. Ponsford (1982) gives wind data for a semi-submersible but usually model experiments need to be undertaken for each rig due to differences in the number of legs and deck geometry.

Catenary Mooring Systems

The wind coefficients depend on the angle of attack and wind forces and moments can be expressed as:-

$$X_w = \tfrac{1}{2}\rho c_{xw}(\alpha) A_T V_w^2$$
$$Y_w = \tfrac{1}{2}\rho c_{yw}(\alpha) A_L V_w^2 \qquad (15.5)$$
$$N_w = \tfrac{1}{2}\rho c_{mw}(\alpha) A_L L_{OA} V_w^2$$

V_w is the wind speed in knots and α is the angle of attack of the wind. Current coefficients also depend on the angle of attack and the forces and moments can be expressed as:

$$X_c = \tfrac{1}{2}\rho c_{xc}(\beta) L_{WL} T V_c^2$$
$$Y_c = \tfrac{1}{2}\rho c_{yc}(\beta) L_{WL} T V_c^2 \qquad (15.6)$$
$$N_c = \tfrac{1}{2}\rho c_{mc}(\beta) L_{WL} T V_c^2$$

V_c is the current speed in knots and β is the angle of attack of the current. Current coefficients for a semi-submersible are non-dimensionalised with respect to the longitudinal and transverse underwater area A_{WL} and A_{WT} rather than $L_{WL} T$ which is used for monohulls.

Mean wave drift forces and moments can also be calculated from available coefficients and typical data is given in Wise et al (1975) for a general cargo ship which was converted to a drill ship.

The mean wave drift forces and moments are assumed to be proportional to the square of the significant wave height and depend on the angle of attack γ:-

$$X_{\bar{\xi}} = \tfrac{1}{2}\rho\, C_{x\bar{\xi}}(\gamma)\, L_{pp}\, \bar{\xi}^2_{1/3}$$
$$Y_{\bar{\xi}} = \tfrac{1}{2}\rho\, C_{y\bar{\xi}}(\gamma)\, L_{pp}\, \bar{\xi}^2_{1/3} \qquad (15.7)$$
$$N_{\bar{\xi}} = \tfrac{1}{2}\rho\, C_{m\bar{\xi}}(\gamma)\, L_{pp}\, \bar{\xi}^2_{1/3}$$

A. K. Brook

A comparison has been discussed in Brook et al (1983) of the wind and current forces which are calculated using the different available model test data for the wind and current forces and moments on an oil tanker. The results of the comparison are shown in Figures 15.5 and 15.6 and it can be seen that there are significant differences in the magnitude of the forces and moments for both current and wind effects.

The OCIMF current coefficients which are given in OCIMF (1977) for the longitudinal force indicate, as shown in Figure 15.5, that at low angles of attack the longitudinal force changes sign and becomes a lift force. Recent correlation between the observed and predicted behaviour of a tanker at a single point mooring in a tidal stream suggests that this lift does not occur in reality until the angle of attack is in excess of 60 degrees, whereas the OCIMF data indicates that it may occur at angles of about 15 degrees in very shallow water.

The effect of these differences in the predicted wind and current forces and moments will be studied using the static analysis.

The current forces and moments which are given in equation (15.6) are those imposed on a restrained ship. If the ship is free to surge, sway and yaw the current speed needs to be replaced by the speed of the ship relative to the water and the angle of attack needs to be evaluated on the basis of the surge and sway components of the relative velocity - $\tan^{-1}(v/u)$. The current forces and moments which act on a moving ship are essentially represented as non-linear damping terms and it can be seen from the current yaw moment expression in equation (15.6) that if the ship is restrained in surge and sway but is constrained to yaw, no damping would be present because the yaw moment expression is independent of the rate of turn. Ratcliffe et al (1975) discusses how this effect can be incorporated since the sway velocity of the ship at a longitudinal position x on the ship is $v + rx$. The following term needs to be added to N_c:

$$\int_{-L/2}^{L/2} \tfrac{1}{2}\rho c_D T \; x \; (rx+v) \; | \; rx+v \; | \; dx \qquad (15.8)$$

Catenary Mooring Systems

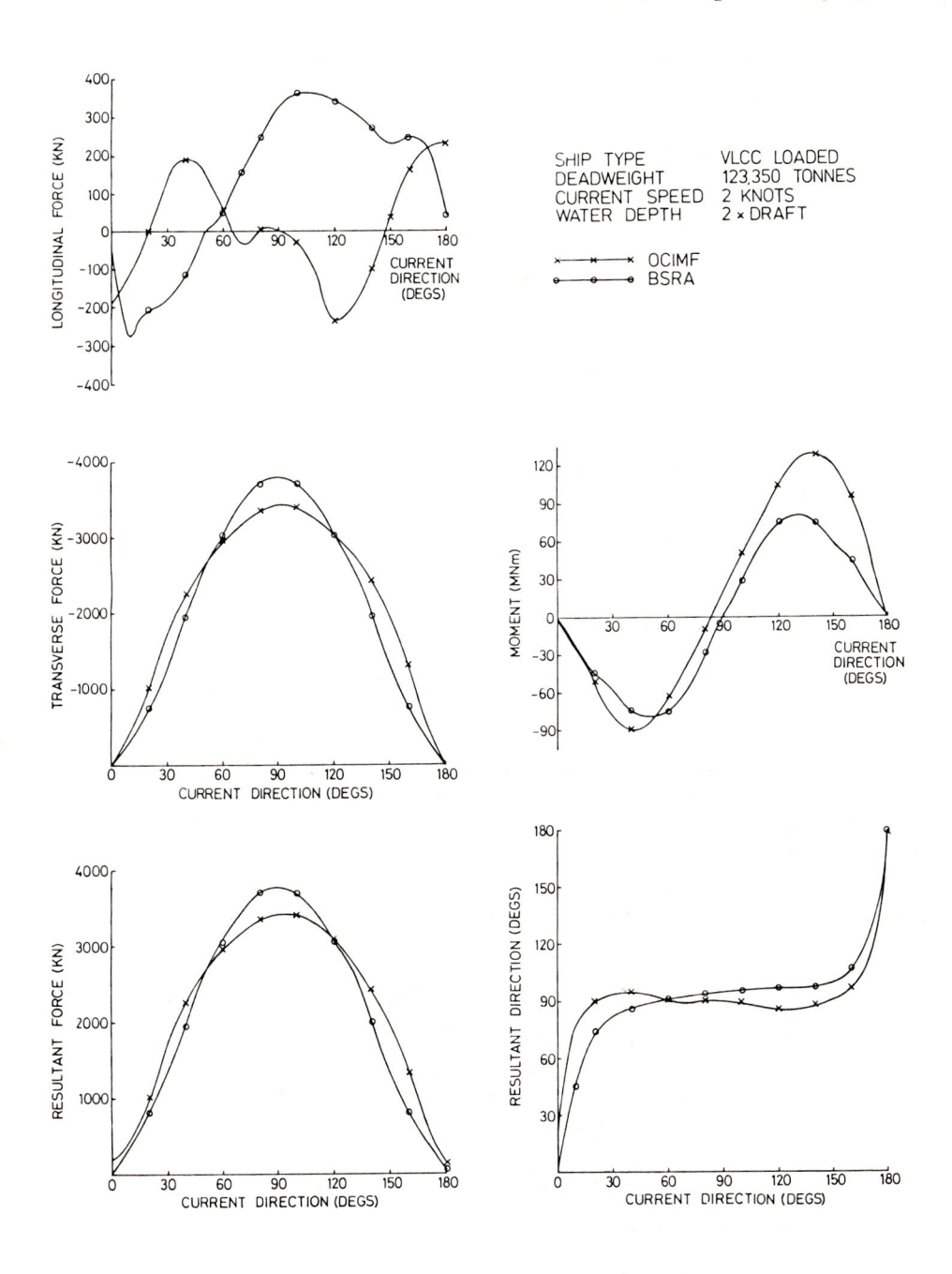

Figure 15.5　Comparison of OCIMF and BSRA Current Forces and Moments

A. K. Brook

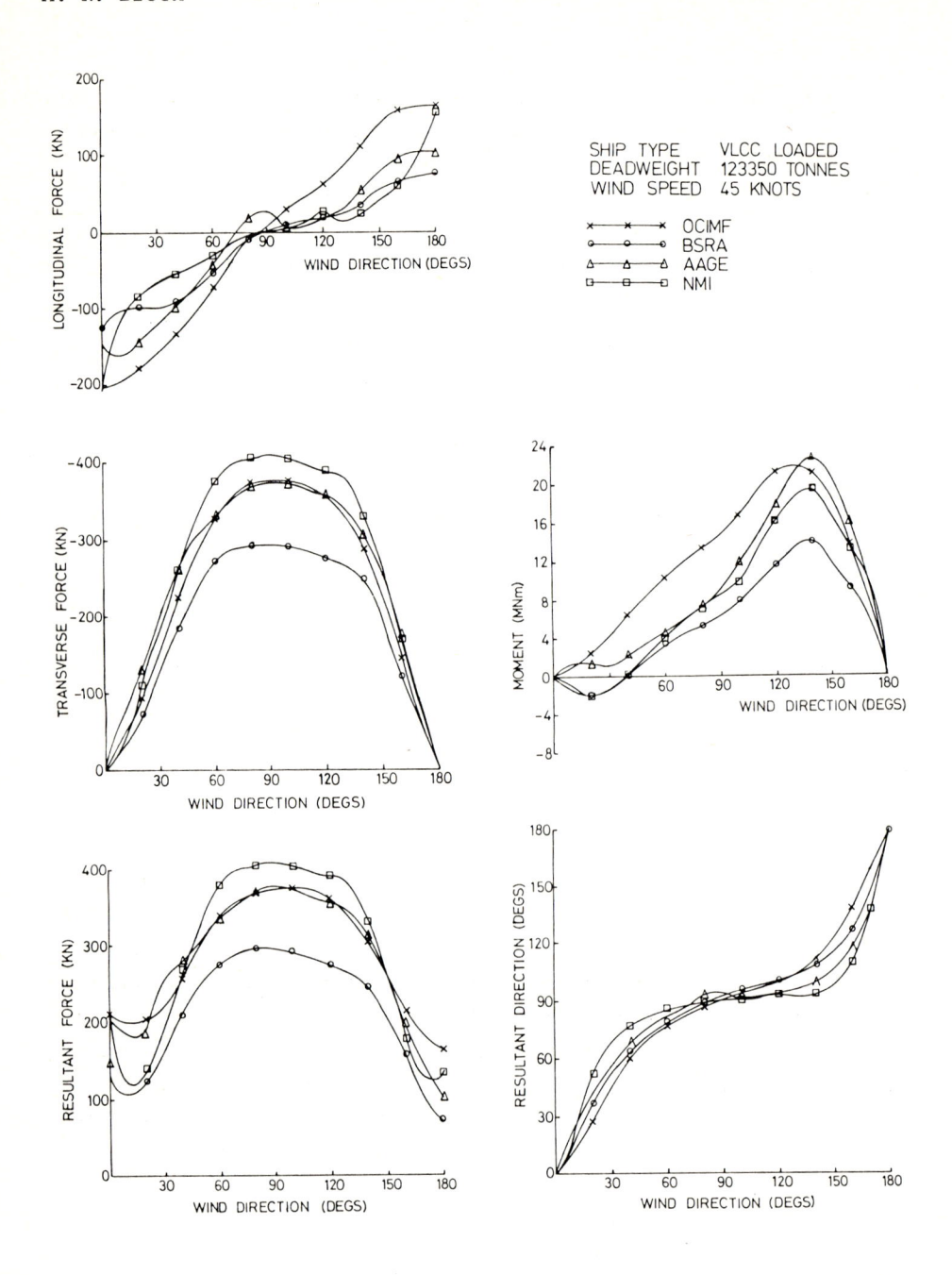

Figure 15.6 Comparison of OCIMF, BSRA, AAGE and NMI Wind Forces and Moments

Catenary Mooring Systems

(b) Irregular Forces and Moments

The time varying irregular forces and moments which are included in this study arise from wind gusting and slow varying wave drift action.

The wind and wave forces and moments can be developed with reference to a wind or wave spectrum which varies with frequency. The commonest wind spectrum is attributed to Davenport (1967), which has been modified slightly by Harris (1973) to:-

$$S_w(\omega) = \frac{4kV_w^2 q}{\omega(2+q^2)^{5/6}} \qquad (15.9)$$

where $q = \frac{286\omega}{V_w}$ and k is a turbulence factor which depends on the roughness of the terrain.

Typical wave spectra are the ITTC and Jonswap which take the form:

ITTC $\quad S_\xi(\omega) = \frac{A}{\omega^5} e^{-B/\omega^4}$

JONSWAP $\quad S_\xi(\omega) = \frac{A}{\omega^5} e^{-B/\omega^4} \gamma^{-a}$

where $a = \exp\alpha(\omega-\mu)^2$

The constants A, B, α, μ, γ are related to the significant wave height and average wave period.

An irregular time series of forces and moments can be constructed from regular forces and moments which are calculated over a range of frequencies by summing a series of sine waves and introducing a random phase angle. The wind or wave spectrum is introduced through the relationship between the amplitude of a since component and the spectral ordinate at a particular frequency:

$$a(\omega_i) = \sqrt{2S(\omega_i)d\omega}$$

A. K. Brook

In the case of wind gusting the process is used to generate an irregular time series of wind speed, rather than forces and moments and this wind speed can be used in equation (15.5) to calculate the wind forces and moments. If the wind spectrum is divided into N-1 equal intervals of $d\omega$ and N random phase angles $\varepsilon_i(\omega_i)$ are generated between 0 and 2π the wind speed at time t can be expressed as:

$$V(t) = V_w + \sum_{i=1}^{N} a_i(\omega_i) \cos(\omega_i t + \varepsilon_i) \tag{15.11}$$

where

$$a_i(\omega) = \sqrt{2S_w(\omega_i) d\omega}$$

Newman (1974) has shown that a similar procedure involving a double sum can be used to construct slow varying drift forces and moments. The surge force is given by:

$$X_{\bar{\xi}}(t) = \sum_{i=1}^{N} a_i T_i \sum_{j=1}^{N} a_j \cos((\omega_j - \omega_i) t - (\varepsilon_j - \varepsilon_i))$$

where

$$a_i(\omega_i) = \sqrt{2S_\xi(\omega) d\omega} \tag{15.12}$$

$a_i^2(\omega_i) T_i(\omega_i)$ is the surge force in regular waves at a particular frequency. ε_i are N random phase angles between 0 and 2π. Similar expressions can be given for the sway force and yaw moment.

15.5 Calculation of Mooring Forces and Moments

The tensions which occur in mooring lines which are attached either between the vessel and the sea-bed or for certain applications between the vessel and a moored buoy need to be calculated. The shape of each mooring line is assumed to be a catenary since the effects of current, wind, waves and the inertias of cables have not been included. It can be shown that under these conditions a cable always lies in a vertical plane and hence the following equations need to be solved for each cable to determine the horizontal and vertical forces which are acting on the ship:

$$\frac{d}{ds}(T\sin\phi) = w \qquad \frac{d}{ds}(T\cos\phi) = 0$$

$$\frac{dz}{ds} = (1+\varepsilon)\sin\phi \qquad \frac{dy}{ds} = (1+\varepsilon)\cos\phi$$

(15.13)

Catenary Mooring Systems

The relationship between the tension and the extension of the cable can be expressed as $T = F(\varepsilon)$ which is non-linear for synthetic materials and linear for wires and chains and for catenary mooring applications wires and chains can be assumed to be inextensible.

It is desirable that the mooring system should be designed so that vessel movements do not cause all of the chain to lift off the sea-bed since once this occurs small movements of the vessel cause significant increases in mooring line tensions. From the catenary equations it can be shown that if the upper end of a chain is located at a vertical distance η and a horizontal distance ξ from the origin shown in Figure 15.7 the horizontal component of tension when some chain lies on the sea-bed at the upper end can be obtained implicitly from:-

$$\frac{P}{h+\eta}(D + \xi - \ell) + \sqrt{P^2 + 2P} = \ln(1 + P + \sqrt{2P + P^2}) \qquad (15.14)$$

where

$$T_H = \frac{w(h+\eta)}{P}$$

The length of cable which is not on the sea-bed is given by:-

$$\ell_s = (h+\eta)\sqrt{1 + \frac{2T_H}{w(h+\eta)}} \qquad (15.15)$$

The vertical component of tension which acts on the ship is given by:

$$T_V = -\omega \ell_s \qquad (15.16)$$

The cable is lifted entirely off the sea bed when $\ell_s = \ell$ and it can be shown that the tension at the upper end is given by:

$$T = \frac{w}{2(h+\eta)}(\ell^2 + (h+\eta)^2) \qquad (15.17)$$

and the horizontal movement is given by:

$$\xi = -D + \frac{(\ell^2 - (h+\eta)^2)}{2(h+\eta)} \ln \frac{(\ell+h+\eta)^2}{\ell^2 - (h+\eta)^2} \qquad (15.18)$$

A. K. Brook

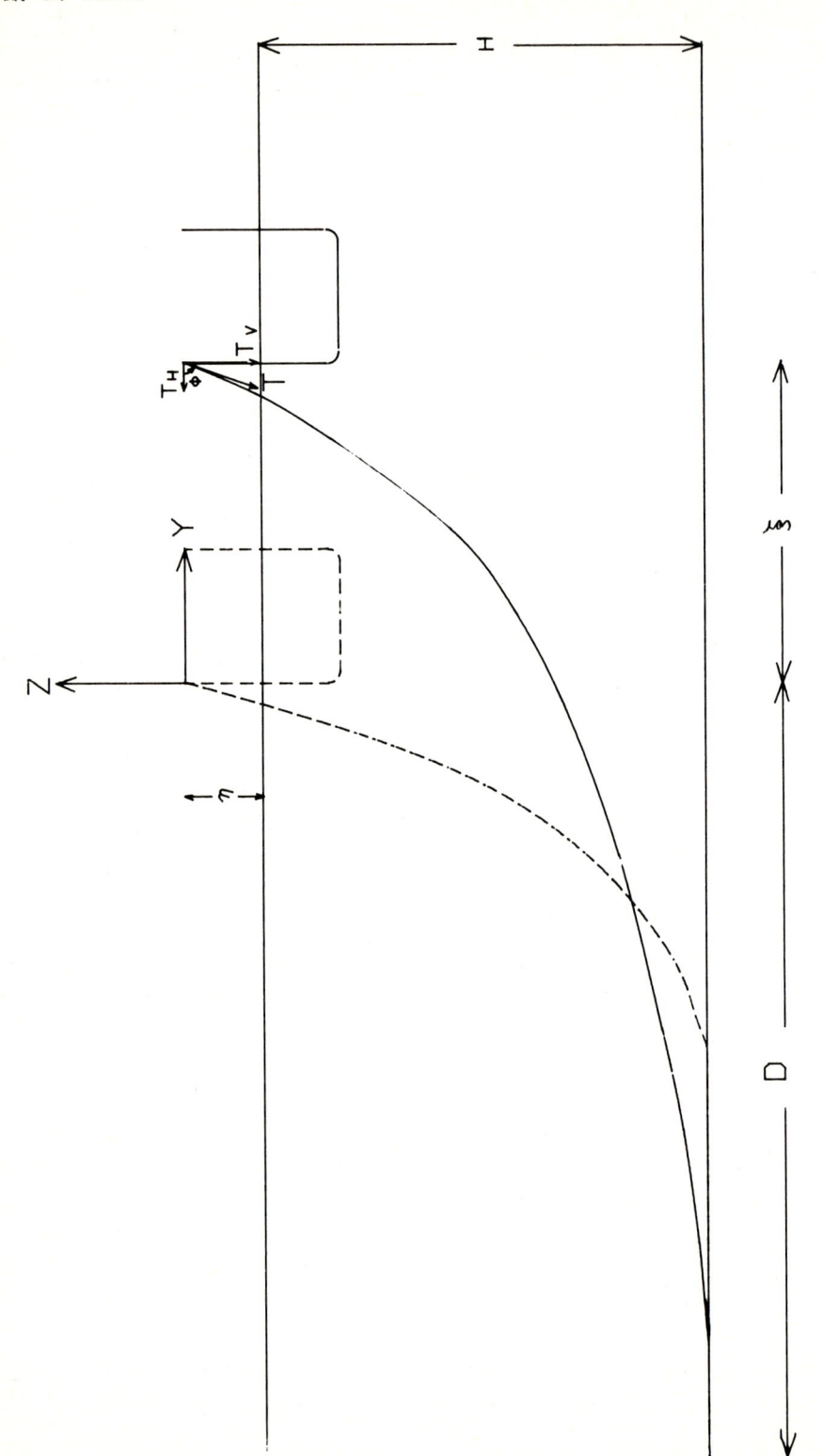

Figure 15.7 Cable Co-ordinate System

Catenary Mooring Systems

Where no cable is on the sea-bed it can be shown that the horizontal component of tension T_H can be determined implicitly from:-

$$\xi + D = \frac{T_H}{w} \ln \frac{t-1}{t+1}$$

where (15.19)

$$t = \sqrt{1 + \frac{4T_H^2}{w^2(\ell^2 - (h+\eta)^2)}}$$

The vertical component of tension is given by:

$$T_V = -\frac{w}{2}(\ell + (h+\eta)t) \qquad (15.20)$$

Hence the tension in the cable can be determined from:-

$$T = \sqrt{T_H^2 + T_V^2} \qquad (15.21)$$

Figure 15.8 shows a typical relationship between the tension in the cable and the horizontal movement of the upper end ξ. It can be seen that when some cable is on the sea-bed the relationship is highly non-linear but that once all of the cable lifts off the sea-bed the relationship is nearly linear and hence equation (15.19) can be replaced by a linear relationship which depends on the values of T_H and ξ at the points when all the cable lifts off the sea-bed and when the cable breaks.

From Figure 15.8, can be seen that the practice of linearising the mooring restoring forces is clearly not valid and will lead to large errors in many cases even when movements of the vessel are small about the equilibrium position.

The solution for a single cable can be generalised to any number of cables located around a vessel where the horizontal movement ξ_i in the vertical plane of each cable can be determined from the surge, sway and yaw motion of the vessel at a given time t. The total horizontal forces in directions parallel and perpendicular to the centreline of the vessel together with the cable restoring moment can be calculated from geometric considerations.

If N cables are attached to a vessel the surge and sway forces and yaw moments acting on the vessel due to the cables are given by:

A. K. Brook

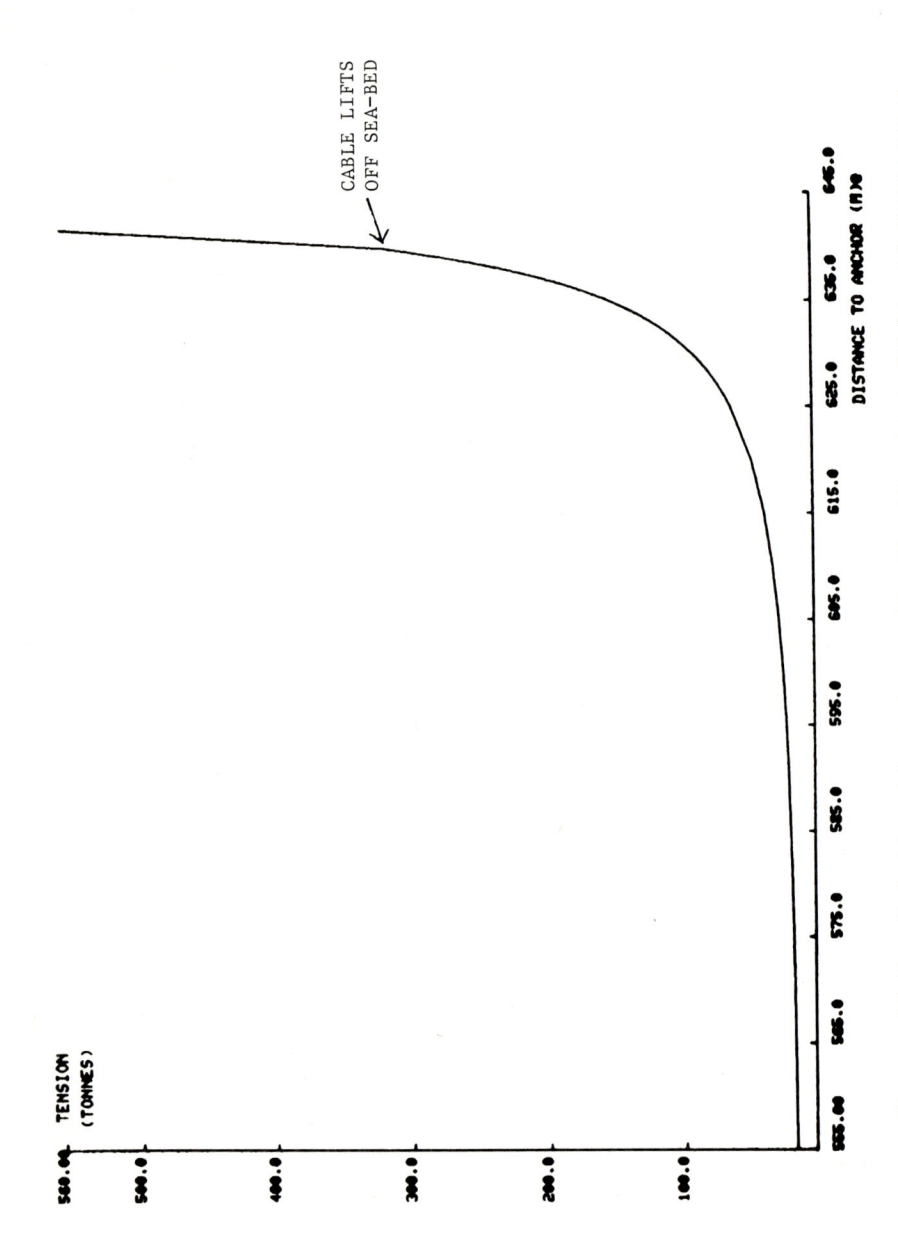

Figure 15.8 Relationship between the Tension in a Cable and Horizontal Movement of the Upper End

Catenary Mooring Systems

$$X_M = \sum_{i=1}^{N} T_{Hi}(\xi_i) \cos(\psi + \mu_i)$$

$$Y_M = \sum_{i=1}^{N} T_{Hi}(\xi_i) \sin(\psi + \mu_i) \quad (15.22)$$

$$N_M = \sum_{i=1}^{N} T_{Hi}(\xi_i) \{\ell_{mi} \sin(\psi + \mu_i) + t_{mi} \cos(\psi + \mu_i)\}$$

where T_{Hi} is the horizontal component of tension in cable i, μ_i is the initial angle of cable i relative to the vessel's centreline and ℓ_{mi} and t_{mi} are the longitudinal and transverse positions of the mooring point as shown in Figure 15.4.

Where the mooring lines from the ship are attached to a buoy similar expressions can be developed to equations (15.19) and (15.20) which also take into account the elasticity of the cable from which the tension and hence the forces and moments which act on the vessel can be calculated.

15.6 Static Analysis

A static analysis is required to determine the mooring loads which give a static balance when the vessel is subject to steady environmental conditions. Hence from equation (15.1) the following relationships need to be satisfied:-

$$X_w + X_c + X_{\dot{\xi}} + X_M = 0$$

$$Y_w + Y_c + Y_{\dot{\xi}} + Y_M = 0 \quad (15.23)$$

$$N_w + N_c + N_{\dot{\xi}} + N_M = 0$$

Where the number and position of the cables has been fixed it can be seen from equations (15.22) and (15.23) that if more than three mooring lines are used, an infinite number of different combinations of horizontal tensions T_{Hi} are possible in order to achieve a static balance with the given environmental conditions. For example if ten cables are used to moor a ship a number of combinations of tensions can be assigned to seven of the cables and the required tensions in the remaining three cables can be calculated to give a static balance with the environment. These values can be used as initial conditions for a dynamic simulation.

A. K. Brook

15.7 Response of Vessel to Wind Gusting and Wave Drift Action

The response of a vessel to wind gusting and slow varying wave drift effects can be calculated by solving equations 15.1 and 15.2 numerically. The wind gusting and wave drift effects can be calculated from the methods described in Secion 15.4. Typical results are shown in Figure 15.9 for a semi-submersible with the mooring configuration shown in Figure 15.2.

15.8 Conclusions

A mathematical model has been discussed which represents the response of a multiple moored vessel to wind, wave drift and current effects. The environmental forces and moments which act on a ship or semi-submersible have been discussed and practical methods given from which these can be determined for steady and irregular conditions. The mooring forces which act on a vessel have been described and expressions have been given from which these can be calculated based on catenary equations. The equations of motion can be solved numerically and typical results have been presented for the response of a semi-submersible to wind gusting and wave drift effects.

Acknowledgements

The author would like to thank the Chairman and Council of the British Ship Research Association for permission to publish this paper.

Catenary Mooring Systems

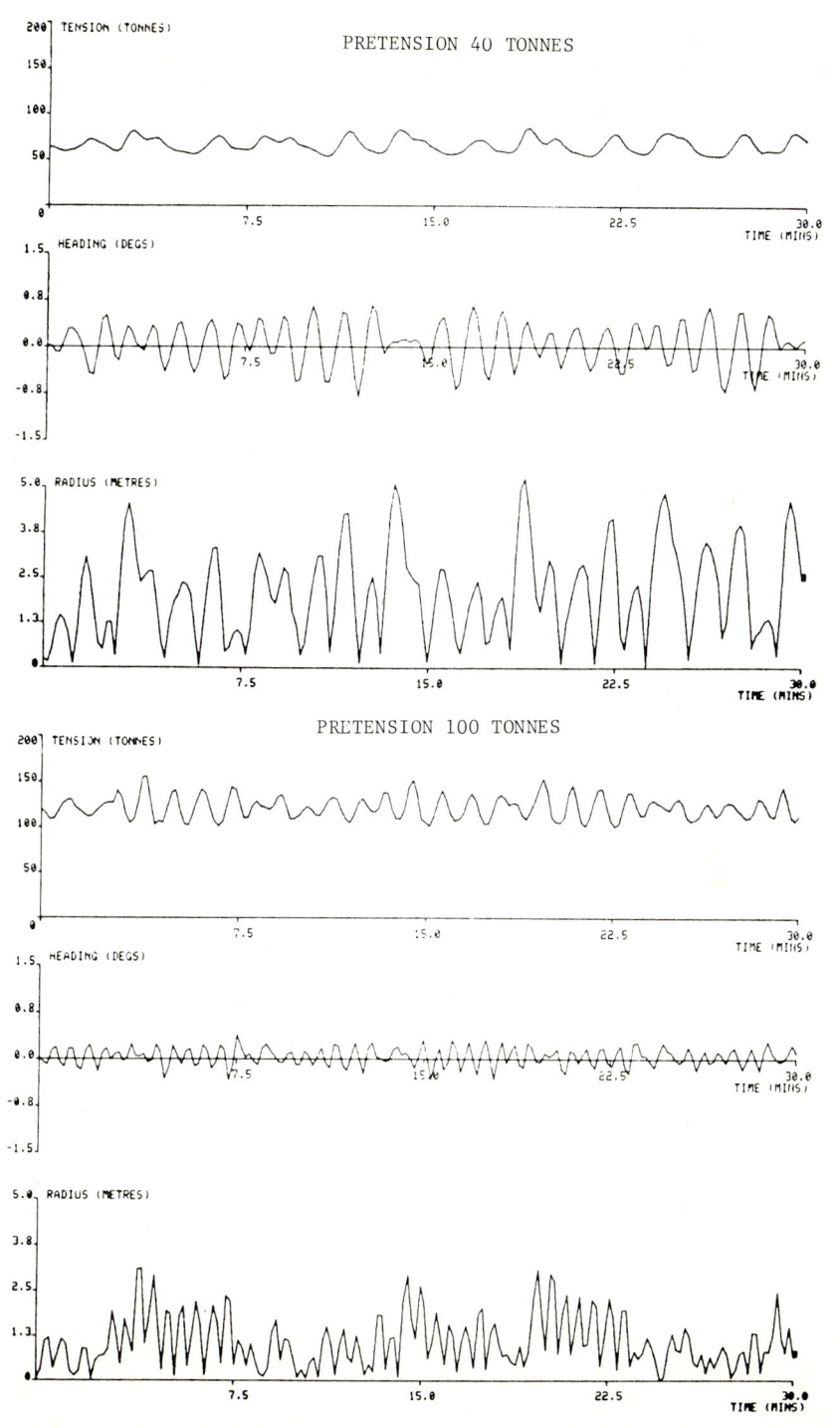

Figure 15.9 Response of Semi-Submersible to Wind Gusting and Slow Wave Drift Effects - 12 Mooring Lines (Vw 60 knots; Vc 1 knot; $\bar{\zeta}_{1/3}$, α, β, γ 90°)

References

Aage, C., 1971
Wind Coefficients for Nine Ship Models. Hydro-Og Aerodynamisk Laboratorium Aerodynamics Section report no. A-3.

Brook, A. K. and Byrne, D., 1983
The Dynamic Behaviour of single and multiple moored vessels. RINA Spring meeting.

B.S.R.A., 1973
Research investigation for the improvement of ship mooring methods. Fourth report (Wind and current data for various classes of ship). Report NS386.

Davenport, A. G., 1967
Gust loading factors. ASCF, Vol. 95, ST3.

Gould, R. W. F., 1967
Measurement of wind forces on a series of models of merchant ships. NPL Aero Rep. 1233.

Harris, R. I., 1973
The nature of wind. Paper no. 3 from Modern Design of Wind Sensitive Structures, CIRA.

Newman, J. N., 1974
Second order slowly varying forces on vessels in irregular waves. Int. Symp. Dynamics of Marine Vehicles and Structures in Waves, London.

Oil Companies International Marine Forum, 1977
Prediction of wind and current loads on VLCC's.

Ponsford, P. J., 1982
Wind-tunnel measurements of aerodynamic forces and moments on a model of a semi-submersible offshore drilling rig. NMI R34.

Ratcliffe, A. T. and Clarke, D., 1981
Development of a Comprehensive simulation model of a single point mooring system. Trans. RINA.

Wise, D. A. and English, J. W., 1975
Tank and wind tunnel tests for a drill-ship with dynamic positioning control. Offshore Technology Conference.

16. SOME PROBLEMS INVOLVING UMBILICALS, CABLES AND PIPES

D.G. Simmonds,
School of Mathematical Sciences and Computer Studies,
RGIT, Aberdeen.

16.1 Introduction

During recent years a number of companies have grown up in the Aberdeen area designing and developing diving bells, mobile diving units (MDU's), remotely operated inspection vehicles etc. to meet the increasing need for subsea inspection of pipelines and other offshore installations.

The school of mathematics and computer studies at RGIT is quite regularly consulted by local companies about the calculation of hydrodynamic forces on submerged objects and, in particular, umbilicals and pipes. These calculations usually involve the prediction of configurations, tensions, and bending stresses under the steady forces of weight and current.

The problems encountered so far have been modelled as two-point boundary-value systems of ordinary differential equations. Figures 16.1 to 16.5 illustrate typical types of problems that have been encountered.

16.2 The Statics of Cables and Pipes

Equations for the static equilibrium of 'cables' (i.e. models which ignore bending) and 'pipes' (i.e. models in which bending moments and shear forces are included) may be regarded as classical. They occur in many equivalent forms in different texts and publications and are simply summarised here:-

i) Two Dimensional Cable

Figure 16.6 shows a cable element ds along which the tension increases by dT and the inclination by $d\theta$ under the effect of external forces Wds (weight, bouyancy), Gds (normal drag), Fds (tangential drag). For static equilibrium

$$(T+dT)\sin d\theta = Gds + W\cos\theta ds \text{ , normally}$$
$$-T + (T+dT)\cos d\theta = W\sin\theta ds - Fs \text{ , tangentially}$$

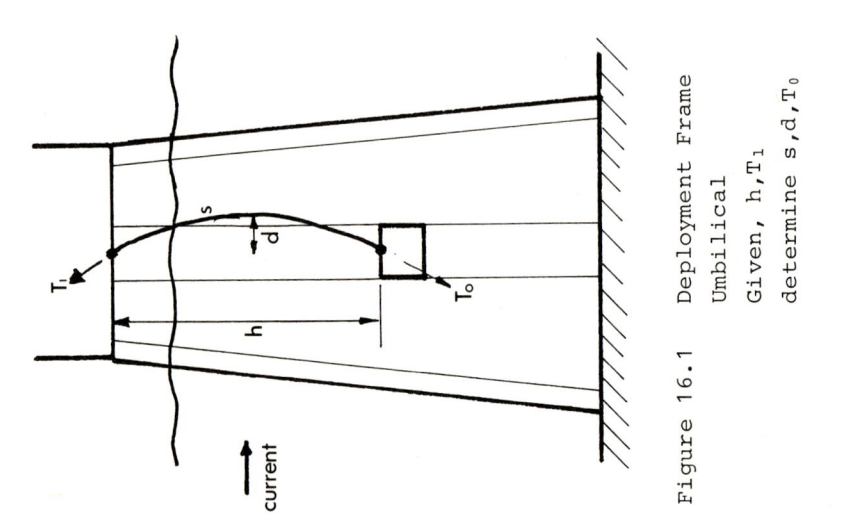

D. G. Simmonds

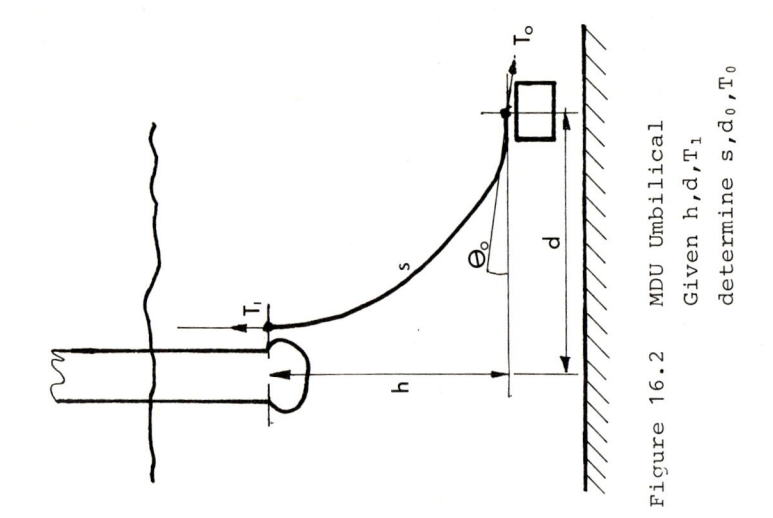

Figure 16.2 MDU Umbilical
Given h,d,T₁
determine s,d₀,T₀

Figure 16.1 Deployment Frame Umbilical
Given, h,T₁
determine s,d,T₀

Umbilicals, Cables and Pipes

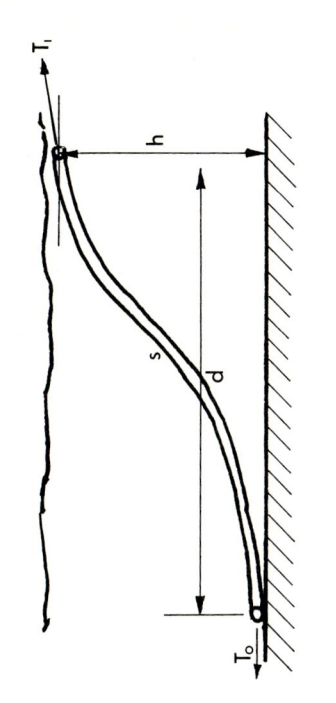

Figure 16.4 Pipelaying
Given $\theta_0(=0)$, $M_0(=0)$, d, h, θ_1
determine T_1, T_0 bending, moment diagram etc.

Figure 16.3 Mooring Cable
Given h, θ_0 and $T_1\cos\theta_1$
determine d, T_0, T_1, s

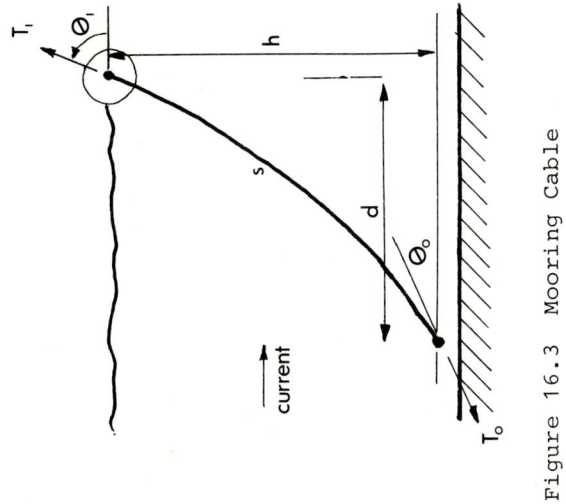

D. G. Simmonds

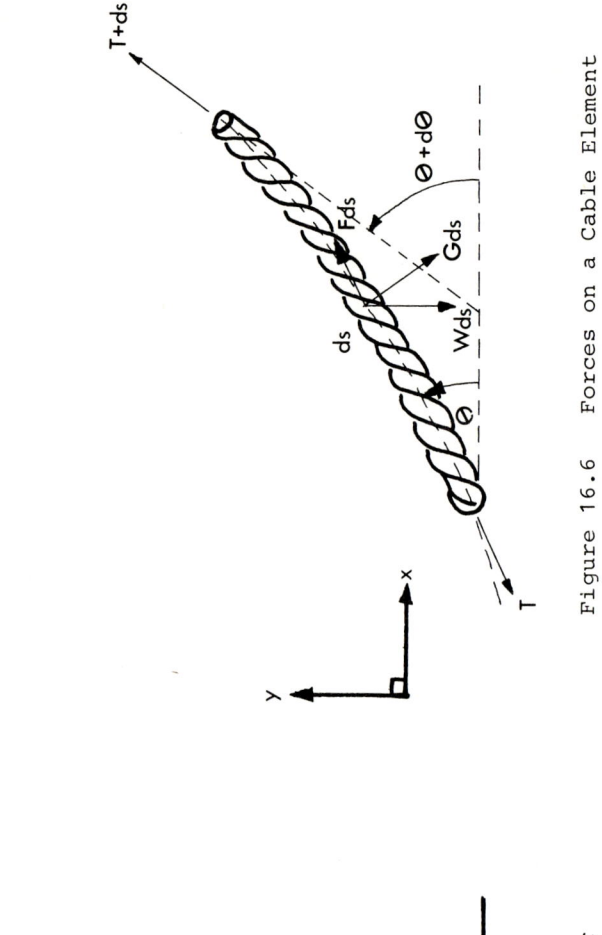

Figure 16.6 Forces on a Cable Element

Figure 16.5 Marine Riser
Given M_0, θ_0, M_1, h
determine d, T_1, s etc

Umbilicals, Cables and Pipes

$$T \frac{d\theta}{ds} = G + W\cos\theta$$

$$\frac{dT}{ds} = W\sin\theta - F \qquad (16.1)$$

ii) Three Dimensional Cable

θ is defined as the inclination to the horizontal, as before, and W, G, F are all as before and still lie in the same vertical plane. The equations (16.1) thus remain unaltered. The horizontal orientation of the element is described by the 'azimuth' angle ϕ, measured from the x-axis, and the horizontal external force per unit length is Hds (see figure 16.7).
Hence

$$T \frac{d\theta}{ds} = G + W\cos\theta$$

$$\frac{dT}{ds} = W\sin\theta - F$$

$$T \frac{d\phi}{ds} = \frac{-H}{\cos\theta} \qquad (16.2)$$

iii) Two Dimensional Pipe

Figure 16.8 is basically the same as Figure 16.6 with the addition of bending and shear forces to allow for bending stiffness. For static equilibrium

$(T+dT)\sin d\theta + Q - (Q+dQ)\cos d\theta = Gds + W\cos\theta ds$, normally
$-T + (T+dT)\cos d\theta + (Q+dQ)\sin d\theta = W\sin\theta ds - Fds$, tangentially

and

$-M + (M+dM) - Qds + (Gds+W\cos\theta ds)\alpha ds = 0 \qquad , (0 < \alpha < 1)$

taking moments about the upper end.

$$\frac{dT}{ds} = -\frac{QM}{EI} + W\sin\theta - F$$

$$\frac{dQ}{ds} = \frac{TM}{EI} - W\cos\theta - G$$

$$\frac{dM}{ds} = Q$$

$$\frac{d\theta}{ds} = \frac{M}{EI} \qquad (16.3)$$

D. G. Simmonds

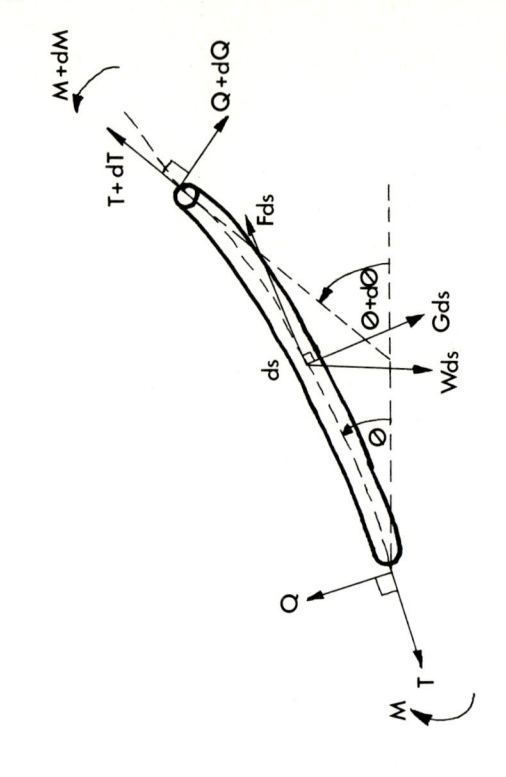

Figure 16.7 A three-dimensional cable element

Figure 16.8 Forces on a pipe element

Umbilicals, Cables and Pipes

The fourth equation expresses the standard relationship between bending moment and radius of curvature.

Equations (16.3) often occur in small-deflection form. Thus if θ is assumed small enough to be replaced by dy/dx and for ds to be replaced by dx then (16.3) are easily reduced to

$$\frac{d^2}{dx^2}\left(EI\frac{d^2y}{dx^2}\right) = T\frac{d^2y}{dx^2} - G \qquad (16.4)$$

the familiar small-deflection beam equation. A vertical beam version is also readily obtained by setting θ = π/2 - φ and assuming φ small enough to be replaced by dx/dy and ds replaced by dy. Equations (16.3)

$$\frac{d^2}{dy^2}\left(EI\frac{d^2x}{dy^2}\right) - T\frac{d^2x}{dy^2} - W\frac{dx}{dy} = G \qquad (16.5)$$

which may be used, for example, to determine the steady-state deflection of a marine riser.

16.3 Hydrodynamic Forces

The primary problem of cable modelling lies in the way in which the hydrodynamic forces, arising from the flow of water past the immersed cable, are expressed through the external forces F, G, and H. There is much literature available on the subject, dating back to the earlier part of this century, and offering a variety of possibilities to the modeller. Comprehensive reviews of the models available are given by Casarella and Parsons (1970) and Every (1981).

A thin smooth cylinder of diameter D which is stationary in a uniform steady flow of fluid normal to its axis will experience a drag force, R, per unit length where

$$R = \tfrac{1}{2} \rho D C_D v|v| \qquad (16.6)$$

In this formula ρ is the density of the fluid, v is the velocity of the flow, and C_D is a coefficient which must be determined experimentally. The formula is well supported by experimental work over many years (e.g. Relf and Powell, 1917, Bursnall and Loftin, 1951). The drag defined by (16.6) is usually termed pressure drag because it arises from the pressure difference caused by boundary layer separation,

D. G. Simmonds

and it is thus also dependent on the pattern of flow around the cylinder. This dependence is usually expressed through the C_D value which varies with Reynolds number typically as shown in Figure 16.9. Umbilicals usually operate in sub-critical flows.

There is much published data on the effects of roughness on C_D value, e.g. Miller (1976), Achenback (1971), the main effect being the onset of critical flow at progressively lower C_D values as roughness increases (see Figure 16.9).

When the cable is inclined to the flow at an angle θ the most obvious approach is to resolve the hydrodynamic forces into normal and tangential components and assume independence. This corresponds to

$$F = \tfrac{1}{2} \rho D C_{D_t} v^2 \cos\theta |\cos\theta|$$
$$G = \tfrac{1}{2} \rho D C_{D_n} v^2 \sin\theta |\sin\theta| \tag{16.7}$$

with separate drag coefficients for the separate components. These formulae were investigated by Relf and Powell (1917) and found to be quite reasonable. Wilson (1960) also supports this approach and suggests that

$$0.008 < C_{D_t} < 0.01$$

More recently Ferriss (1981) uses the same approach taking $C_{D_t} = 0.01$

Although all the experimenters mentioned before have found good correlation between the normal force, G, as given in (16.7), and observation, the tangential force, F, has proved more difficult to validate largely because of the difficulties of measuring tangential forces accurately. Attempts have been made to extend the conceptual models of tangential drag. Eames (1967) suggests decomposing the normal force, R, of (16.6) into a pressure drag $(1-\mu)R$ and a friction drag μR. With the cable inclined to the flow he argues that pressure drag acts in the direction of the flow and is independent of angle (see Figure 16.10). In the Eames model, then, we have

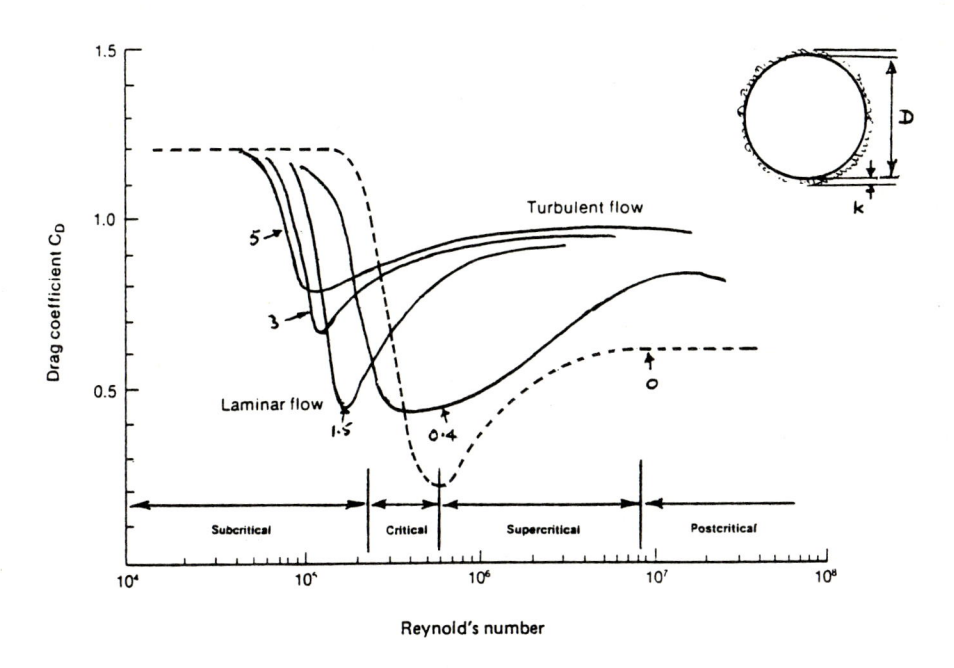

Figure 16.9   The dependence of C_D on Reynolds number and surface roughness (figures on curves are values of $k/D*10^3$)

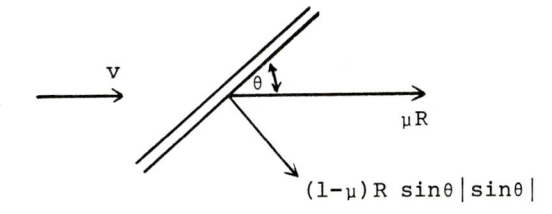

Figure 16.10   Pressure drag and friction drag on an inclined cable

D. G. Simmonds

$$F = \mu R \cos\theta$$
$$G = (1-\mu)R \sin\theta |\sin\theta| + \mu R \sin\theta \qquad (16.8)$$

and only one drag coefficient, C_D, is specified. At cable angles less than 30° experimental data show the friction force model of Eames to be inadequate and he has suggested adding a term $R\nu \sin\theta \cos\theta$ to the F of (16.8). Other experimenters have found difficulty in correlating the $\mu R \sin\theta$ term in the G of (16.8) with observation.

Whicker (1957) proposed hydrodynamic forces of the form

$$F = aR \cos\theta |\cos\theta| + bR \cos\theta$$
$$G = dR \sin\theta |\sin\theta| + eR \sin\theta \qquad (16.9)$$

determining the coefficients a,b,d,e to fit observations and Springsten (1967) proposed Fourier expansions for F and G in terms of sines and cosines of the cable angle θ, the coefficients being determined by curve fitting to observed data.

Friction drag is small and has proved difficult to measure with sufficient accuracy to properly validate the models that attempt to account for it. The model in (16.7) is widely used and appears to be as good as any, offering the scope to adjust C_{D_n}, C_{D_t} separately according to flow or cable angle. Choo and Casarella (1970) have proposed a model for the variation of C_{D_t} with Reynolds number and cable angle.

Many authors feel that concern to model the relatively small effects of friction drag (unless there are long sections of cable at small inclination to the flow) is unrealistic when there is another phenomenon known to have a far greater effect on cable behaviour. This phenomenon, of vortex shedding behind a cylinder in a uniform flow will occur at moderate velocities and at a frequency described by the Strouhal relationship. If the frequency is close to one of the cable natural frequencies then the cable may start to oscillate at this frequency and control the vortex shedding to reinforce its oscillation. Many experimenters have observed and measures this 'strumming' effect and have attributed to it massive increases in the drag coefficient:-

Umbilicals, Cables and Pipes

Grimminger (1945)	-	220%
Wieselsberger (1921)	-	140%
Dale & McCandless (1968)	-	200%
Diana & Falco (1971)	-	210%
Skop & Griffin (1977)	-	250%

Many of the above experimenters have also published empirical formulae which may be used to predict the drag coefficients of strumming cables but we will not review them here.

In the following sections the model (16.7) is used exclusively. The various empirical formulae for the variation of C_D with Reynolds number, roughness, strumming frequency are not included. If some idea of the effects of strumming is required then this is obtained by 'doing a run' with a C_D value of 2 instead of 1.2.

16.4 Analytical Solutions

16.4.1 Solutions of Equations (16.1)

If G and F can be ignored so that the cable is in equilibrium under weight and buoyancy forces alone, i.e. in static fluid, then equations (16.1) reduce to

$$T \frac{d\theta}{ds} = W \cos\theta$$

$$\frac{dT}{ds} = W \sin\theta \qquad (16.10)$$

The addition of the geometrical relationships

$$\frac{dx}{ds} = \cos\theta$$

$$\frac{dy}{ds} = \sin\theta \qquad (16.11)$$

yields a system that may be integrated analytically introducing four constants of integration - A,B,C,D say. In this solution, the classical catenary solution, any of the variables T,s,θ,x,y may be expressed in terms of any one of the others. The following are typical of the relationships that can be found:-

D. G. Simmonds

$$T \cos\theta = A$$
$$s = (A/W) \tan\theta + B$$
$$y = (A/W) \cosh\left[W(x-C)/A\right] + D$$
$$T^2 = A^2 + (s-B)^2 W^2$$
$$T = W(y-D) \tag{16.12}$$

The first of these expresses the fact that the horizontal component of tension is always constant and the last shows that T is a linear function of y if W is constant, which leads to the result

$$T_1 - T_0 = W(y_1 - y_0) \tag{16.13}$$

i.e. the difference in tension between two points is the weight of cable between those points. This is often a useful check on calculations.

The catenary solution is well-validated, much-used, and even in cases where F and G cannot be ignored will provide a useful first approximation.

Some integration of equations (16.1) is possible with the hydrodynamic force terms as in (16.7) included but under certain restrictions. Thus ignoring tangential drag and assuming a uniform steady current v in the positive x direction we have

$$T \frac{d\theta}{ds} = W(\cos\theta + \alpha \sin^2\theta) \quad ; \quad \frac{dx}{ds} = \cos\theta$$
$$\frac{dT}{ds} = W \sin\theta \quad ; \quad \frac{dy}{ds} = \sin\theta \tag{16.14}$$

where

$$\alpha = \tfrac{1}{2} \rho D C_D v^2 / W \tag{16.15}$$

and

$$\sin\theta \geq 0 \Rightarrow 0 \leq \theta \leq \pi \tag{16.16}$$

With these restrictions integration yields, as in the catenary case,

$$T = W(y - K_4)$$

Umbilicals, Cables and Pipes

and

$$T = K_1 \left| \frac{\cos\theta - \tfrac{1}{2}\alpha - \beta}{\cos\theta - \tfrac{1}{2}\alpha + \beta} \right|^{\frac{1}{2\alpha\beta}} \tag{16.17}$$

where

$$\beta^2 = 1 + \tfrac{1}{4}\alpha^2 \tag{16.18}$$

This result is mentioned by many others, e.g. Dove (1950), although in a slightly different form. Again it may be used for checking numerical solutions or providing starting approximations.

Equations (16.14) may be integrated completely if the further restriction that W=0, i.e. the cable is neutrally buoyant, is included. In this case the equations reduce to

$$T \frac{d\theta}{ds} = \gamma \sin^2\theta$$

$$\frac{dT}{ds} = 0 \tag{16.19}$$

where

$$\gamma = \tfrac{1}{2} \rho D C_D v^2 \tag{16.20}$$

and integration yields:-

$$T = T_0 \quad , \text{ constant along cable}$$
$$s = A_1 - \frac{T_0}{\gamma} \cot\theta$$
$$x = A_2 - \frac{T_0}{\gamma} \csc\theta$$
$$y = \frac{T_0}{\gamma} \ln(\tan\theta/2) + A_3 \tag{16.21}$$

Partial integrations of the cable equations are often used to tabulate non-dimensionalised forms of T,s,x,y (called cable functions) against θ. Wilson (1960) for example shows how this may be done for equations (16.1) with the full hydrodynamic terms of (16.7) included. Such tables are still used by designers.

D. G. Simmonds

16.4.2 Solutions of Equations (16.2)

Similar solutions are possible for equations (16.2) with the inclusion of the geometrical relations

$$\frac{dx}{ds} = \cos\theta \cos\phi$$

$$\frac{dy}{ds} = \cos\theta \sin\phi$$

$$\frac{dz}{ds} = \sin\theta \tag{16.22}$$

Thus with F ignored we obtain

$$T = W(z-K_4)$$

as before and ignoring G and H yields a catenary solution.

A solution analogous with (16.17) can be obtained by ignoring F and supposing that a uniform current v is flowing in the x direction. The components of this velocity in the F,G,H directions will be V_F, V_G, V_H where:-

$$V_F = v \cos\phi \cos\theta$$

$$V_G = v \cos\phi \sin\theta$$

$$V_H = -v \sin\phi \tag{16.23}$$

The resultant velocity normal to the cable is $(V_G^2 + V_H^2)^{\frac{1}{2}}$ which produces a normal pressure drag with components

$$G = \tfrac{1}{2} \rho\, D\, C_D\, V_G\, (V_G^2 + V_H^2)^{\frac{1}{2}}$$

$$H = \tfrac{1}{2} \rho\, D\, C_D\, V_H\, (V_G^2 + V_H^2)^{\frac{1}{2}} \tag{16.24}$$

Substituting (16.23) and (16.24) into (16.2) yields

$$T \frac{d\theta}{ds} = W \alpha \cos\phi \sin\theta\, (\cos^2\phi \sin^2\theta + \sin^2\phi)^{\frac{1}{2}} + W\cos\theta$$

$$\frac{dT}{ds} = W\sin\theta$$

$$T \frac{d\phi}{ds} = \frac{W \alpha \sin\phi\, (\cos^2\phi \sin^2\theta + \sin^2\phi)^{\frac{1}{2}}}{\cos\theta} \tag{16.25}$$

Umbilicals, Cables and Pipes

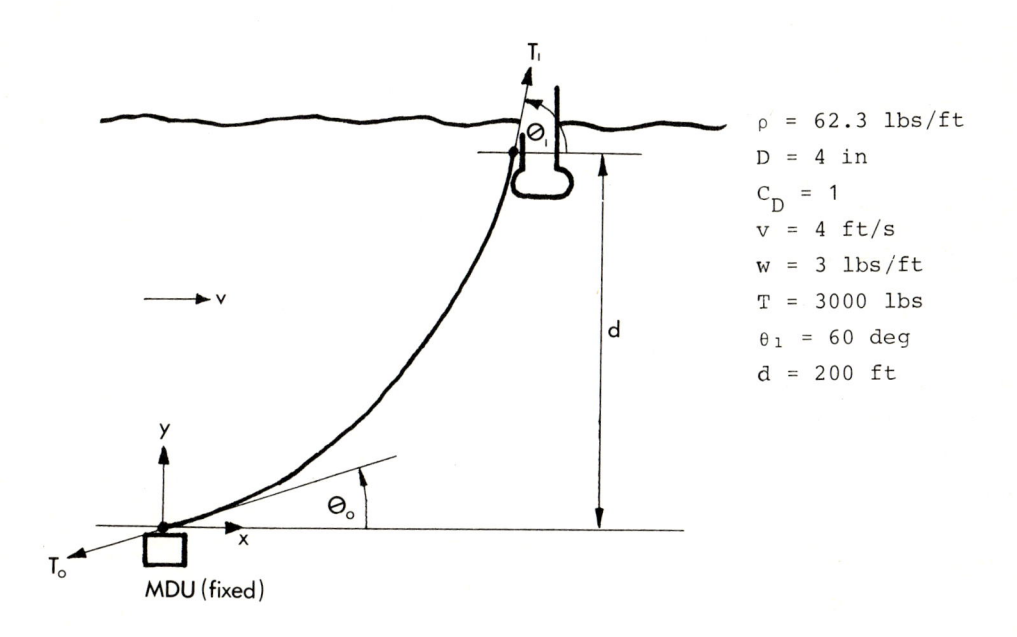

ρ = 62.3 lbs/ft
D = 4 in
C_D = 1
v = 4 ft/s
w = 3 lbs/ft
T = 3000 lbs
θ_1 = 60 deg
d = 200 ft

Figure 6.11   A typical umbilical problem: Calculate T_0, θ_0

D. G. Simmonds

where α is as defined in (16.15).

Partial integration is now possible if we ignore the normal weight component, $W\cos\theta$, supposing it is small in comparison with G. Integration yields

$$\tan\theta = K_1 \sin\phi \qquad (16.26)$$

and

$$T = K_2 \left| \frac{\sqrt{1+K_1^2 \sin^2\phi} - \cos\phi}{\sqrt{1+K_1^2 \sin^2\phi} + \cos\phi} \right|^{\frac{1}{2\alpha(1+K_1^2)^{\frac{1}{2}}}} \qquad (16.27)$$

(see Simmonds, 1983 for details). This partial solution has been used for checking numerical solutions of (16.2).

A neutrally buoyant solution analogous with (16.21) is also obtainable. Setting W=0 in (16.25) leads to the full solution

$$\tan\theta = K_1 \sin\phi$$

$$T = T_0 \quad \text{(constant along cable)}$$

$$s = K_2 - \frac{T_0}{\gamma(1+K_1^2)^{\frac{1}{2}}} \cot\phi$$

$$s = K_2 - \frac{T}{\gamma(1+K_1^2)^{\frac{1}{2}}} \sqrt{K_1^2 \cot^2\theta - 1}$$

$$x = \frac{T_0}{\gamma(1+K_1^2)^{\frac{1}{2}}} \{K_3 - (\csc^2\phi + K_1^2)^{\frac{1}{2}}\} \qquad (16.28)$$

16.4.3 Solutions of Equation (16.3)

The author is not aware of any analytical solutions of (16.3) apart from the small-deflection versions given in (16.4) and (16.5).

16.5 Typical Problems and Numerical Solutions

Figure 16.11 shows a typical umbilical problem as posed by an Aberdeen engineering company involved in the design of diving systems. The concern here is to calculate the horizontal force $T_0 \cos\theta_0$ on the MDU.

Umbilicals, Cables and Pipes

The solution, using analytical solution (16.17) ignoring tangential drag, runs as follows:-

With the figures given then, $T_0 = 3000 - 600 = \underline{2400}$

$$\alpha = \tfrac{1}{2} \rho D C_D v^2/W = 1.7306 \qquad \therefore \beta = 1.0835$$

(16.17) becomes

$$T = K_1 \left| \frac{\cos\theta - 1.3724}{\cos\theta + 0.7946} \right|^{0.2667}$$

Given $\theta = 60°$ when $T = 3000$ yields

$$3000 = K_1 \left| \frac{0.5 - 1.3724}{0.5 + 0.7946} \right|^{0.2667}$$

$$K_1 = 3333.03$$

So θ_0 is given by

$$2400 = 33333.03 \left| \frac{\cos\theta_0 - 1.3724}{\cos\theta_0 + 0.7946} \right|^{0.2667}$$

$$\theta_0 = 30.7°$$

This solution can offer no information on the length of cable or horizontal excursion of the MDU, for example, without numerical intergration. The neutrally-buoyant solution (16.21) does provide a complete solution if the wet weight of the cable can be regarded as negligible. From the value of α this seems unlikely but calculation yields a $T_0\cos\theta_0$ value of 2142 lbs compared with 2064 lbs above. The horizontal excursion obtained is 158 ft and the total length of cable is 256 ft.

The real inadequacy of all the analytical solutions is the assumption that $v, W, D,$ and C_D are constant. In the problems actually encountered the fluid flow is usually specified as a current profile and some sections of the cable may be different from others, e.g. they have buoyancy-collars fitted. Numerical integration of the model equations is thus essential.

D. G. Simmonds

The first one or two problems occurring at RGIT were tackled by trial-and-error. The usual routine was to estimate T_0 and θ_0 and then integrate (16.1) numerically to see if the cable turned up at the required attachment point or with the required angle, changing the estimates suitably if not. Even with the addition of an optimisation routine to automate the changing of estimated values this method proved tedious and unreliable. The arrival of further problems involving pipes as well as umbilicals and a greater variety of boundary-value configurations stimulated the production of software with greater flexibility and more reliable solution techniques.

The equations (16.1) and (16.11) together form a system of four ordinary differential equations which, if a solution is required in the s-range $[s_0, s_1]$ with a total of four values being specified at $s = s_0$ and $s = s_1$, lead to a simple two-point boundary-value problem. The values specified might be x_0, y_0, T_1, θ_1 or y_0, y_1, T_1, θ_1 (as Figure 16.11) or any combination. Using s as the independent variable is not always convenient, for although we may set $s_0 = 0$, the value of s_1 (the amount of umbilical payed out) may not be known. So the boundary-value problem is sometimes best tackled by rewriting equations (16.1) and (16.11) with x or y or θ or T as the independent variable. The computer packages developed have this flexibility built in. As far as solution techniques are concerned, the literature (see Morrison et al, 1964 and Holt, 1964) suggests that shooting methods are the easiest to implement with multiple-shooting methods or finite difference methods being used in cases of failure through instability. Ferriss (1981) has described a different approach using a selected points method.

In the event, the programs were written in FORTRAN and make use of the NAG subroutine D02ADF. This uses a standard Runge-Kutta-Merson code to integrate from both boundary points using supplied estimates of the 'free' boundary value. The two numerical solutions are compared at a 'matching point' and suitable changes made to the free boundary values using a modified Newton integration. The matching point is set at one of the boundary points initially but if this fails to produce convergence D02ADF is called again with one of eight interior points or the other boundary point as the matching point until all ten points have been tried.

Umbilicals, Cables and Pipes

Four programs have been developed based on the model equations (16.1), (16.2) and (16.3) described before, and also the equations for a three dimensional bending element which have not been stated but may be found in Love (1952). All the programs produce complete numerical details of configurations as well as drawings of the configurations, including plan and elevation drawings in three dimensional cases and bending moment or tension diagrams where required. Testing with standard analytical solutions and cable function tabulations published by Wilson (1960) has provided some validation of the programs since these solutions and tabulations have themselves been validated with experimental data.

The following examples illustrate the application of the programs:-

Example (i)

An MDU is at a depth of 50m, its umbilical being attached to a platform 50m above the mean sea level (MSL). The horizontal excursion of the MDU is 100m. The current is a uniform 1m/s and the wind speed is 2m/s. Cable properties are uniform with

$$w = 3 \text{ kg/m (dry)} \quad ; \quad D = .035\text{m}$$
$$C_{D_n} = 1 \quad ; \quad C_{D_t} = 0 \quad ; \quad \rho = 1000 \text{ kg/m}^3 \quad ; \quad g = 9.81 \text{ m/s}^2$$

Determine configurations for a variety of cable lengths.

Figures (16.12), (16.13), (16.14) and (16.16) show typical results.

Example (ii)

A 'cage' containing a remotely controlled vehicle (RCV) is lowered 300m from the surface to the sea bed where the RCV leaves the cage to carry out inspection work. Given the various cable properties, current profile, and drag characteristics for the cage plus RCV determine the lateral displacement of the cage at touchdown.

D. G. Simmonds

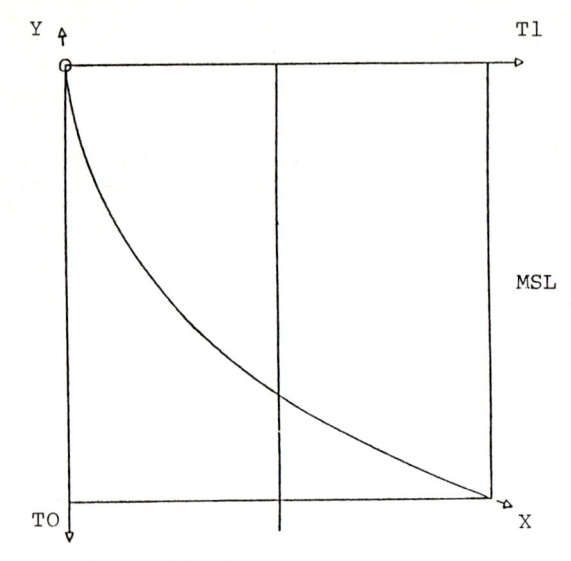

```
T0  = 198.    FREE
T1  = 450.    FIXED
TH0 = 7.      FREE
TH1 = 69.     FREE
S0  = 0.      FIXED
S1  = 150.    FREE
Y0  = 0.      FIXED
Y1  = 100.    FIXED

CABLE LENGTH = 150.
X LIMITS   0.   100.
Y LIMITS   0.   100.
```

Figure 16.12

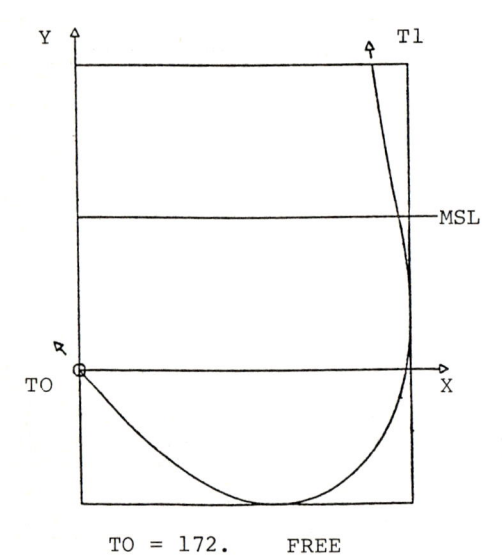

```
T0  = 172.    FREE
T1  = 424.    FREE
TH0 = -48.    FREE
TH1 = 98.     FREE
X0  = 0.      FIXED
X1  = 100.    FIXED
Y0  = 0.      FIXED
Y1  = 100.    FIXED

CABLE LENGTH = 250.
X LIMITS   0.    112.
Y LIMITS -44.    100.
```

Figure 16.13

```
T0  = 138.    FREE
T1  = 390.    FREE
TH0 = -40.    FREE
TH1 = 90.     FREE
X0  = 0.      FIXED
X1  = 100.    FIXED
Y0  = 0.      FIXED
Y1  = 100.    FIXED

CABLE LENGTH = 200.
X LIMITS   0.    100.
Y LIMITS -20.    100.
```

Figure 16.14

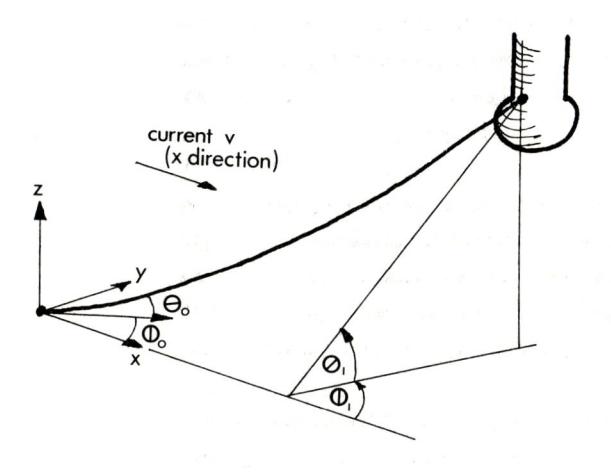

Figure 16.15 A three-dimensional version of Figure 16.11

D. G. Simmonds

```
T0  = 198.      FREE
T1  = 450.      FIXED
TH0 = 7.        FREE
TH1 = 69.       FREE
S0  = 0.        FIXED
S1  = 150.      FREE
Y0  = 0.        FIXED
Y1  = 100.      FIXED

CABLE LENGTH = 150.
X LIMITS    0.   100.
Y LIMITS    0.   100.
```

```
T0  = 172.      FREE
T1  = 424.      FREE
TH0 = -48.      FREE
TH1 = 98.       FREE
X0  = 0.        FIXED
X1  = 100.      FIXED
Y0  = 0.        FIXED
Y1  = 100.      FIXED

CABLE LENGTH = 250.
X LIMITS    0.   112.
Y LIMITS  -44.   100.
```

```
T0  = 138.      FREE
T1  = 390.      FREE
TH0 = -40.      FREE
TH1 = 90.       FREE
X0  = 0.        FIXED
X1  = 100.      FIXED
Y0  = 0.        FIXED
Y1  = 100.      FIXED

CABLE LENGTH = 200.
X LIMITS    0.   100.
Y LIMITS  -20.   100.
```

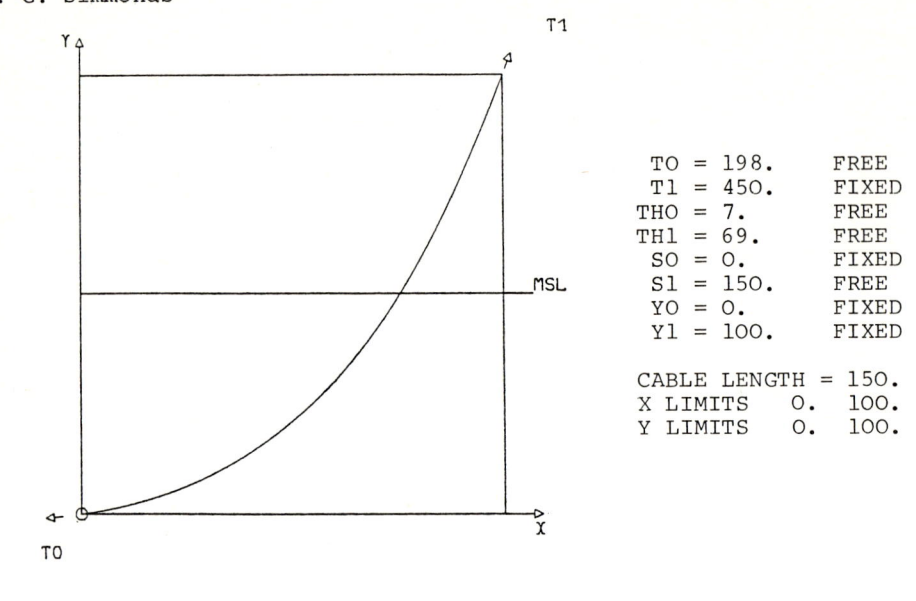

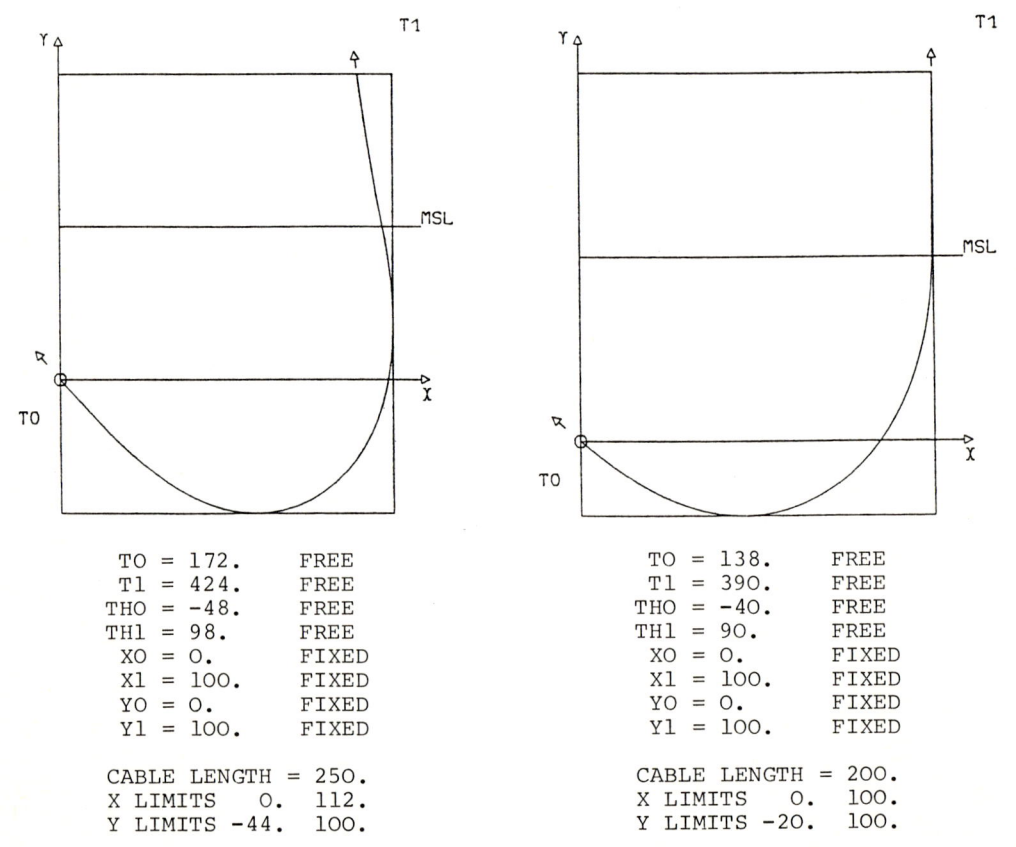

Figure 16.16

Umbilicals, Cables and Pipes

The horizontal drag, F, on the cage (at sea bed level), and its buoyant weight, W, are calculated from the information provided and hence the bottom tension $T_0 = (F^2+W^2)^{\frac{1}{2}}$ and angle $\theta_0 = \pi-\arctan(W/F)$. We thus have a boundary value problem with S_0 (=0), T_0, θ_0 specified at the cage and x_1 (=0) at the point of suspension (vertical distance, y, being the independent variable).

Example (iii)

Figure 16.15 shows a cable in a cross current. With the data as specified in Figure 16.11 and with $\phi_1 = 60°$. Figure 16.16 shows the three drawings produced by the three dimensional cable program.

The independent variable is z and the six fixed boundary values are

$x_0 = 0$ $y_0 = 0$ $s_0 = 0$
$T_1 = 3000$ $\theta_1 = 60$ $\phi_1 = 60$

System equations (16.2) and (16.22).

Example (iv)

A firm in charge of laying small diameter pipe on the sea bed using a 'J-lay' method (i.e. pushing the pipe down vertically from the barge) were interested to know what the maximum bending stresses in the pipe and side forces on the barge would be during the operation. Figure 16.17 shows the determination of the position of the 'touchdown' point for a given length (108m) of pipe using the boundary values

$x_0 = 0$ $y_0 = 0$ $\theta_0 = 0$
$M_1 = 0$ $y_1 = 84$ $\theta_1 = 90$

Figure 16.18 shows corresponding bending moment and tension diagrams.

Example (v)

Figure 16.19 illustrates the typical configuration when pipelaying using an 'S-lay' method with the pipe fed over a 'stinger' at the rear of the barge.

In the particular problem tackled (from the CASTORO II in the Forties

D. G. Simmonds

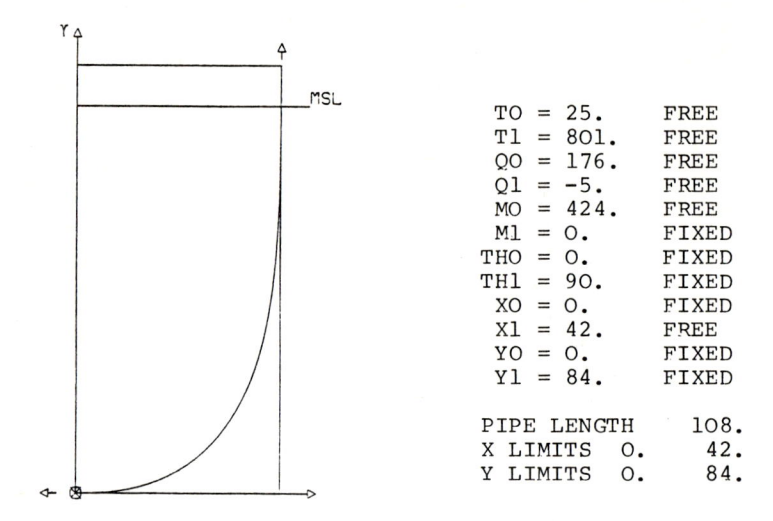

```
T0  = 25.     FREE
T1  = 801.    FREE
Q0  = 176.    FREE
Q1  = -5.     FREE
M0  = 424.    FREE
M1  = 0.      FIXED
TH0 = 0.      FIXED
TH1 = 90.     FIXED
X0  = 0.      FIXED
X1  = 42.     FREE
Y0  = 0.      FIXED
Y1  = 84.     FIXED

PIPE LENGTH   108.
X LIMITS  0.   42.
Y LIMITS  0.   84.
```

Figure 16.17

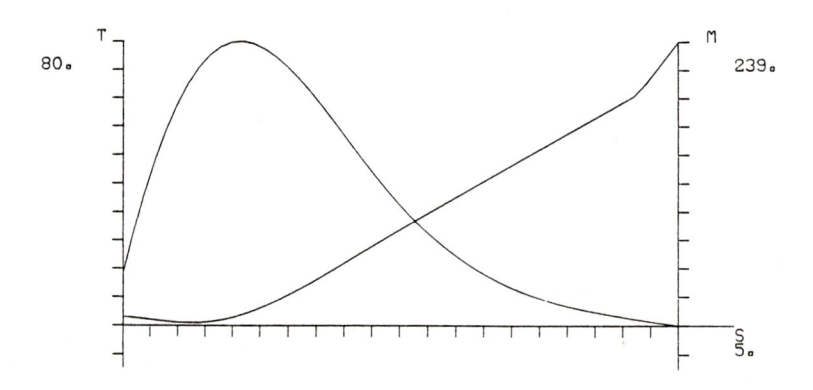

Figure 16.18

Umbilicals, Cables and Pipes

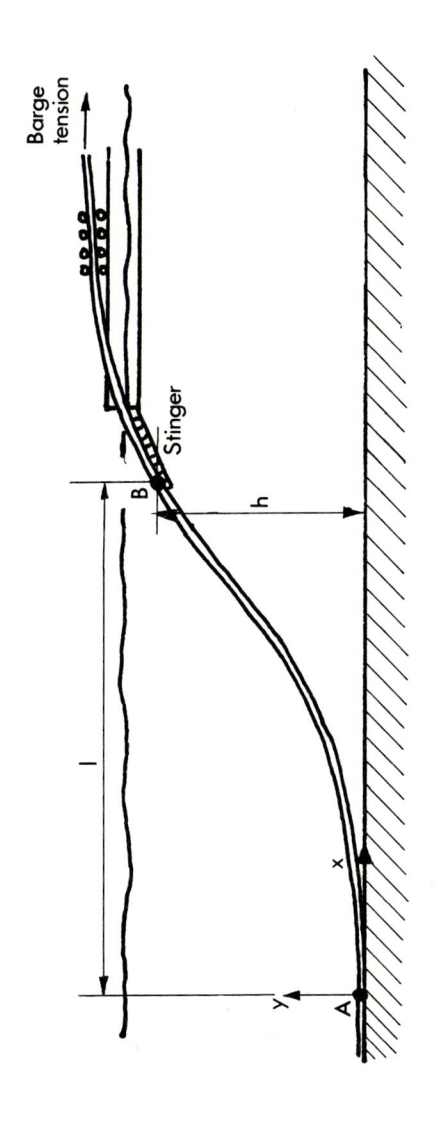

Figure 16.19   S-lay pipelaying

D. G. Simmonds

Field) the data specified was

$h = 98m$　　　　　$l = 200m$　　　　　$\theta = 26.35°$ (stinger angle)
$W = 802.9$ kg/m　　$I = 3.743*10^{-3} m^4$　　$D = .9398m$

and the six fixed boundary values were

$M_0 = 0$　　　$\theta_0 = 0$　　　$s_0 = 0$　　　$y_0 = 0$　　at A
$\theta_1 = 26.35$　$y_1 = 98$　　　　　　　　　　　　　　at B

with x as independent variable.

Obtaining estimates for $T_0, Q_0, T_1, Q_1, M_1, s_1$ proved somewhat tedious as it became clear that quite good values were going to be needed to get the Newton iteration to converge on a pair of 'matching' Runge-Kutta solutions. A rough calculation based on an assumed deflection shape yielded the values

$T_0 = 30836$　　$Q_0 = -5461$
$T_1 = 40429$　　$Q_1 = 12152$　　$M_1 = -340469$　　$s_1 = 321$

and with these estimates the results of Figure 16.20 were obtained. Bending moment and tension variation is shown in Figure 16.21.

16.6　Final Comments

We have found that modelling the steady-state cable and pipe problems that we have encountered so far has not presented any great difficulties and design engineers have found the information provided by the software to be quite useful. But there are some reservations:-

i) Although the software has been checked against analytical and semi-analytical solution and uses hydrodynamic force terms, true validation is sadly lacking. Predicted values for deflections, tensions, bending moments, and so on have not been compared with observed values in the sorts of problems described above because usually detailed observations of what actually happens are not made. The best we can say is that there have been no complaints!

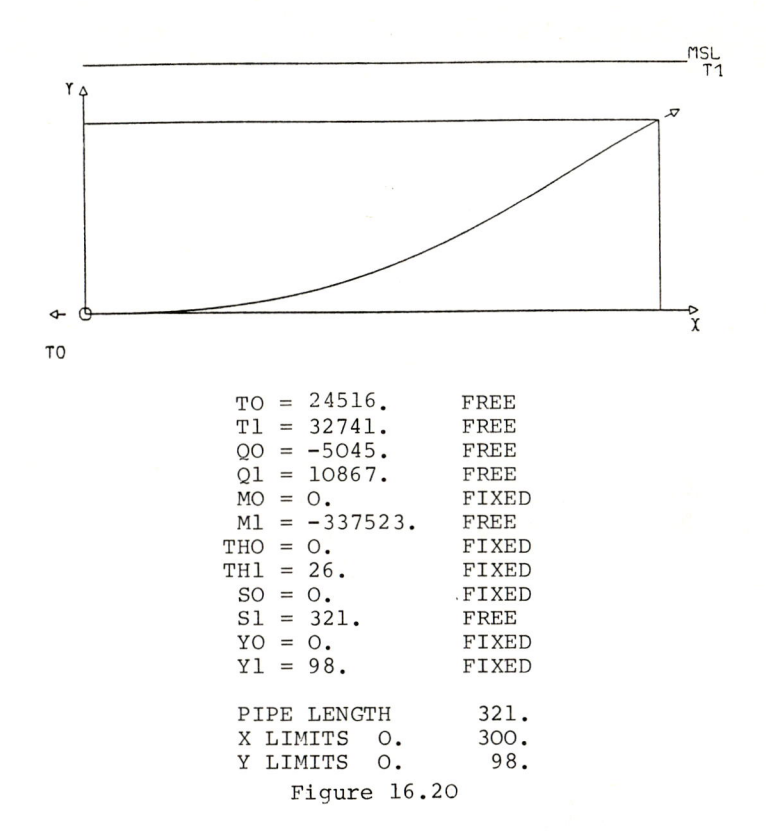

Umbilicals, Cables and Pipes

```
TO  = 24516.      FREE
T1  = 32741.      FREE
Q0  = -5045.      FREE
Q1  = 10867.      FREE
M0  = 0.          FIXED
M1  = -337523.    FREE
TH0 = 0.          FIXED
TH1 = 26.         FIXED
S0  = 0.          FIXED
S1  = 321.        FREE
Y0  = 0.          FIXED
Y1  = 98.         FIXED

PIPE LENGTH      321.
X LIMITS    0.   300.
Y LIMITS    0.    98.
```

Figure 16.20

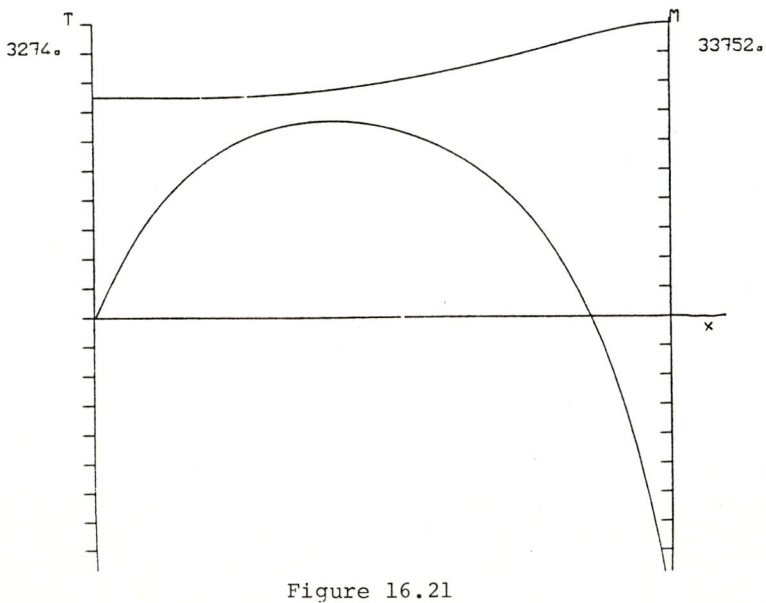

Figure 16.21

D. G. Simmonds

ii) Although sometimes some thought is needed to get good enough estimates of free values to ensure good convergence, on the whole this has not been a problem. We feel that perhaps we have been rather lucky in this respect and that some effort should be made to compare the efficiency of simple shooting methods with the various other methods available when applied to cable/pipe models.

We have made no mention of dynamic models of umbilical or umbilical/submersible systems in this paper but problems requiring such models have arisen also at RGIT. A vibration model applied to a cable system is described by Simmonds (1982) and current activity is centred around the development of computer programs that will predict unsteady behaviour.

Umbilicals, Cables and Pipes

References

Achenback, E., 1971
'Influence of surface roughness on the cross flow around a circular cylinder', Journal of Fluid Mechanics, Vol. 46, Part 2, pp 321-225.

Bursnall, W.J. and Loftin, L.K., 1951
'Experimental investigation of the pressure distribution about a yawed circular cylinder in the critical Reynolds number range', NACA TN 2463.

Casarella, Mario J. and Parsons, Michael., 1970
'Cable Systems Under Hydrodynamic Loading', Marine Technology Journal, Vol. 4, No. 4, July.

Choo, Y. and Casarella, M.J., 1971
'Hydrodynamic resistance of towed cables', Jnl. Hydronautics, Oct. Vol. 5, No. 4.

Dale, J.R., McCandless, J.M. and Holler, R.A., 1968
'Water Drag Effects of Flow Induced Cable Vibrations', ASME Paper 63-WA/FE-47.

Diana, G. and Falco, M., 1971
'On the forces transmitted to a vibrating cylinder by a blowing fluid (Experimental study and analysis of the phenomenon)', Mechanica. (Jnl. Italian Assn. for theoretical and applied mechanics), 6, pp 9-22, March.

Dove, H.L., 1950
'Investigations on Model Anchors', Trans. Inst. Nav. Arch., Vol. 92, pp 351-362.

Eames, M.C., 1967
'Steady-State Theory of Towing Cables', Defence Research Establishment Atlantic (Canada) Report 67/5, and TransactionsRINA, Vol. 110. No 2, April.

Every, M.J., 1981
"A brief survey of the steady drag data available for application to unfaired umbilical cables', BHRA Project RP 21708, June.

Ferriss, D.H., 1981
'Numerical determination of the three-dimensional equilibrium configuration of an underwater umbilical subjected to steady hydrodynamic loading', NPL Report DNACS 50/81, October.

Grimminger, G., 1945
'The effect of rigid guide vanes on the vibration and drag of a towed circular cylinder', David Taylor Model Basin Report 504, April.

Holt, J.F., 1964
'Numerical Solution of Nonlinear Two-Point Boundary Problems by Finite Difference Methods', Communications of the ACM, Vol. 7, No. 6, June.

Love, A.E.H., 1952
'A Treatise on the Mathematical Theory of Elasticity', 4th Edition, Cambridge University Press.

Miller, B.L., 1976
'The hydrodynamic drag of roughened circular cylinders', NPL Report Mar. Sci R147, OT-R-7602, April.

Morrison, D.D., Riley, J.D. and Zancanaro, J.F., 1962
'Multiple Shooting Method for Two-Point Boundary Value Problems', Communications of the ACM, Vol. 5, Part 12, pp 613-14.

Relf, E.F. and Powell, C.H., 1917
'Tests on Smooth and Stranded Wires in Air and Water', Advisory Committee for Aeronautics, Great Britain, Reports and Memoranda (New Series) No. 307, January.

Simmonds, D.G., 1982
'The Response of a Simple Pendulum with Newtonian Damping', Journal of Sound and Vibration, Vol. 84(3), pp 453-461.

Simmonds, D.G., 1983
'The Numerical Determination of the Steady-State Configurations of Subsea Cables', School of Mathematics Internal Report, RGITMR 83-1.

Skop, R.A., Griffin, O.M. and Ramberg, S.E., 1977
'Strumming Predictions for the Seacon II experimental mooring', Proc. 9th Offshore Technology Conf. Houston, Texas, Paper OTC 2884.

Springsten, G.B. Jr., 1967
'Generalized Hydrodynamic Loading Fuctions for Bare and Faired Cables in Two-Dimensional Steady-State Cable Configurations', U.S. Naval Ship Research and Development Center, Report 2424.

Whicker, L.F., 1957
'The Oscillatory Motion of Cable Towed Bodies', University of California, Report Series 82, Issue 2.

Wieselsberger, C., 1921
'New data for the laws of fluid resistance', NACA Tech Note 84, April.

Wilson, Basil W., 1960
'Characteristics of Anchor Cables in Uniform Ocean Currents', The A & M College of Texas, Technical Report No. 204-1, April.

17. MATHEMATICAL MODELLING IN OFFSHORE CORROSION

A. Turnbull,
Division of Materials Application,
National Physical Laboratory,
Teddington, Middlesex. TW11 OLW.

17.1 Introduction

Sea water can influence the long-term performance of structures or components in the North Sea in a number of ways through environment assisted fracture, general corrosion, corrosion under heat transfer conditions, crevice corrosion and corrosion in concrete. It is essential that adequate data and predictive methods are available to ensure reliable design in relation to fatigue performance, cathodic protection and materials selection.

17.1.1 Environment Assisted Fracture (fatigue)

This is the major factor determining structural life and involves the development of a crack from an initial defect at the tow of a weld and its subsequent propagation until failure of the structural component. The majority of the fatigue life in offshore structures is involved in the slow growth of the crack and the rate of growth depends on the interaction of the cyclic stresses or strains at the crack tip with the local environment within the crack. Fatigue life prediction offshore is made difficult by the time-dependent (or frequency-dependent) nature of crack growth (which limits the amount of data obtainable), the problem of reproducing offshore conditions in a laboratory and the sensitivity of crack growth to a wide range of experimental and operational variables. The proper obtention and application of fatigue data requires improved understanding of the environmental fatigue process and the development of quantitative models of fatigue crack growth is arguably the most effective method of developing that understanding. However, there are essentially two basic problems in developing a crack growth model:

(a)   mathematical modelling of the chemical/electrochemical environment at the crack tip;

(b)   modelling of corrosion fatigue crack growth rates in relation to the reactions within this crack tip environment.

A. Turnbull

In this paper current progress in mathematical modelling of the electro-chemical conditions within cracks will be described and in view of the importance of fatigue to structural integrity, will be considered most extensively.

17.1.2 General Corrosion/Corrosion Under Heat Transfer

The main concern with general corrosion and corrosion under heat transfer is simply to reduce it to an acceptable level through the application of a cathodic protection scheme. This usually involves the use of a sacrificial anode system which is essentially a galvanic couple with the corrosion of the anode conferring protection on the steel structure, or an impressed current system in which a cathodic current is applied to a structure with the cathodic potential controlled by suitably positioned reference electrodes (operating in a feedback system).

The major problem in design of a cathodic protection system is to ensure adequate current and potential distribution but this is made difficult by complex anode-cathode geometries (particularly in jacket structures), mutual interference from multiple anodes, shielding and environmental factors such as water resistivity, velocity, oxygen content and temperature. Adequate consideration of all these variables increasingly requires computer techniques and a short description of the underlying physical basis for the calculations is included in this paper.

17.1.3 Crevice Corrosion

Crevice corrosion offshore is generally associated with fasteners, often of stainless steel, and solution to the problem usually relates to proper materials selection. The tension-leg platform currently being constructed for deep-water application also has crevices inherent in the design of the complex anchor assembly. In this case the materials are structural steels and the primary concern is whether cathodic protection can be achieved in deep, "tight" crevices. Mathematical modelling as a basis for materials selection will be described briefly and methods to predict cathodic protection in crevices will be discussed.

Mathematical Modelling in Offshore Corrosion

17.1.4 Corrosion in Concrete

Corrosion of steel reinforcements in concrete structures offshore depends on depassivation of the steel principally through the ingress of chloride ion and oxygen through the concrete cover and the attainment of a threshold chloride ion concentration. The transport of these species to the steel surfaces depends on pore structure of the concrete, and the presence of cracks or their inducement because of volume expansion due to corrosion product or due to applied stress. Mathematical modelling of the multitude of processes involved in complex but some preliminary work has been initiated and will be described briefly.

17.2 General Mass Transport Theory

Although the models developed for the varied corrosion processes described differ in detail they all derive from the same fundamental equations describing the mass transport of species in electrolyte solutions. For dilute solutions (few models of relevance consider concentrated solutions) the important relationships are as follows:-

The flux of each dissolved species is given by

$$J_i = C_i V \quad - \quad D_i \nabla C_i \quad - \quad z_i \frac{D_i}{R'T} F V_i \nabla \phi \qquad (17.1)$$

$$\text{convection} \qquad \text{diffusion} \qquad \text{ion-migration}$$

where v is the fluid velocity, C_i is the concentration of species i with diffusion coefficient D_i, ∇C_i is the concentration gradient, $-\nabla \phi$ is the electric field, z_i is the charge, R' is the gas constant, T is the absolute temperature and F is the Faraday constant.

The mass conservation of species is described by

$$\frac{\partial C_i}{\partial t} = -\nabla J_i + R_i \qquad (17.2)$$

where $\frac{\partial C_i}{\partial t}$ is the rate of change of concentration with time and R_i represents the rate of production (or depletion) of species by chemical reaction in solution (e.g. hydrolysis reaction).

A. Turnbull

In addition, the solution is electrically neutral to a good approximation

$$\sum_i z_i c_i = 0 \qquad (17.3)$$

The current density in an electrolyte solution can be expressed as

$$i = F \sum_i z_i J_i \qquad (17.4)$$

The analysis of electrochemical systems by means of the above differential equations requires in addition a statement of the geometry of the system and of conditions existing at the boundaries of the system.

These four equations provide a consistent description of transport processes in electrolytic solutions and their physical significance is worth repeating. The first states that species in solution can move by convection, diffusion and ion migration. The second is a material balance for a species. The third is the condition of electroneutrality and the fourth states that the flux of a charged species constitutes an electric current.

17.3 Mathematical Modelling of the Electrochemistry in Cracks

The objective of research into the electrochemistry within cracks is to determine the rate of metal dissolution and the rate of production of hydrogen atoms at the crack tip (the primary factors in accelerated failure) for a range of experimental and operational conditions and to correlate these rates with observed crack growth. However, the rates of these reactions cannot be predicted from experiment in the bulk environment because the solution composition and electrode potential in the crack can vary markedly due to restricted mass transport. Furthermore direct measurement of reaction rates in a crack is not readily possible. Experimental measurement of the local environment and potential is possible (though difficult because of the restricted geometry) but may not be sufficient to predict electrode reaction rates if these are mass transport limited. Mathematical modelling is thus a necessary as well as a useful complement to experimental measurement.

However, mathematical modelling of the electrochemistry in fatigue cracks is complex and requires consideration of crack geometry, chemical and electrochemical reactions, and mass transport. The only realistic

Mathematical Modelling Offshore Corrosion

models currently developed are those of Turnbull (1982,1983,1984), Ferriss (1984), Ferriss and Turnbull (1983) and these form the basis of the following sections.

17.3.1 Definition of Crack Geometry

The specimen geometry used predominantly in this analysis was the compact tension type (Figure 17.1(a)) since the majority of crack propagation data for offshore structure steels in sea water was obtained using this test-piece (Scott et al, 1983). For this specimen geometry the crack can be considered reasonably to be trapezoidal in shape with a half-width defined by

$$h(x,t) = h_0(t) + x \tan \theta(t) \qquad (17.5)$$

where x is the distance from the crack tip (x = 0), $h_0(t)$ is the half-width of the crack tip and $\theta(t)$ is the crack angle. The displacement at the crip tip, $2h_0(t)$, is represented by the crack tip opening displacement and is based on the Dugdale model of plasticity with the assumption of small scale yielding and is given by the following expressions for plane-strain conditions (McCartney, 1980) viz.

$$\text{unloading} \quad \text{CTOD}(t) = \frac{1}{3} \frac{K^2_{max}}{E''\sigma_y} \left\{ 1 - 0.5 \left[1 - \frac{K(t)}{K_{max}} \right]^2 \right\} \qquad (17.6)$$

$$\text{loading} \quad \text{CTOD}(t) = \frac{1}{3} \frac{K^2_{max}}{E''\sigma_y} \left\{ 1 - 0.5(1-R)^2 + 0.5 \left[\frac{K(t)}{K_{max}} - R \right]^2 \right\} \qquad (17.7)$$

where R = minimum load/maximum load, $E'' = E'(1-\mu^2)$ where E' is Young's modulus, μ is Poisson's ratio and σ_y is the yield strength. The stress intensity factor will vary with time according to

$$K = K_m + \delta K \sin 2\pi f t \qquad (17.8)$$

where f is the cyclic frequency.

The crack angle $\theta(t)$ is given by

$$\tan \theta(t) = \frac{\delta(t) - \text{CTOD}(t)}{2a} \qquad (17.9)$$

where a is the crack length as measured from the loading line and $\delta(t)$

A. Turnbull

Figure 17.1

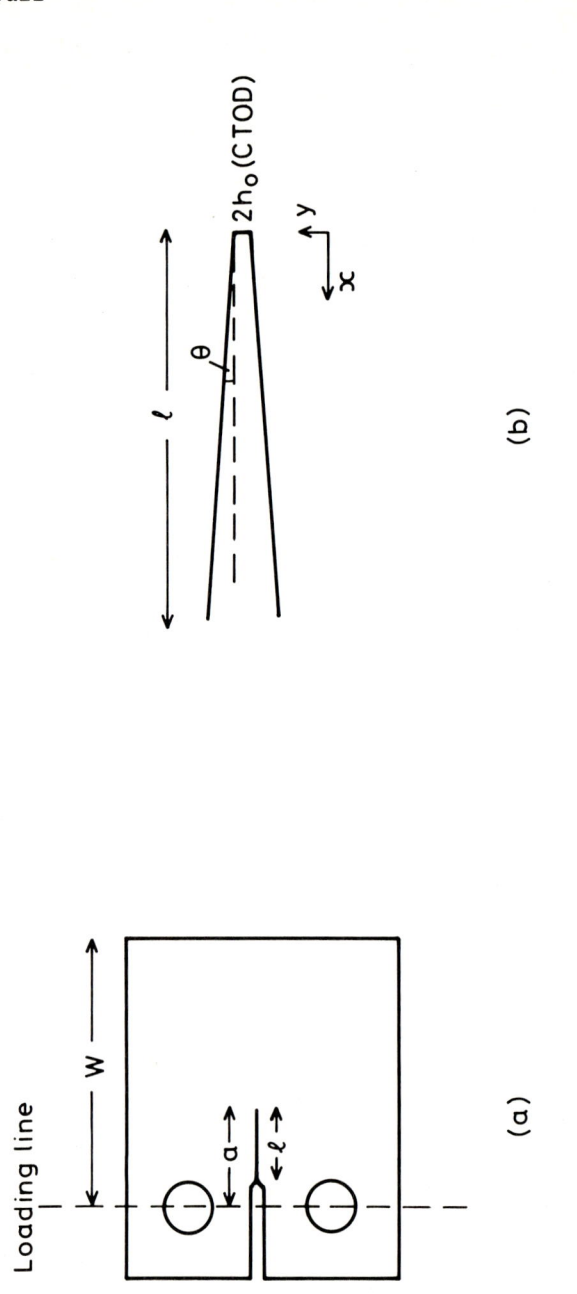

(a) Compact tension specimen (b) crack geometry assumed in model calculations

Mathematical Modelling in Offshore Corrosion

is the displacement at the loading line defined by Tada et al (1973).

$$\delta(t) = \frac{K(t) W V_2 (1-\mu^2)}{\sqrt{a}\, FE'} \qquad (17.10)$$

where W is the distance of the loading line to the base of the specimen and V_2 is a function of a/W and can be evaluated from the data provided in Tada et al (1973). F is a function of a/W and for $0.3 \leqslant a/W \leqslant 0.7$ is given by

$$F\{\tfrac{a}{W}\} = 29.6 - 185.5\{\tfrac{a}{W}\} + 655.7\{\tfrac{a}{W}\}^2 - 1017\{\tfrac{a}{W}\}^3 + 638.9\{\tfrac{a}{W}\}^4$$

$$(17.11)$$

17.3.2 Electrochemical Reactions

A range of chemical and electrochemical reactions are possible in cracks depending on the applied potential and composition of the bulk environment. In modelling crack electrochemistry the strategy at NPL has been to develop models of increasing complexity with respect to the nature and number of reactions occurring with the crack.

The first stage of the work has been to develop a model to describe the electrochemistry in a corrosion fatigue crack in BS 4360 50D steel cathodically protected in 3.5% NaCl solution (Ferriss & Turnbull, 1983, Turnbull, 1984). Under cathodic protection metal dissolution can be neglected and the important reactions within the crack for a steel in 3.5% NaCl are then

$$H_2O + e^- \rightarrow H + OH^- \qquad (17.12)$$

$$H_2O \underset{k_{-2}}{\overset{k_2}{\rightleftharpoons}} H^+ + OH^- \qquad (17.13)$$

$$O_2 + 2H_2O + 4e^- \rightarrow 4OH^- \qquad (17.14)$$

The first reaction represents reduction of water producing hydrogen atoms on the metal surface (which can subsequently permeate the steel), and hydroxyl ions in solution. The second reaction is the ionisation of water where k_2 and k_{-2} are the rate constants for the forward and reverse reactions respectively. The third reaction is reduction of oxygen producing hydroxyl ions. The electrochemical reactions will occur both at the crack tip and on the crack walls and the rates of

A. Turnbull

these reactions are given below:-

Water reduction:-

$$i_{H_2O} = k(H^+) \exp\{-\frac{\beta(H^+)FE}{R'T}\} \quad (17.15)$$

where i_{H_2O} is the current density for water reduction k is a rate constant, with a complex dependence on hydrogen ion concentration, β is the transfer coefficient also dependent on the hydrogen ion concentration (Turnbull & May, 1984). F is Faraday's constant, E is the electrode potential, R' is the gas constant and T is the temperature.

Oxygen reduction:-

$$i_{O_2} = k'[O_2] \exp\{-\frac{\beta'FE}{R'T}\} \quad (17.16)$$

i_{O_2} is the current density for oxygen reduction, k' is a rate constant and $[O_2]$ is the oxygen concentration in the crack.

In future work it is intended to include the buffering reaction of sea water with equations (17.12-17.14) for cathodic protected steel viz.

$$OH^- + HCO_3^- \underset{k_{-3}}{\overset{k_3}{\rightleftharpoons}} CO_3^{2-} + H_2O \quad (17.17)$$

$$Ca^{2+} + CO_3^{2-} \rightarrow CaCO_3\downarrow \quad (17.18)$$

where the second reaction represents calcareous deposition.

Subsequently, allowance for metal dissolution and hydrolysis will be added viz.

$$Fe \rightarrow Fe^{2+} + 2e^- \quad (17.19)$$

$$Fe^{2+} + H_2O \rightleftharpoons FeOH^+ + OH^- \quad (17.20)$$

Hence each successive stage of modelling will involve the inclusion of further chemical and electrochemical reactions and the mathematical method is required to be sufficiently robust to accommodate these additional features.

Mathematical Modelling in Offshore Corrosion

17.3.3 Mass Transport Equations

Mass transport of a dissolved molecular or ionic species in a corrosion fatigue crack can occur due to advection, diffusion and ion-migration. Advection is the fluid flow resulting from the cyclic displacement of the crack walls, while diffusion and ion migration are a consequence of electrode reactions on the walls and tip of the crack, which by generating or removing ionic species create concentration and potential gradients. Mass transport and continuity equations are required for each of the dissolved species in the crack (H^+, OH^-, O_2) in addition to sodium and chloride ions which, although not consumed or produced, will undergo ionic migration to maintain electrical neutrality.

It is assumed in this model that transport in the through-thickness direction can be neglected. This is appropriate for real cracks in structures but in standard laboratory specimens such as the compact tension type access via the crack sides is possible though a deficiency of the test method. The model is, therefore, only relevant to the latter when the crack sides are sealed or the thickness dimension is much greater than the crack depth ℓ (measured from the notch root). The numerical solution of the unsteady mass transport equations in two-dimensions (x and y) would prove to be too costly but the narrow geometry of the crack permits an approach involving the averaging of the species concentrations in the y-direction, i.e. from wall to wall. In this way the two-dimensional problem is reduced to a one dimensional problem. The essential assumption is that the concentration is approximately uniform across the width of the crack which has been shown to be reasonable (Turnbull, 1981)).

The fluid velocity (averaged across the crack at any position in the crack), has the form (Turnbull, 1983).

$$v(x,t) = - \frac{x(\dot{h}_o(t) + \frac{x}{2} \dot{\theta}(t))}{h(x,t)} \qquad (17.21)$$

where $\dot{h}_o(t)$ and $\dot{\theta}(t)$ are respectively dh_o/dt and $d\theta/dt$. As θ is small, $\tan \theta$ has been approximated by θ. In the above expression for the fluid velocity it has been assumed that inertial terms could be neglected. Secomb (1978) showed that the fluid flow in a channel with fluctuating walls was viscosity dominated for $\alpha^2 \ll 1$ where α is the Womersley number and is defined by

A. Turnbull

$$\alpha = h_m \left[\frac{2\pi f}{\nu}\right]^{\frac{1}{2}} \tag{17.22}$$

where h_m is the mean half-width of the channel and ν is the kinematic viscosity. Although the crack geometry in this analysis varies from the parallel-sided channel considered by Secomb (1978), the Womersley number is likely to provide a good approximation for other geometries and on this basis it can be concluded that inertial effects on the fluid flow can be neglected for frequencies $\leq 1\,Hz$ ($h_m \cong 10^{-3}\,cm$).

Defining C_1, C_2, C_3, C_4, C_5 as the concentration of hydrogen ions, hydroxyl ions, sodium ions, chloride ions and oxygen molecules respectively, the species mass transport equations are described by

$$J_1 = C_1 v - D_1 \frac{\partial C_1}{\partial x} - D_1 \frac{FC_1}{RT} \frac{\partial \phi}{\partial x} \tag{17.23}$$

$$J_2 = C_2 v - D_2 \frac{\partial C_2}{\partial x} + D_2 \frac{FC_2}{RT} \frac{\partial \phi}{\partial x} \tag{17.24}$$

$$J_3 = C_3 v - D_3 \frac{\partial C_3}{\partial x} - D_3 \frac{FC_3}{RT} \frac{\partial \phi}{\partial x} \tag{17.25}$$

$$J_4 = C_4 v - D_4 \frac{\partial C_4}{\partial x} + D_4 \frac{FC_4}{RT} \frac{\partial \phi}{\partial x} \tag{17.26}$$

$$J_5 = C_5 v - D_5 \frac{\partial C_5}{\partial x} \tag{17.27}$$

In these equations D_1 to D_5 are the diffusion coefficients of the relevant species and ϕ is defined as the difference in potential between the external surface and a point along the crack ($\phi = E^{ext} - E(x)$; at the crack mouth $\phi = 0$ by definition). The mass conservation of each species is given by

$$\frac{\partial C_1}{\partial t} - C_1 \frac{\partial v}{\partial x} = -\frac{\partial J_1}{\partial x} + k_2 - k_{-2} C_1 C_2 + \frac{D_1 \theta}{h} \frac{\partial C_1}{\partial x} + \frac{D_1 F}{RT} \frac{\theta}{h} C_1 \frac{\partial \phi}{\partial x} \tag{17.28}$$

$$\frac{\partial C_2}{\partial t} - C_2 \frac{\partial v}{\partial x} = -\frac{\partial J_2}{\partial x} + k_2 - k_{-2} C_1 C_2 + \frac{k(C_1)}{hF} \exp\left\{-\frac{\beta(C_1) FE^{ext}}{RT}\right\}$$

$$\exp\left\{\frac{\beta(C_1) F\phi}{RT}\right\} + \frac{i' C_5}{hF} \exp\left\{\frac{B' F\phi}{RT}\right\} + D_2 \frac{\theta}{h} \frac{\partial C_2}{\partial x}$$

$$- D_2 \frac{F}{RT} \frac{\theta}{h} C_2 \frac{\partial \phi}{\partial x} \tag{17.29}$$

Mathematical Modelling in Offshore Corrosion

$$\frac{\partial C_3}{\partial t} - C_3 \frac{\partial v}{\partial x} = - \frac{\partial J_3}{\partial x} + \frac{D_3 \theta}{h} \frac{\partial C_3}{\partial x} + \frac{D_3 F}{RT} \frac{\theta}{h} C_3 \frac{\partial \phi}{\partial x} \quad (17.30)$$

$$\frac{\partial C_4}{\partial t} - C_4 \frac{\partial v}{\partial x} = - \frac{\partial J_4}{\partial x} + \frac{D_4 \theta}{h} \frac{\partial C_4}{\partial x} - \frac{D_4 F}{RT} \frac{\theta}{h} C_4 \frac{\partial \phi}{\partial x} \quad (17.31)$$

$$\frac{\partial C_5}{\partial t} - C_5 \frac{\partial v}{\partial x} = - \frac{\partial J_5}{\partial x} - \frac{i' C_5}{4fH} \exp\left\{\frac{\beta' F \phi}{RT}\right\} + D_5 \frac{\theta}{h} \frac{\partial C_5}{\partial x} \quad (17.32)$$

It is assumed that the solution is electrically neutral so that

$$C_1 - C_2 + C_3 - C_4 = 0 \quad (17.33)$$

In order to solve equations (17.23) to (17.32) a set of initial and boundary conditions are required viz.

$$t = 0 \quad C_1 = C_1(x,0), \quad C_2 = C_2(x,0), \quad C_3 = C_3(x,0),$$
$$C_4 = C_4(x,0), \quad C_5 = C_5(x,0), \quad \phi = \phi(x,0) \quad (17.34)$$

The form of the initial conditions at $t = 0$ is somewhat arbitrary and was chosen to give a relatively rapid convergence to the "steady" periodic solution which is the most practically relevant.

Adoption of more practical starting conditions such as

(a) all concentrations are at their bulk value, and

(b) concentrations are at the steady-state value of a static crack prior to load cycling gave rise to impractical computing times.

This reflects the physical situation in which two days may be necessary for "steady" conditions in the crack to be achieved. A calculated initial profile (essentially a "guess-estimate") gave more rapid attainment of the "steady" periodic concentrations without affecting the final result.

The boundary conditions for $t > 0$ were as follows

$$x = \ell \quad C_1 = C_1^\infty, \quad C_2 = C_2^\infty, \quad C_3 = C_3^\infty, \quad C_4 = C_4^\infty, \quad C_5 = C_5^\infty, \quad \phi = 0$$
$$(17.35)$$

$$x = 0, \quad J_1 = 0 \quad (17.36)$$

A. Turnbull

$$J_2 = \frac{k^*}{F} \exp\left\{-\frac{\beta^* F E^{ext}}{RT}\right\} \exp\left\{\frac{\beta^* F\phi}{RT}\right\} + \frac{i'C_5}{F} \exp\left\{\frac{\beta' F\phi}{RT}\right\} \quad (17.37)$$

$$J_3 = 0 \quad (17.38)$$

$$J_4 = 0 \quad (17.39)$$

$$J_5 = \frac{i'C_5}{4F} \exp\left\{\frac{\beta' F\phi}{RT}\right\} \quad (17.40)$$

In the above equations $i' = k' \exp\left\{\frac{\beta' F E^{ext}}{RT}\right\}$

The expression for the water reduction kinetics at the crack tip (equation 17.33) differs from that in equation (17.25) and was designed to allow an assessment of the effect of enhanced crack tip reaction rates on local conditions at the crack tip.

17.3.4 Numerical Method

To facilitate in the numerical solution the mass transport and conservation equations were non-dimensionalised. The resulting system, together with the charge neutrality equation, is a non-linear set of parabolic differential equations. The problem thus defined is complicated by the frequent domination of the total transport by the advection terms due to crack width variation. When applied to problems of this type, conventional finite difference techniques frequently produce spurious spatial oscillations in the computed solution. For this reason an unconventional finite difference scheme was used which combines the effects of advection and ion-migration and represents local spatial variations by expressions involving exponential functions. In outline the calculation proceeds as follows. An estimate is made to the electrode potential at the new time level and the partial differential equation (pde) representing the change in profile of any one species is treated whilst the other species concentrations are assumed known. Each species is treated in turn before a new estimate is made for the potential distribution. This process is repeated until the results converge at which point the time variable is incremented and the calculation continued.

This procedure was adopted instead of simultaneous solution of all equations for two main reasons. Firstly, and most important, additional

Mathematical Modelling in Offshore Corrosion

chemical reactions can be added to the model if necessary in a convenient way. Secondly the sub-problem for each species is linear and involves only the solution of a tri-diagonal system of algebraic equations. A description of the technique for four species in a parallel-sided crack is given in Ferriss (1984).

17.3.5 Examples of Results

It is not within the scope of this paper to discuss the results of the numerical study in detail but it is useful to demonstrate examples of the output and to discuss some implications of the work. The parameters of most relevance to crack growth are the long-time "steady" (periodic) values at the crack tip and examples of the variation of the crack tip pH and potential during a cycle are shown in Figure 17.2. The fluctuation in potential during a cycle can be significant (depending on the input parameters) but the pH variation for a wide range of input data was invariably small (< 0.1 pH units). In addition the pH variation during a cycle was not always in harmony with the stress intensity factor as shown by the example in Figure 17.2. The variation in pH and potential drop along the crack are shown in Figures 17.3 and 17.4. It can be observed that the most significant variation in pH and potential occurs immediately at the crack mouth and that oxygen reduction has an important influence on the concentration profile. The role of oxygen is interesting since evaluation of the oxygen concentration along the crack shows that, at long times, reaction is confined to less than 10% of the crack, very close to the mouth. Nevertheless this reaction at the mouth is sufficient to locally increase the pH and potential drop.

The pH in the crack under cathodic protection conditions is predicted to be alkaline (Figure 17.5) with values in the range 11-13. In view of the production of OH^- ions in the crack which increases with decreasing potential this high pH is not surprising. It is noticeable however that the pH is affected by crack depth when all other parameters are held constant. This is true also of the potential drop which increases with increasing crack depth (Figure 17.6). This dependence on crack depth of crack tip pH and potential, and hence the rate of production of hydrogen atoms, is of considerable significance for corrosion fatigue crack growth. The conventional concept of fracture mechanics is that the effect of crack depth on corrosion fatigue growth can be characterised entirely the stress intensity factor ($K \propto \sigma\sqrt{a}$ where a is the depth).

A. Turnbull

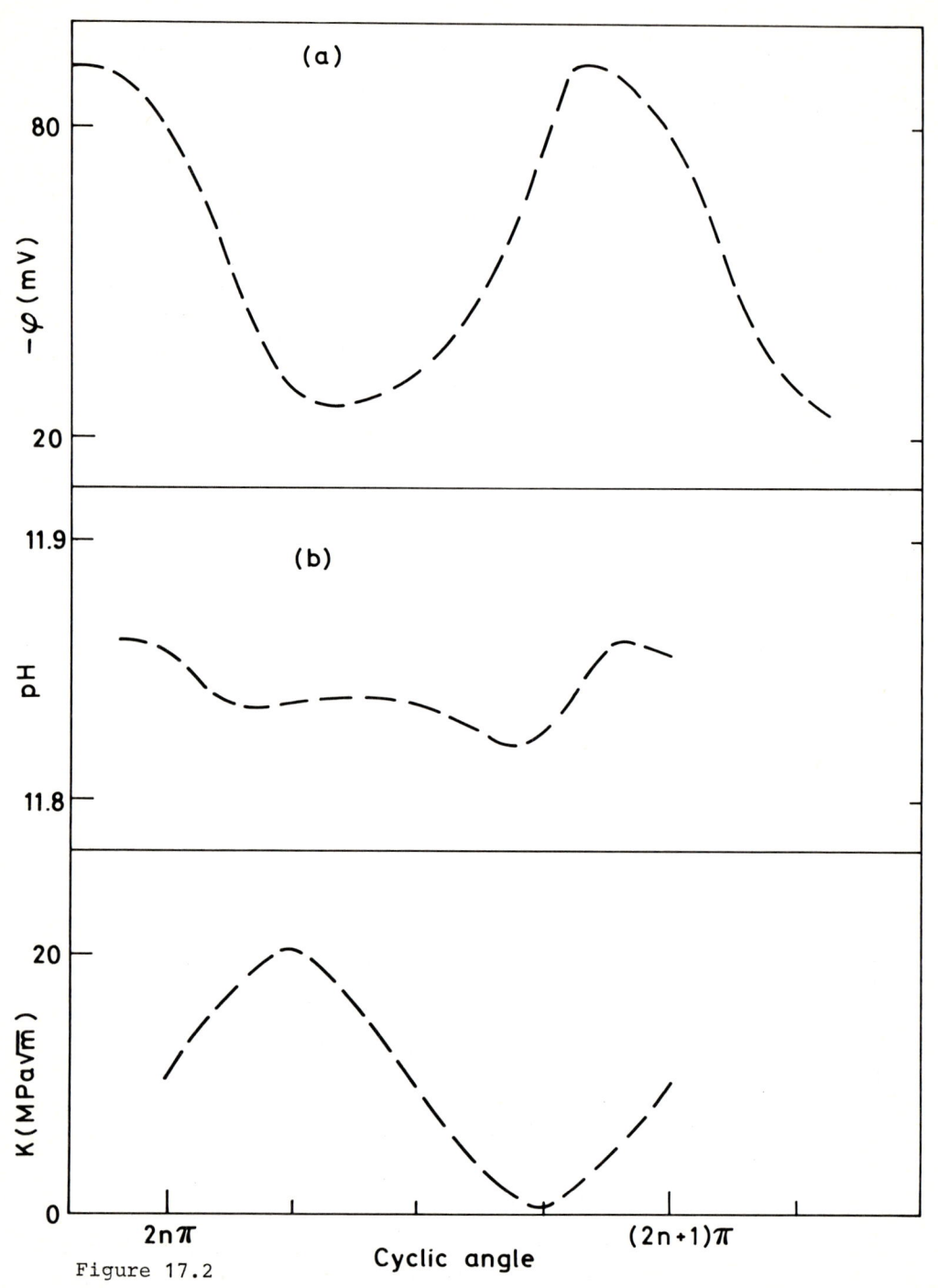

Figure 17.2
Cyclic variation of crack tip potential (a) and pH values (b) ("steady" periodic values) for $\Delta K = 20$ MPa$\sqrt{m}$, R = 0.01, f = 0.1 Hz, l = 2.0 cm, E^{ext} = -IV(SCE) and T = 5°C. BS 4360 50D steel in 3.5% NaCl at pH 6

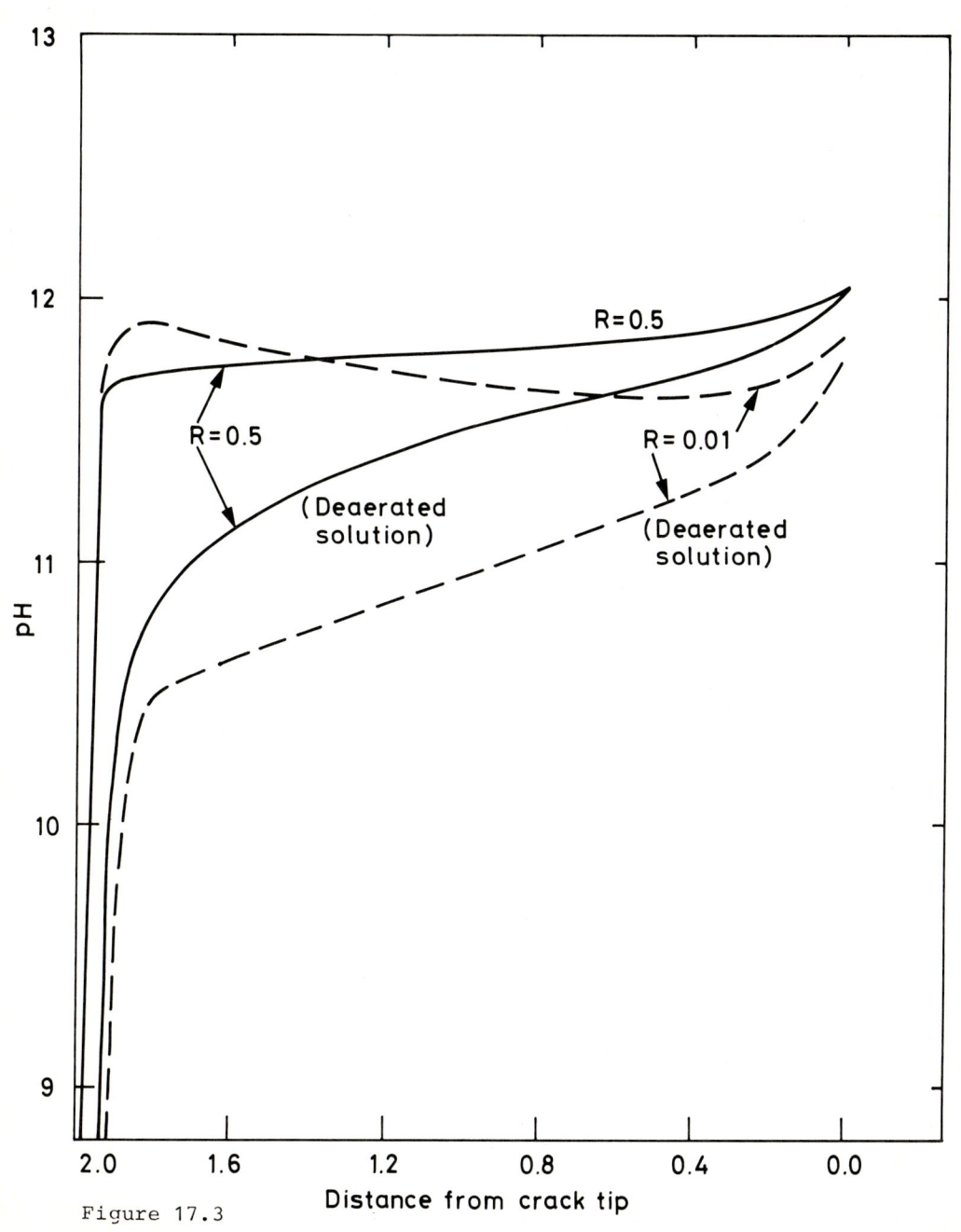

Mathematical Modelling in Offshore Corrosion

Figure 17.3

Predicted variation of pH with distance from the crack tip ("steady" periodic value at $2\pi ft = 2n\pi$), $\Delta K = 20$ MPa$\sqrt{m}$, $f = 0.1$ Hz, $l = 2.0$ cm, $E^{ext} = -1$V(SCE), T=5°C, BS 4360 50D steel in 3.5% NaCl at pH 6

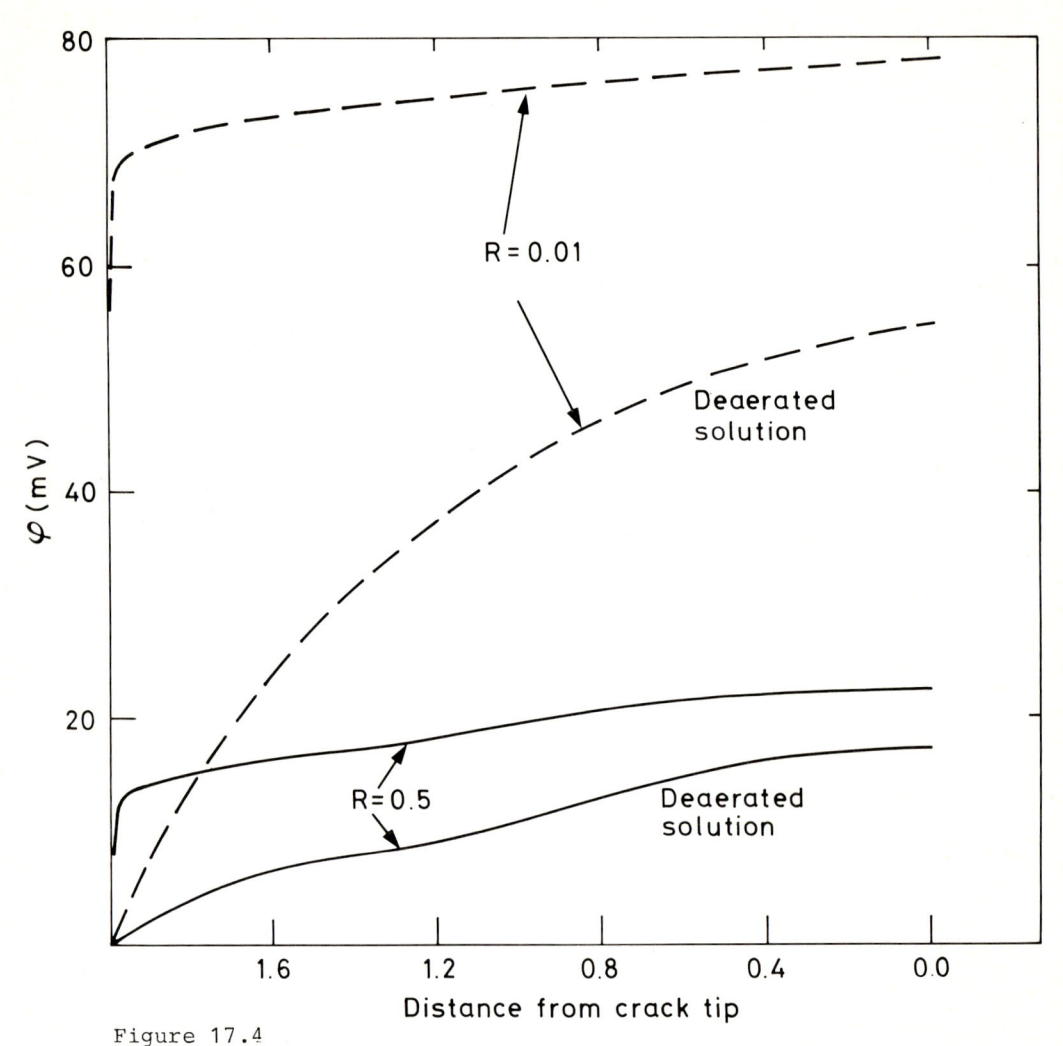

A. Turnbull

Figure 17.4
Predicted variation of the potential drop with distance from the crack tip ("steady" periodic value at $2\pi ft = 2n\pi$), $\Delta K=20$ MPa$\sqrt{m}$, $f=0.1$ Hz, $l=2.0$ cm, $E^{ext}=-1$V(SCE), T=5°C, BS 4360 50D steel in 3.5‰ NaCl at pH

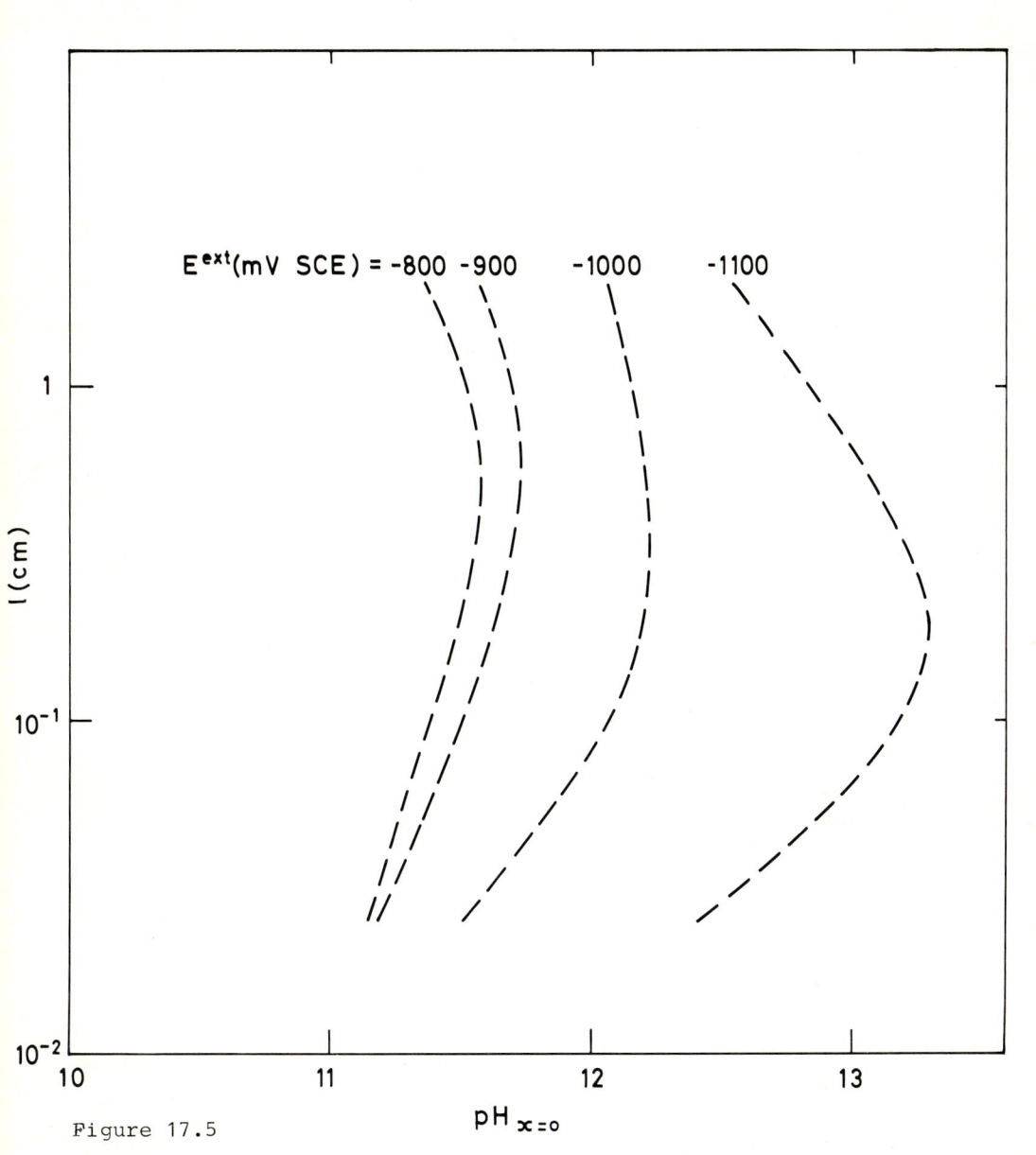

Mathematical Modelling in Offshore Corrosion

Figure 17.5

Predicted variation of the crack tip pH ("steady" periodic value at $2\pi ft = 2n\pi$) with crack depth and external potential for constant $\Delta K = 20$ MPa$\sqrt{m}$, R=0.5, T=5°C, f=0.1 Hz ; BS 4360 50D steel in 3.5‰ NaCl at pH 6

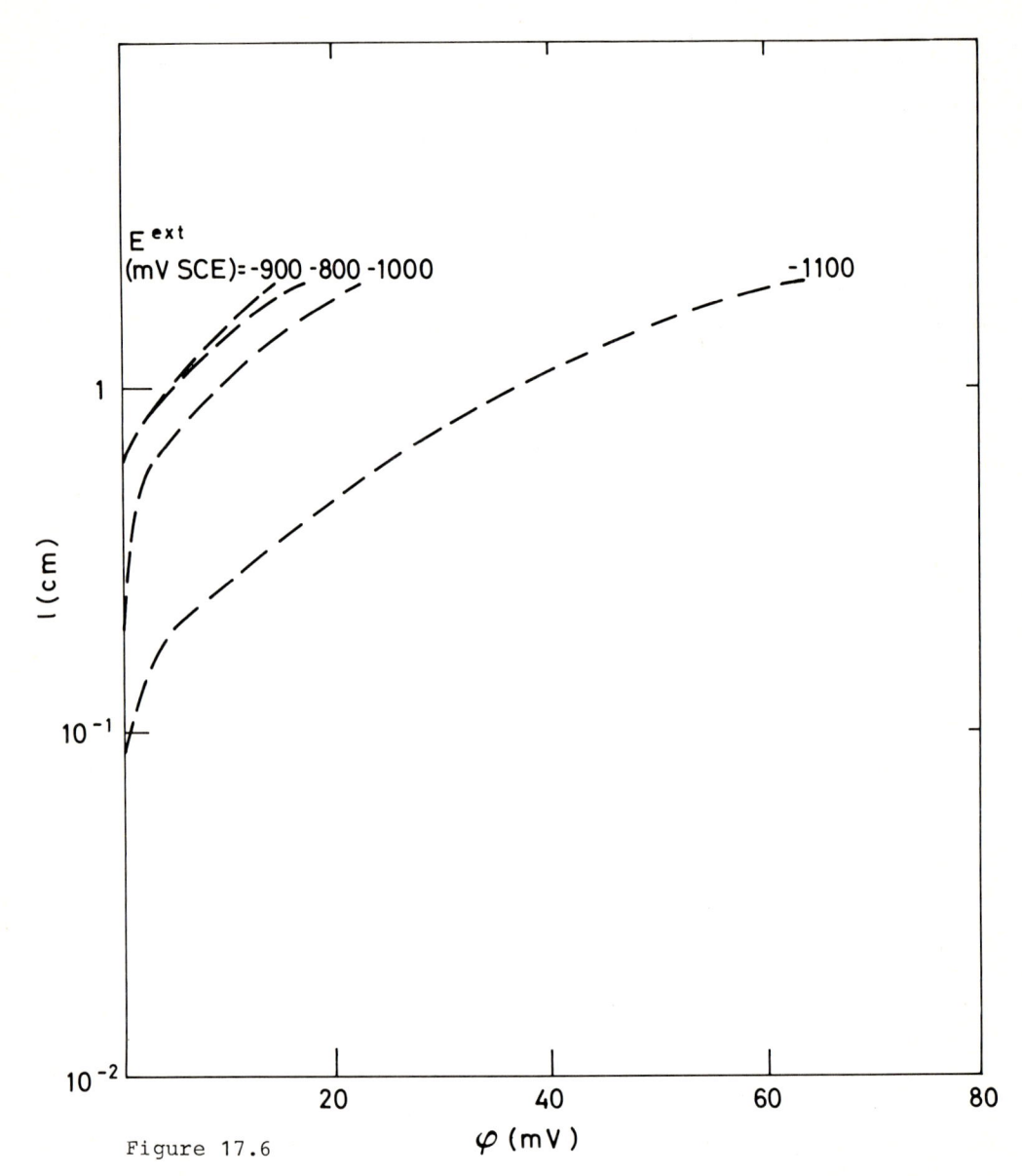

A. Turnbull

Figure 17.6
Predicted variation of the potential drop at the crack tip ("steady" periodic value at $2\pi ft = 2n\pi$) with crack depth and external potential for constant $\Delta K = 20$ MPa$\sqrt{m}$, $R = 0.5$, $T = 5°C$, $f = 0.1$ Hz; BS 4360 50D steel in 3.5% NaCl at pH 6

Mathematical Modelling in Offshore Corrosion

This enables laboratory data determined from simple crack geometries to be translated to more complex geometries based on the unifying concept of the stress intensity factor. The predictions of this mathematical study that electrode reaction rates and hence crack growth are influenced by crack depth because of differences in mass transport undermines the basis of fracture mechanics in this application.

17.4 Mathematical Modelling Cathodic Protection Offshore

The physical basis underlying the principles of cathodic protection as conventionally used in offshore practice is comparatively straightforward.

The current in an electrolyte solution is due to the motion of charged particles and can be expressed as (Newman, 1973).

$$i = F \sum_i z_i J_i \tag{17.41}$$

Now

$$J_i = C_i v - D_i \nabla C_i - \frac{z_i D_i}{RT} C_i \nabla \phi \tag{17.42}$$

Hence

$$i = Fv \sum_i z_i C_i - F \sum_i z_i D_i C_i - F^2 \nabla \phi \sum \frac{z_i^2 D_i}{RT} C_i \tag{17.43}$$

The first term will be zero by virtue of electroneutrality and when there are no concentration variations in the solution this equation reduces to

$$i = - \kappa \nabla \phi \tag{17.44}$$

where

$$\kappa = F^2 \sum_i z_i^2 \frac{D_i}{RT} C_i \tag{17.45}$$

is the conductivity of the solution. This is an expression for Ohm's Law, valid for electrolytes in the absence of concentration gradients.

Now charge conservation requires that

$$\nabla \cdot i = - \frac{\partial q}{\partial t} \tag{17.46}$$

where q is the charge density. For steady-state systems $\partial q / \partial t = 0$ and insertion of (17.41) into (17.46) yields

A. Turnbull

$$\nabla \cdot (\kappa \nabla \phi) + F \sum_i z_i \nabla \cdot (D_i \nabla C_i) = 0 \qquad (17.47)$$

In the absence of concentration gradients and with a uniform value of the conductivity κ this reduces to

$$\nabla^2 \phi = 0 \qquad (17.48)$$

i.e. the potential satisfies Laplace's equation in a region of uniform composition.

The boundary conditions may involve a statement of the voltage or voltage gradient at the anode and cathode surface.

This equation has subsequently been applied to simple geometries such as the circular corrosion cell of McCafferly (1977) using analytical methods, but extension to more complex geometries with account of variable polarisation parameters has required the use of finite difference (Doig & Flewitt (1979)), finite element (Munn (1982)) or boundary element methods (Warne (1982)). The paper by Warne gives an excellent summary of the techniques available and discusses future developments. Warne argues that boundary element techniques are the best suited to cathodic protection studies. Irrespective of the technique the accuracy of the predictions requires improved information on the polarisation characteristics of reactions at the metal surface when calcareous scale is formed and when marine growth is present.

17.5 Mathematical Modelling of Crevice Corrosion and Protection

A range of mathematical models have been developed to describe the solution composition and/or potential inside crevices (Turnbull, 1982(a), 1982). The most significant of these have been the model developed for passive systems (protective oxide film on metal) by Oldfield and Sutton (1978) and that for steels in the active state (no protective oxide film) developed by Turnbull and Thomas (1982).

In the former (Oldfield and Sutton, 1978), mass transport of material in the crevice was not considered in a rigorous way. Rather, in a complex model, a series of approximations were invoked aimed at obtaining a reasonable assessment of the oxygen concentration pH and chloride ion concentration within the crevice, these being the factors which determine the breakdown of the passive oxide film. The attainment of the critical composition depended on the crevice dimensions, the passive

Mathematical Modelling in Offshore Corrosion

current and the steel composition. The model has provided a useful basis for predicting the effects of steel composition on crevice corrosion attack and thus as as an assistance to materials selection. The model of Turnbull and Thomas (1982) for steels in the active state was developed for a parallel-sided crevice or crack in which the crevice walls were static and mass transport was by diffusion and ion migration. A wide range of chemical and electrochemical reactions was included; cathodic reduction, dissolution of elements from the steels and to an extent the buffering reactions of sea water.

The mass transport equations can be considered a sub-set of equations (17.23)-(17.32) but with the fluid velocity zero and the crack angle zero. Predictions of solution composition and electrode potential in the crevice were made and a particular feature was the prediction of low potential drops at potentials > - 1V(SCE) even for long "tight" crevices, a factor of considerable importance to cathodic protection which was subsequently verified experimentally (Turnbull and May, 1983).

17.6 Mathematical Modelling of Corrosion in Concrete

Modelling the mass transport and corrosion processes in concrete is not particularly extensive but the essential basis of a model has been set out by Bazant (1977). Bazant's model uses conventional mass transport theory (equations (17.1)-(17.4)) and describes diffusion of oxygen, chloride ions and pore water through the concrete cover; diffusion of ferrous hydroxide near the steel surface; the depassivation of steel due to critical chloride ion concentration; the cathodic and anodic electrode potentials (depending on oxygen and ferrous hydroxide concentrations according to the NERNST equation); the polarisation of electrodes due to changes in concentration of oxygen and ferrous hydroxide; the flow of electric current though the electrolyte in pores of concrete; the mass sinks or sources of oxygen, ferrous hydroxide and hydrated red rust near the electrodes based on Faraday's Law; and the rust production rate based on reaction kinetics.

Clearly the model is very complex even with the assumption of one-dimensionality with perhaps the major difficulty being a lack of detailed knowledge of all the relevant input parameters. Nevertheless the model has been applied to simplified calculations of corrosion rates and times to cracking of concrete cover although no test data for verification of the model were shown.

A. Turnbull

17.7 Conclusions

Mathematical modelling has been applied to a range of corrosion problems associated with the performance of offshore structures. Models of mass transport and electrochemistry in corrosion in fatigue cracks have been developed and applied to prediction of crack growth behaviour relevant to offshore structures.

The numerical technique adopted allows for the addition of further chemical and electrochemical reactions which is necessary as the preliminary model is developed from a consideration of cathodic protection of steel in 3.5% NaCl to include the buffering reactions of sea water and also metal dissolution and hydrolysis.

Mathematical modelling of crack electrochemistry is particularly useful in identifying the variables of most significance to crack tip reaction rates and crack growth but complementary experimental measurement is necessary to verify the predictions of the model and this is the approach adopted at NPL.

Modelling of cathodic protection is largely associated with the solution of Laplace's equation but the primary difficulty is the complex geometry of the structure and anode system and the variable polarisation characteristics of the reactions at the metal surfaces. Of the numerical methods boundary element techniques appear to be most suited to cathodic protection studies.

Modelling of the electrochemistry in crevices has been extensive and models exist which can predict the important role of metal composition on passivity breakdown and on the extent of cathodic protection possible in long, "tight" crevices.

Prediction of the local solution composition and corrosion rates at reinforcing bars in concrete is complex and although a very broad-based model has been set up solution was possible only with some limiting assumptions, and no experimental verification was available.

References

Bazant, Z. P., 1977
ASCE, J. Structural Division, 105 (ST6), 1137.

Doig, P. and Flewitt, P. E. J., 1979
J. Electrochemical. Soc. 126 (12), 2057.

Ferriss, D. H., 1984
Numerical solution of mass conservation equations representing electrochemical reactions in an unsteady corrosion fatigue crack. Proc. 2nd Int. Conf. for Numerical Methods for Non-linear Problems, Barcelona, (Eds. C. Taylor, E. Hinton, D. R. J. Owen), Pineridge Press.

Ferriss, D. H. and Turnbull, A., 1983
Theoretical modelling of the electrochemistry in a corrosion fatigue crack in Numerical solution of time-dependent mass conservation equations involving advection, diffusion and ion-migration. NPL Report DMA(A), 44.

McCafferty, E, 1977
Mathematical analysis of Circular Corrosion Cells having unequal Polarisation Parameters. Naval Research Lab. Report 8107.

McCartney, L. N., 1980
NPL, Private communication.

Munn, R. S., 1982
Materials Performance, 29, August.

Newman, J., 1973
Electrochemical Systems, Prentice-Hall, N.J.

Oldfield, J. W. and Sutton, W. H., 1978
Brit. Corre. J., 13 (1), 13.

Scott, P. M., Thorpe, T. W. and Silvester, D. R. V., 1983
Corr.Sci., 23 (6), 559.

Secomb, T. W., 1978
J.Fluid.Mechs., 88 (2), 273.

Tada, H., Paris, P. and Irwin, G., 1973
The stress analysis of cracks handbook. Del Research Corp.

Turnbull, A., 1981
A theoretical evaluation of the oxygen concentration in a corrosion fatigue crack. NPL Report DMA(A) 31.

A. Turnbull

Turnbull, A., 1982(a)
Reviews in Coatings and Corrosion, 5 (1-4), 43.

Turnbull, A., 1982(b)
Corr.Sci., 22 (9), 877.

Turnbull, A., 1983
A theoretical analysis of the influence of crack dimensions and geometry on mass transport in corrosion fatigue cracks. NPL Report DMA(A) 69.

Turnbull, A., 1984
Mathematical modelling of the electrochemistry in corrosion fatigue cracks 1. Structural steel cathodically protected in 3.5% NaCl. (to be published).

Turnbull, A. and May, A. T., 1983
Materials Perofmrance, 22 (10), 34.

Turnbull, A. and May, A. T., 1984
Electrochemical polarisation studies of BS4360 50D in deaerated 3.5% NaCl of varying pH, using the rotating disc electrode. (to be published).

Turnbull, A. and Thomas, J. G. N., 1982
J.Electrochem,Soc., 129 (7), 1412.

Warne, M. A., 1982
Application of Numerical Analysis Techniques. Proc. of Conf. on Cathodic protection theory and practice - the present status, I.Corr. S.T./NACE, Coventry.

18. FATIGUE CRACK GROWTH PREDICTIONS IN TUBULAR WELDED JOINTS

S. Dharmavasan,
London Centre for Marine Technology,
Department of Mechanical Engineering,
University College London.

18.1 Introduction

The present method of fatigue life prediction for tubular joints is based on stress-life (S/N) curves which are obtained by testing large scale tubular joints and then relating a characteristic stress, known as the hot spot stress (Department of Energy, 1981), to the total number of cycles to failure. It is important, however, to note that these S-N curves are not true S-N curves for the material but 'fictitious' S-N curves relevant to the joint geometry. This means that part of the full analysis of the failure mechanism can be omitted, because the data has come from the full scale test.

The main disadvantages of the S-N curve approach are that it cannot be easily adapted to consider new conditions and cannot give any guidance on what to do if a crack is discovered in service. Welded structures, in general, have crack like defects, and as a result, the initiation life is negligible and all the fatigue life is one of crack propogation. This leads to the second method based on fracture mechanics which can be used for both fatigue life prediction and crack growth rates. Fracture mechanics was originally developed to study the fracture behaviour of brittle materials, but has been generalised to be applied to any situation which involves crack propagation from a nucleated crack.

18.2 Fatigue Crack Growth Behaviour of Tubular Joints

The work at University College London has concentrated on carrying out realistic random load fatigue tests on tubular welded T, Y, and K joints of a reasonable size (Dover and Chaudhury, 1981, Dover and Dharmavasan, 1982). During the course of these tests the crack growth behaviour, especially in the depth direction, was monitored using an alternating current potential drop technique developed at UCL but which is now commercially available. To facilitate an understanding of the stresses which cause the fatigue cracks to grow experimental and finite element stress analyses were also carried out.

S. Dharmavasan

A Y joint is shown in Figure 18.1 and the various terms used are defined. The loading on a tubular joint is normally composed of three main modes - axial, in-plane bending and out of plane bending. The general stress distributions around the welded intersection of a T joint for all three modes of loading are shown in Figure 18.2. It is seen that high stress concentrations exist near the intersection and the distributions are non-uniform.

Further examination of the nature of the stress distribution around the intersection shows that most of the stresses are due to local bending and that the membrane component is relatively small. An example of this feature is shown in Figure 18.3 which is obtained from finite element analysis of a T joint subjected to axial loading. This figure shows the top and bottom surface stresses.

The crack shape evolution for a Y joint subjected to in-plane bend random loading in air, is shown in Figure 18.4. It can be seen that two adjacent semi-elliptical cracks formed, grew separately for some time and then jointed up to form a single semi-elliptical crack. Subsequent crack growth retained an irregularity in the crack front corresponding to these initial separate semi-elliptical cracks. Measurements of these changes in the aspect ratio, from this stress amplitude are shown in Figure 18.5. These indicate that early growth and subsequent large crack growth gave an aspect ratio of approximately 0.1. The model to be presented later will assume a constant aspect ratio of 0.1.

The crack depth averaged 20mm either side of the deepest point is plotted as a function of cycles in Figure 18.6. This crack growth curve shows three distinct phases:-

1. Very low linear growth - initiation phase,
2. Rapidly varying growth rate - transition phase,
3. Linear and more rapid growth - propagation phase.

The crack growth behaviour can be explained qualitatively by comparing the various stages of crack growth with the crack profiles shown in Figure 18.4. The initiation phase is found when the individual defects are growing and this lasts till the depth is about 1.3-1.5 millimetres. Once the individual defects get to this stage they begin to co-alesce, and as this happens the growth rate increases to give the transition phase.

Fatigue Crack Growth Predictions

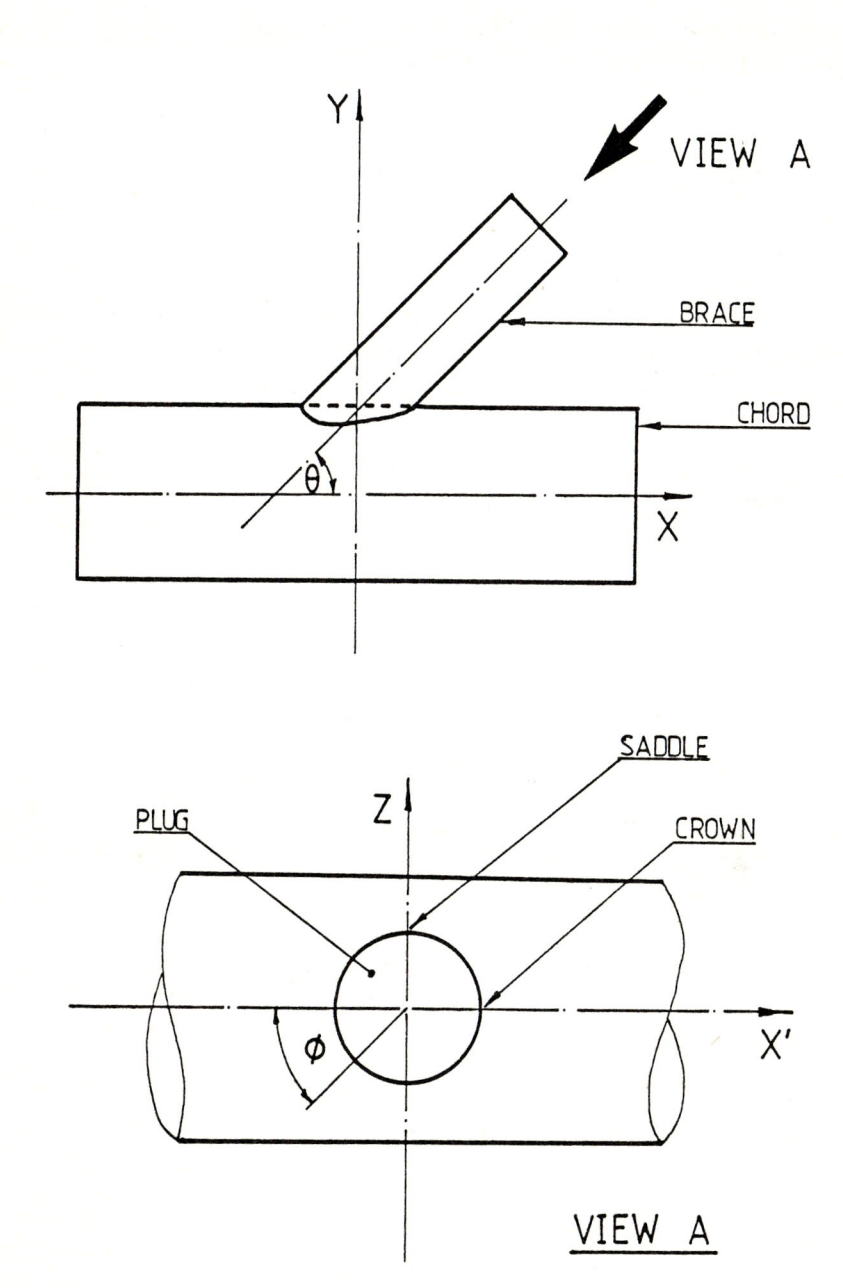

Figure 18.1 A tubular Y joint with definition of terms used

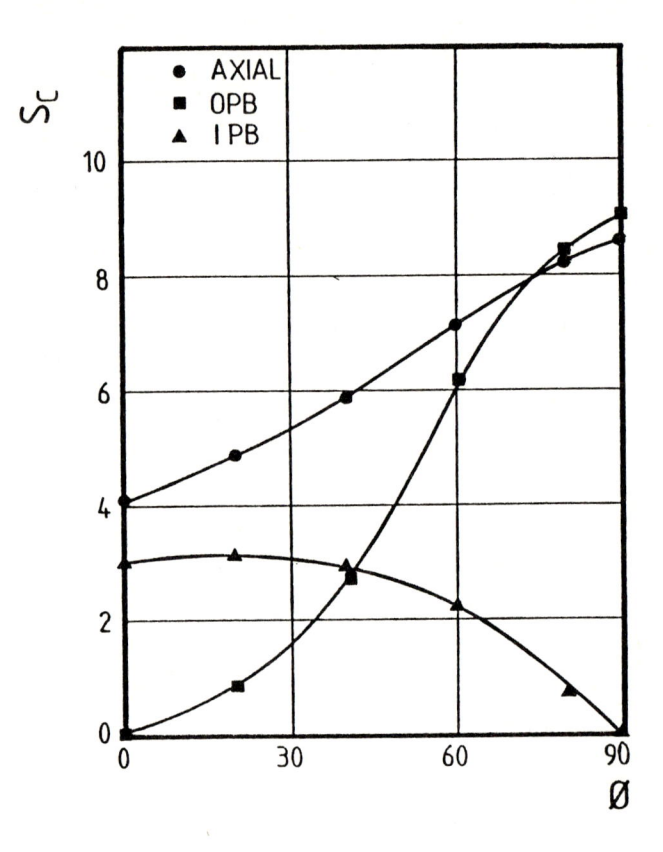

Figure 18.2 The stress distributions around a chord intersection for a T joint subjected to axial, in-plane bend and out-of-plane bend loading

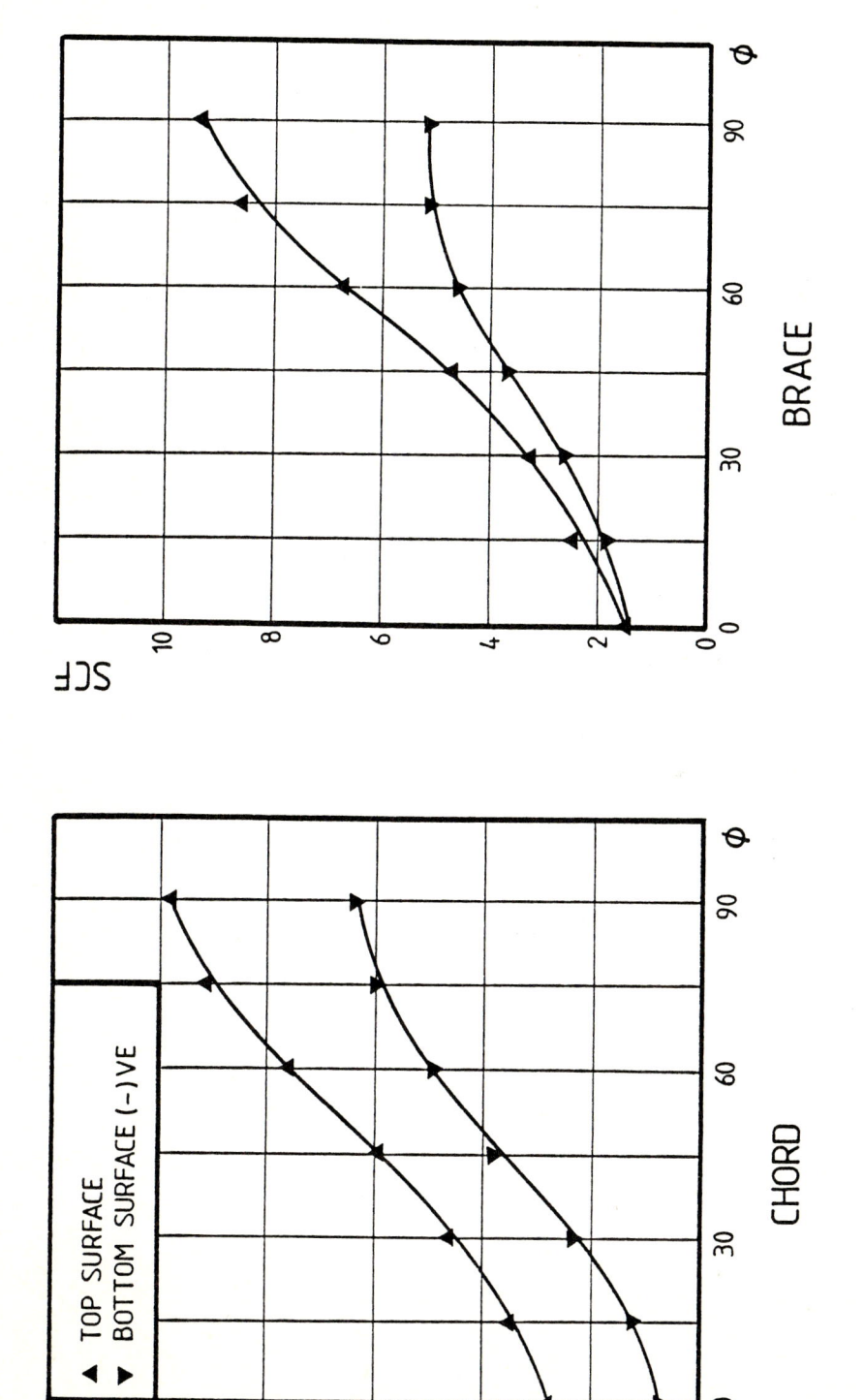

Fatigue Crack Growth Predictions

Figure 18.3 The top and bottom surface stress distributions for T joint subjected to axial loading

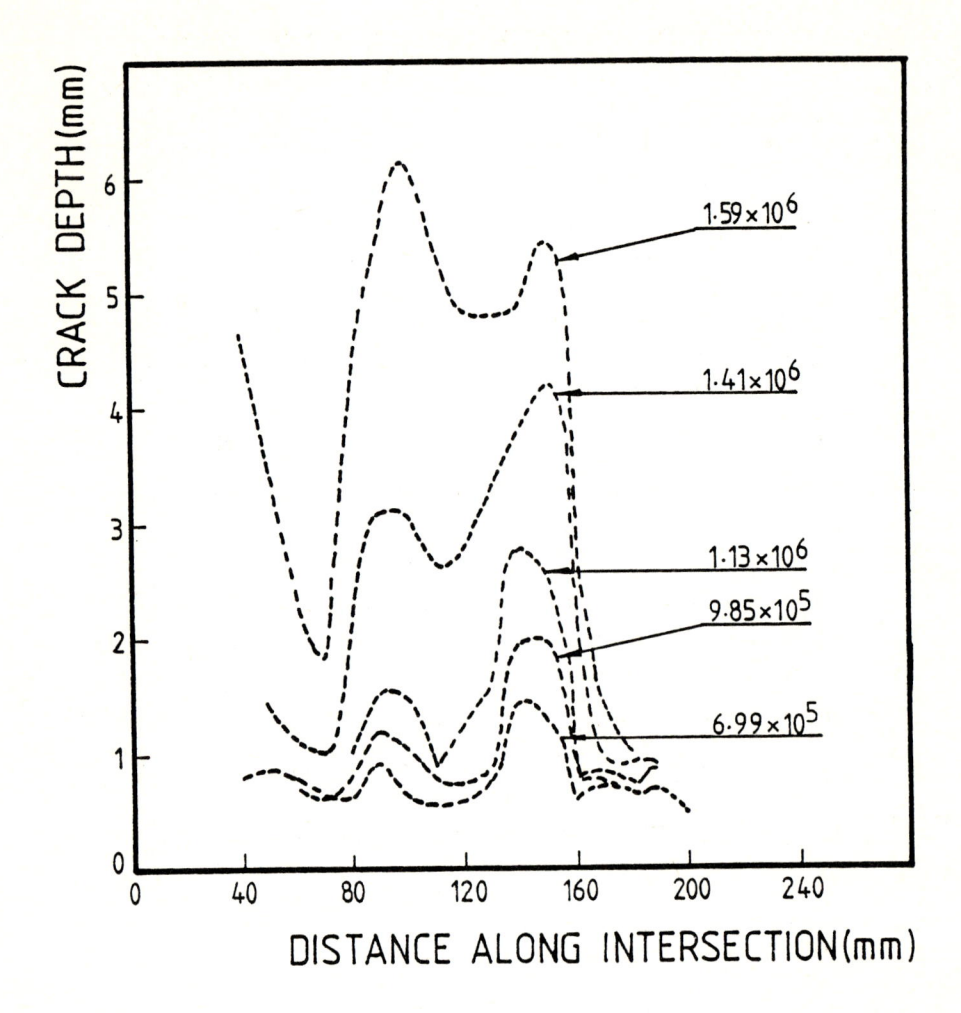

Figure 18.4 Crack shape evolution for a tubular Y joint subjected to in-plane bending

Fatigue Crack Growth Predictions

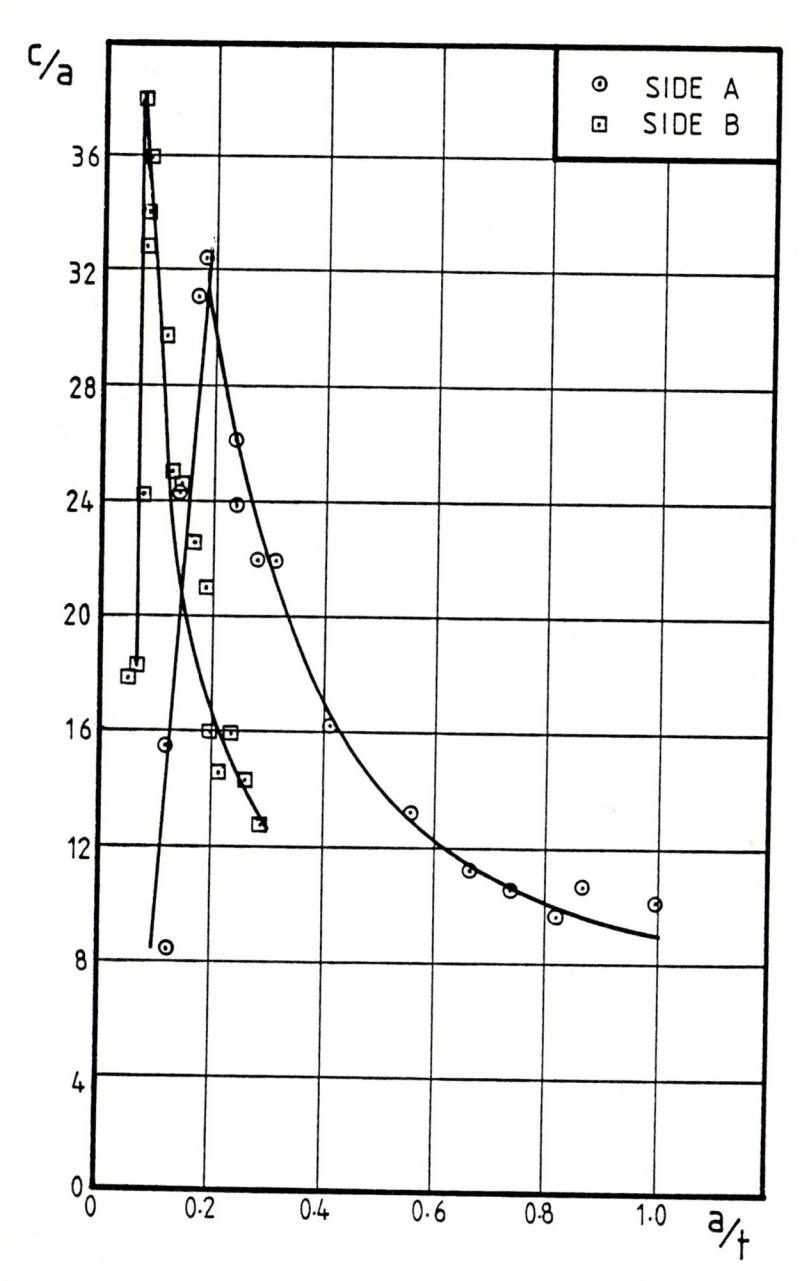

Figure 18.5 The variation of aspect ratio with increasing crack depth for a Y joint subjected in-plane bending

S. Dharmavasan

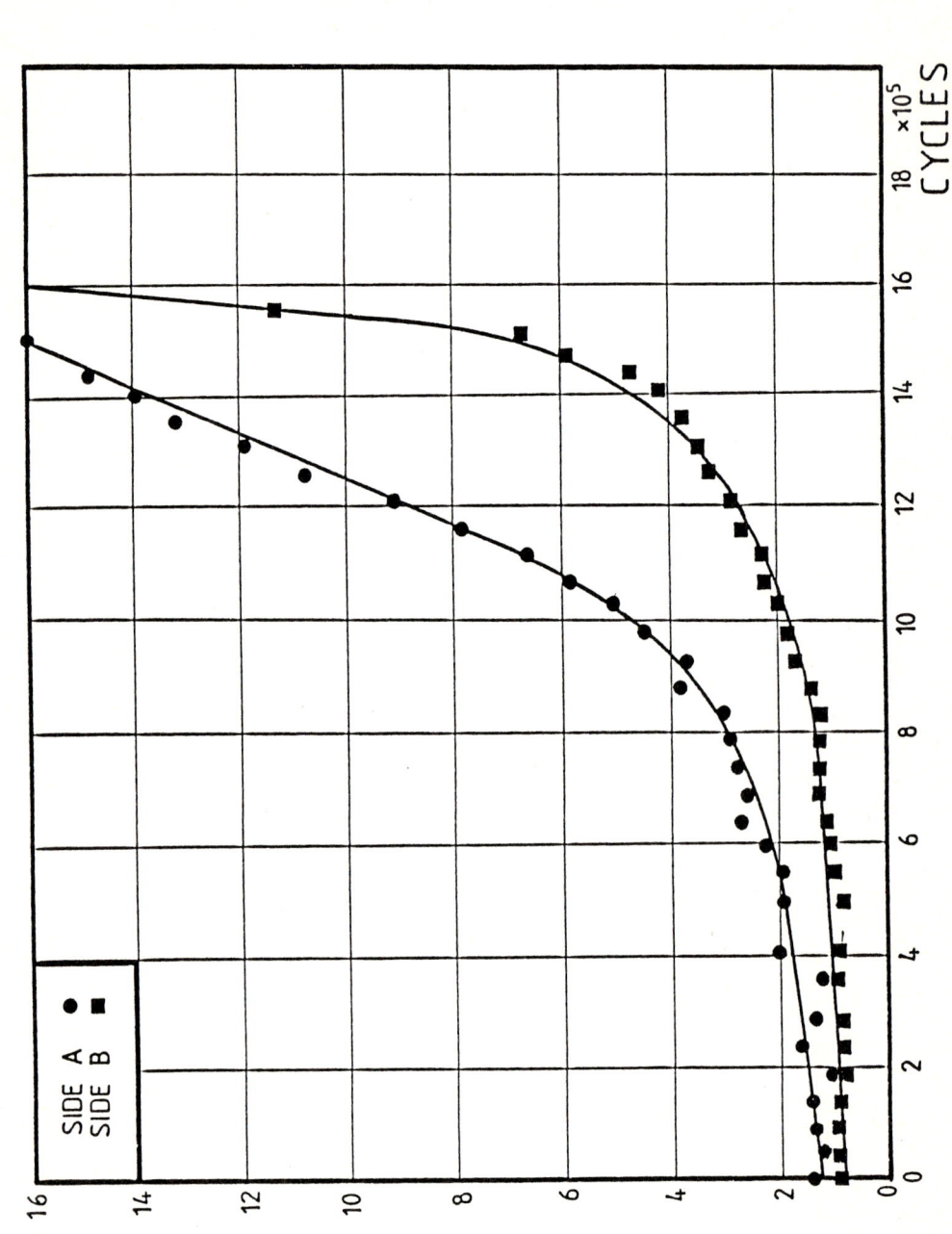

Figure 18.6 Crack depth growth curve for a Y joint subjected to in-plane bend loading

Fatigue Crack Growth Predictions

As soon as these cracks join up to form one large crack the growth rate becomes linear again, albeit at a higher rate. This is the propogation phase.

The Paris Law expresses the crack growth rate as a function of the stress intensity factor as follows:-

$$\frac{da}{dN} = C(\Delta K)^m \qquad (18.1)$$

where C and m are material properties and ΔK is the stress intensity factor range. The stress intensity factor may be expressed as follows:-

$$K = Y_s Y_\sigma \sigma \sqrt{\pi a} = Y \sigma \sqrt{\pi a} \qquad (18.2)$$

where Y_s is a factor dependent on crack shape and path and Y_σ is dependent on the loading, joint geometry and local geometry. Y is generally known as the stress intensity calibration factor and has been determined from the experimental data for the Y joint in question by using equation 18.2 with the following crack growth expression, obtained from specimen tests (5):-

$$\frac{da}{dN} = 4.5 \times 10^{-12} (\Delta K)^{3.3} \quad \text{m/cycle} \qquad (18.3)$$

The results are shown in Figure 18.7 as a plot of Y vs a/t.

Due to the complexity of tubular joint behaviour it has not been possible to derive the stress intensity factor theoretically, but some attempts have been made to propose suitable fracture mechanics models. These models should try to represent accurately the Y vs a/t plot obtained from experiment.

18.3 Theoretical Analysis of Crack Growth

An attempt has been made here, to use Linear Elastic Fracture Mechanics to describe the fatigue behaviour of tubular welded joints. The main assumption that has been made is that weld defects exist at the welded intersections and are similar to cracks and thus, there is no crack initiation as such (i.e. all the life is one of crack propogation).

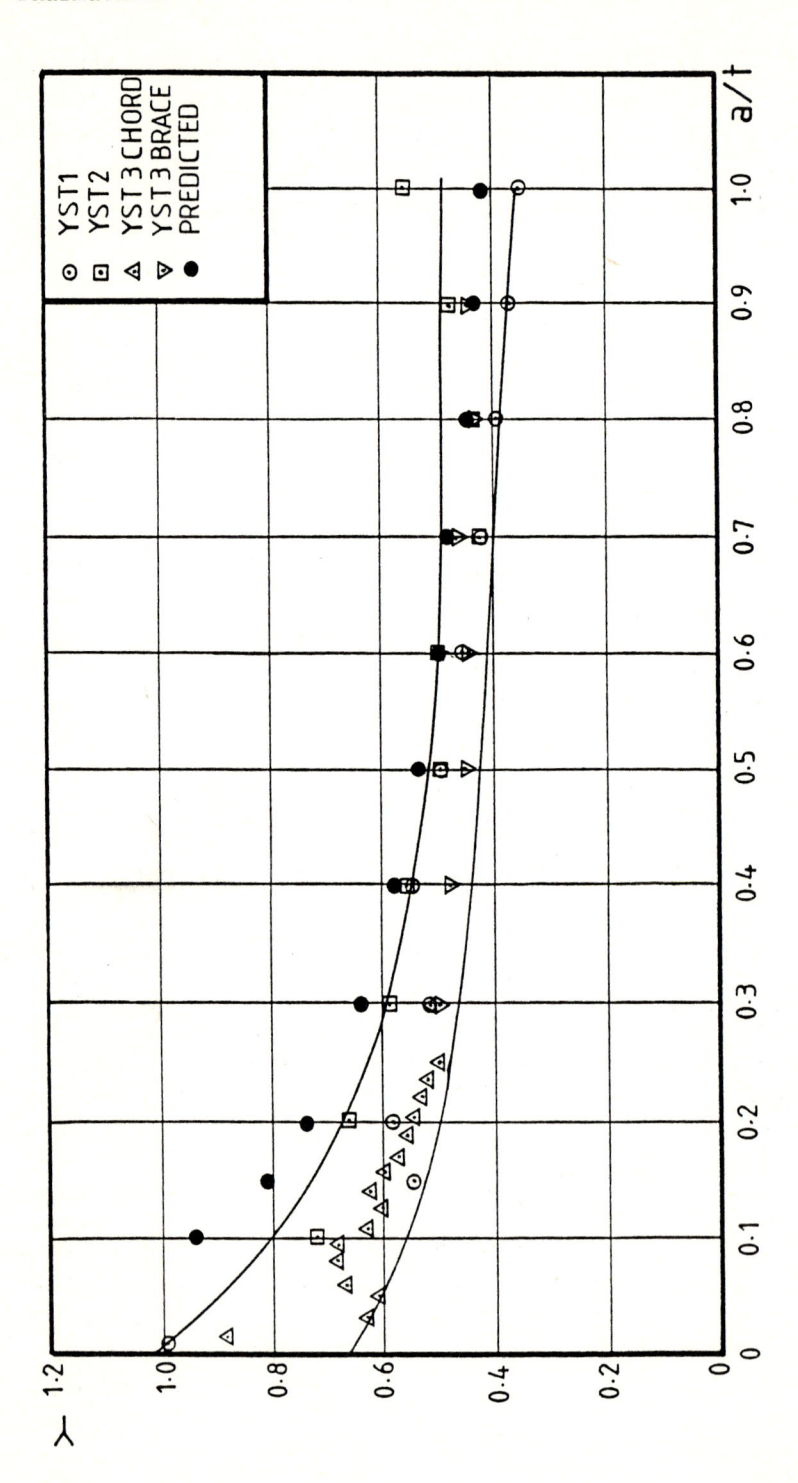

Figure 18.7 Variation of stress intensity calibration factor Y with crack depth for a Y joint subjected to in-plane bending

Fatigue Crack Growth Predictions

At present, it is impossible to derive a theoretical stress intensity factor due to the geometric complexity of tubular welded joints. Thus, in this analysis well known stress intensity factor solutions for simple geometries have been used with suitable correction factors to allow for the complex stress distribution around the welded intersection.

In a review of crack tip stress intensity factors, for semi-elliptic surface cracks, Scott and Thorpe (1982) suggest that for the case of surface cracked thin cylinders quite good results are obtained by using a stress intensity factor derived for a crack in an infinite place. They tested various stress intensity factor solutions by predicting the crack shape and comparing these with experimental data. From this it seemed that the solution proposed by Newman (1973) with a slight modification for a plate in uniform tension and the one by Koterazawa and Minamisaka (1977) for a plate in bending were the most accurate. The two solutions are reproduced below:-

Uniform Tension

$$K_I = [M_f + (E(k)\sqrt{\tfrac{c}{a}} - M_f) (\tfrac{a}{t})^P] \frac{\sigma_m}{E(k)} \sqrt{\pi a} \tag{18.4}$$

$$P_{(\pi/2)} = 1.6 + 3(\tfrac{a}{c})^3 + 8(\tfrac{a}{c})(\tfrac{a}{t})^5 + 0.008(\tfrac{c}{a}) \tag{18.5}$$

$$P_{(0)} = 0.3 + 1.15(\tfrac{c}{a})^{1.3(\tfrac{a}{t})(\tfrac{a}{c})^{0.2}} + 0.8(\tfrac{a}{c})^3 \tag{18.6}$$

$E(k)$ = Elliptic integral of the second kind

$$\simeq [1 + 1.47(\tfrac{a}{c})^{1.64}]^{0.5} \tag{18.7}$$

$$M_{f(\pi/2)} = 1.13 - 0.07(\tfrac{a}{c})^{0.5} \tag{18.8}$$

$$M_{f(0)} = [1.21 - 0.1(\tfrac{a}{c}) + 0.1(\tfrac{a}{c})^4]\sqrt{\tfrac{a}{c}} \tag{18.9}$$

S. Dharmavasan

Pure Bending

$$K_{I_{(\pi/2)}} = M_{f_{(\pi/2)}} [1 - 1.36 (\tfrac{a}{t})(\tfrac{a}{c})^{0.1}] \frac{\sigma_b}{E(k)} \sqrt{\pi a} \qquad (18.10)$$

$$K_{I_{(0)}} = [[M_{f_{(0)}} (1 - 0.3(\tfrac{a}{t}))(1 - (\tfrac{a}{t})^{12})]$$

$$+ [0.394\ E(k) (\tfrac{a}{t})^{12} \sqrt{\tfrac{c}{a}}]] \frac{\sigma_b}{E(k)} \sqrt{\pi a} \qquad (18.11)$$

The results obtained using the stress intensity factors mentioned above are found to be over-conservative. Thus, the stress drop or redistribution of stresses predicted by the formulae is less than that found from experimental data. One of the reasons for this is the presence of cracks in a region of stress concentration which varies around the intersection and the stress intensity factor solutions are strictly valid only for a semi-elliptic surface crack in an infinite place under uniform loading. Thus, it is necessary to apply correction factors to allow for the uniform stress concentration and the position of the crack.

Albrecht and Yamada (1977) define a geometric correction factor F_G which accounts for the non-uniformity of the stress field in a structural detail. This procedure requires the knowledge of the stress distribution of the uncracked body along the line of crack propagation. In the case of tubular joints this would be along the welded intersection. This function is defined as follows:-

$$F_G = \frac{2}{\pi} \sum_{i=1}^{n} \frac{\sigma x_i}{\sigma} [\arcsin \frac{x_i+1}{a} - \arcsin \frac{x_i}{a}] \qquad (18.12)$$

where σ_{x_i} is the stress between x_i and x_{i+1} as defined in Figure 18.8.

Thus, the stresses intensity is expressed as follows:-

$$K = F_G [K_{membrane} + K_{bending}] \qquad (18.13)$$

The results predicted by the use of this corrected stress intensity factor solution are shown in Figures 18.9 and 18.10. The results obtained from an earlier empirical model (Dover and Chaudhury, 1981) are also included. The agreement with experimental results is not as good as for

Fatigue Crack Growth Predictions

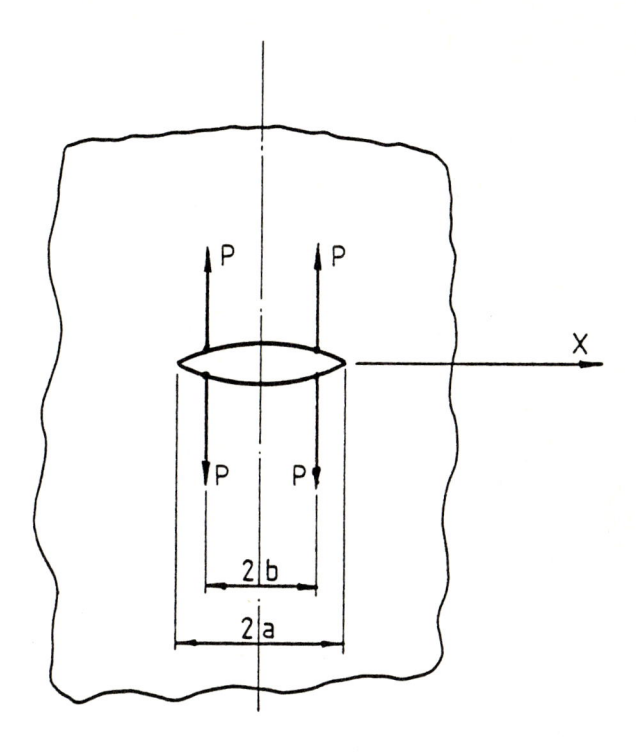

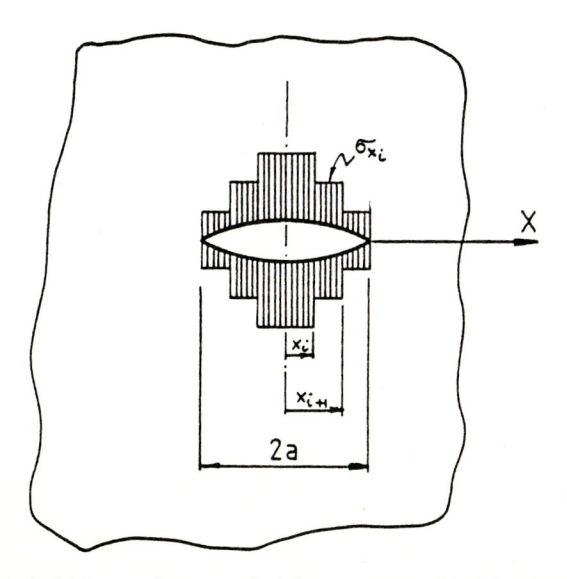

Figure 18.8 Modelling of a variable stress field along a crack for the purpose of calculating K

S. Dharmavasan

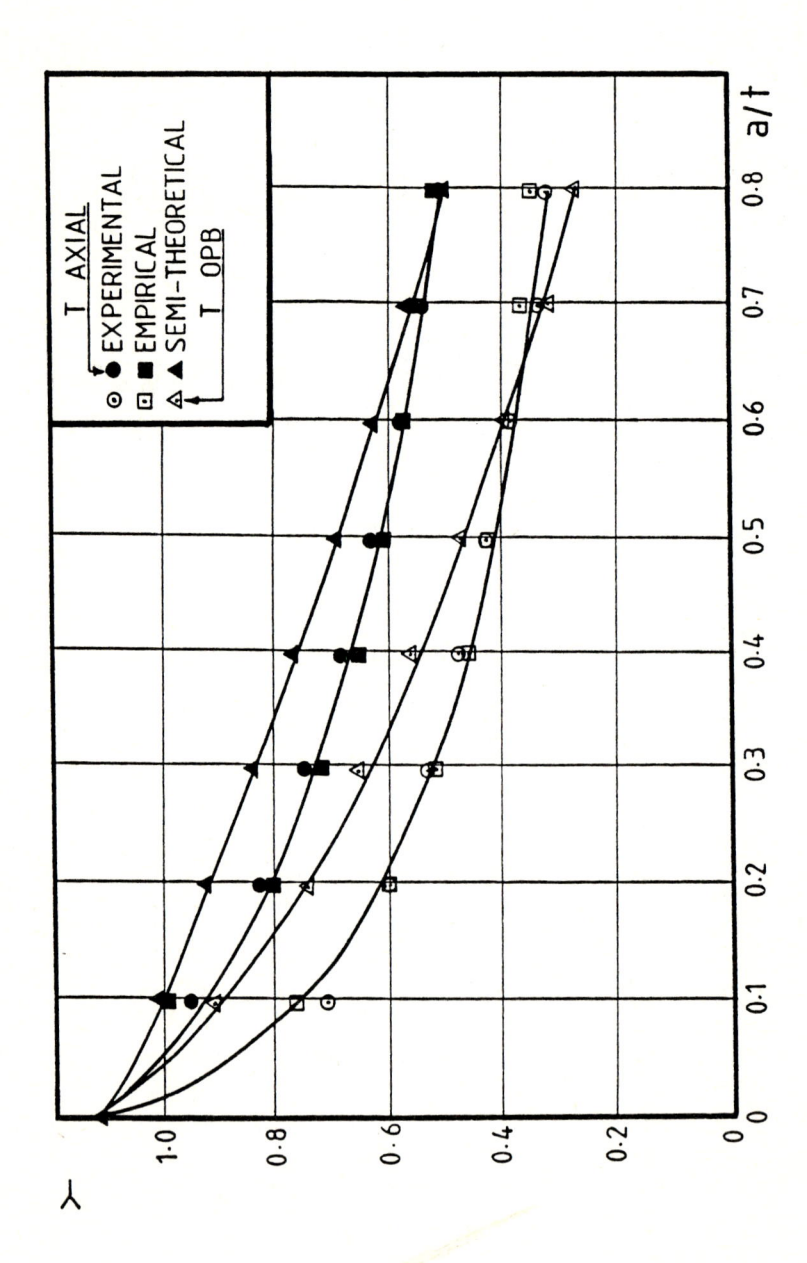

Figure 18.9 Predicted results for a T joint

Fatigue Crack Growth Predictions

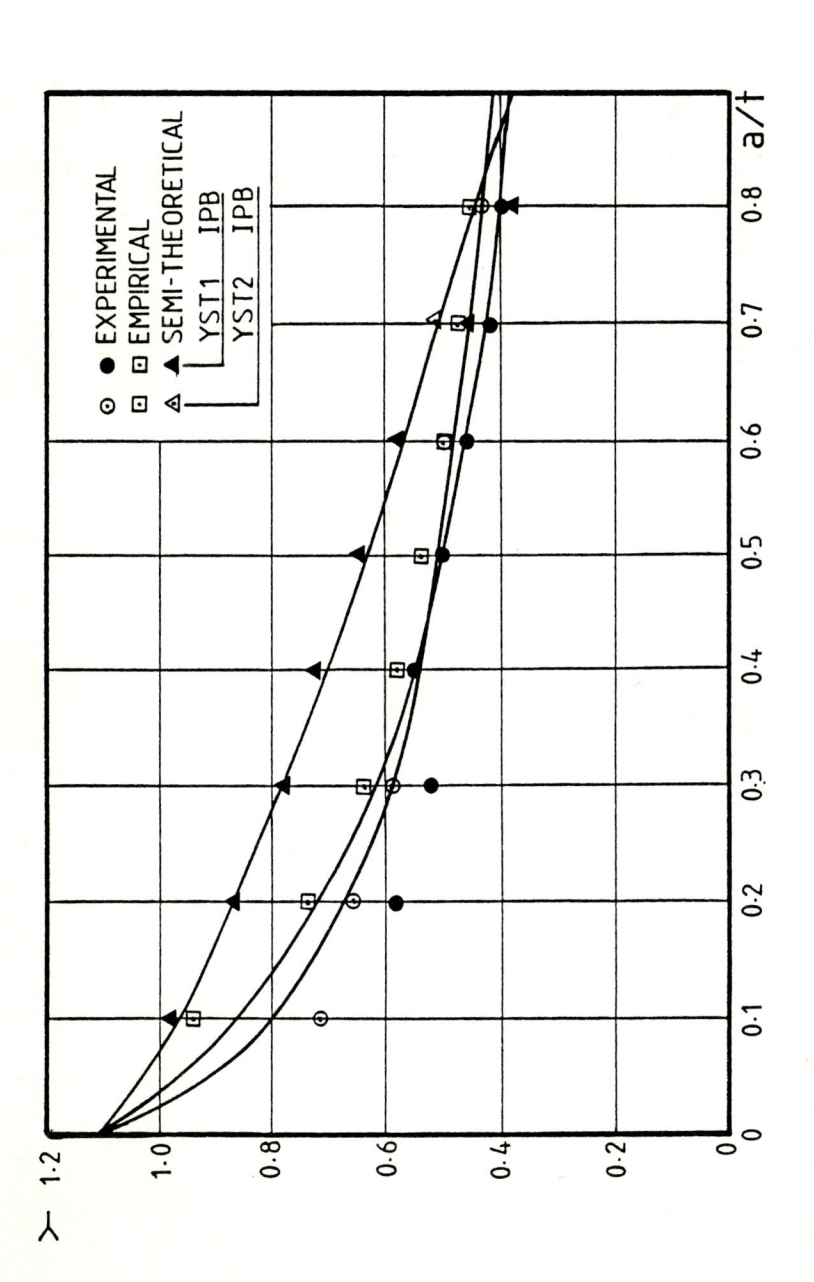

Figure 18.10 Predicted results for a Y joint

the empirical model but nevertheless the general behaviour is predicted and a conservative estimate of fatigue life would be obtained.

18.4 Conclusions

1. The fatigue crack growth behaviour of tubular joints has been defined from experiments.

2. Three distinct regions of crack growth have been identified:
    a. Initial crack growth
    b. Transition region
    c. Crack propagation region

3. A semi-theoretical crack growth model based on Fracture Mechanics has been presented. The results predicted by the model were found to be reasonable and on the conservative side.

4. The various steps for the use of this model are as follows:-

    a. Determine membrane and bending stresses around the intersection

    b. Assume or determine initial and final crack sizes.

    c. Calculate geometric correction factor for the crack size.

    d. Calculate stress intensity factor.

    e. Use Paris Law to calculate growth rate and hence cycles for an incremental change in crack size.

    f. Repeat steps c, d and e till the required size is achieved.

Acknowledgements

The author is grateful to the Science and Engineering Research Council Marine Technology Directorate for financial support of this work.

References

Albrecht, P. and Yamada, K., 1977
Rapid calculation of stress intensity factor. J. of the Structural Division, Proc. of the American Society of Civil Engineers, Vol. 103.

Department of Energy, 1983
Background to proposed new fatigue design rules for steel welded joints in offshore structures. Issue M.

Dover, W. D. and Chaudhury, G. K., 1981
Fatigue crack growth in welded T joints. ICF5, Cannes, Paper 461.

Dover, W. D. and Dharmavasan, S., 1982
Fatigue fracture mechanics analysis of tubular welded Y joints. Offshore Technology Conference, Houston, OTC 4404.

Dover, W. D. and Holdbrook, S. J., 1979
Fatigue crack growth in tubular welded connections. Proc. 2nd Int. Conf. on the Behaviour of Offshore Structures, London.

Koterazawa, R. and Minamisaka, S., 1977
Stress intensity factors of semi-elliptical surface cracks in bending. J.Soc.Mater.Sci.Jap. 26, pp 1-7.

Newman, J. C., 1973
Fracture analysis of surface and through-cracked sheets and plates. Engng.Fract.Mech., 4, pp667-689.

Scott, P. M. and Thorpe, T. W., 1981
A critical review of crack tip stress intensity factors for semi-elliptical cracks. Fatigue of Engineering Materials and Structures, Vol. 4, No. 4, pp291-309.

INDEX

Added mass effects	308, 310
advective terms	3
airpower	194
alongshore transport	221, 225, 230
atmospheric pressure	15
Beach plan shape	219
beaches	111, 126
boundary element techniques	372
boundary value system	340
breakers	125
Cable statics	317, 323, 327
case histories	137
catenary equations	303
catenary solution	323, 333
cathodic protection	313, 360, 365
channels (dredged)	123, 127
Chezy coefficients	94
circulation	3, 42
clam device	187
coastal protection	202
coastline	111, 113
compliant structures	239
concrete corrosion	355
conservation form (of p.d.e's)	84
continental shelf model	3, 9
corrosion fatigue	356
crack growth	385, 392
crack tip stress	387
cracks	356, 357, 358, 365
crevice corrosion	354
cross-isobar angle	31, 35
current	61, 71
current forces	309, 311
Diffusion	355
dimensional results	194
directional spectrum	143, 146

discretisation (of p.d.e.'s)	84
Dover Strait	21
drag coefficients	332
dredging	110, 123
dynamic analysis	208, 282, 303
dynamic programming	89
Economics	1, 235
eddy-viscosity	10, 42, 55, 56, 60, 62, 76
efficiency (of turbines)	90
eigenfunction	58
Ekman's theory	42
electrochemistry	356, 359
energy studies	90
estuarine dynamics	83
experiments (wave energy)	187
Fabrication procedure	242
fatigue life	377, 392
fetch	61
finite difference approximation	59, 84, 282, 364
finite elements (see PAFEC)	138
first order solution	170
flanges and variable radius	263, 268, 279, 284
fluid flow	361
foundation engineering	236
fracture mechanics	385
friction, drag	331, 332
friction, parameterisation of	94
Galerkin method	58
geostrophic wind	23
grain size	220, 227
group velocity	111
groynes	109
Harmonics (waves)	169
heave compensator	241
hose failure	251, 255
hydrodynamic forces	329
hydrodynamics	55, 83, 90

Incipient motion	228, 229
installation	247
interface	169
ion migration	360, 361
Irish Sea	3, 7
irregular wave drift	313
Jack-up barge	241
joints	377
JONSWAP spectrum	143
Linear wave theory	139
littoral drift	219, 220
loading	235
long wave equations	83
Manoeuvring equations	307
marine riser	326
mass transport	361
mathematical modelling	65, 95, 138, 219, 254, 280, 307, 355, 356
mathematical modelling, aims	95
method of lines	104
model calibration	221
modelling (see mathematical)	247
mooring of offshore vessels	305, 306
Non-linear interaction	21, 26
non-linear waves	138
North Sea	3, 6, 10, 15
numerical modelling (see mathematical modelling)	201, 222, 223, 364
Offshore corrosion	353
operations offshore	235
optimal control	85
PAFEC	201, 247, 248
perturbation technique	166
pipe statics	327
pipelaying	325, 345, 347
porosity	211
power take-off	186

pressure drag	329, 331, 326
pure bending	387
Quadratic friction	26
Ray methods	112, 139, 146, 180
Reynolds' number	330, 331
river model	98
rotor dynamics	191
roughness	116
Runge-Kutta Merson method	58, 259, 348
Scour	161
sea model	99
sea walls	201
second order solution	172
sediment	109
sediment transport	223
self-floating platforms	237
semi-submersible	241, 308
separation	180
Severn barrage	83
sheltering	179
shingle beach	219
shooting methods	340
shoreline changes	220, 223
simulation	192
single point mooring buoy	252
southern North Sea	3, 21, 24
spectra, wind and wave (see wave)	313
static analysis	206, 253
statics of cables	323
statics of pipes	323
storm of 1953	105
storm surges	10, 65, 95
stress life curves	377
structural engineering	237
strumming	332
surface current	66, 76
surface hose-string	251, 279
surface wind	35, 71

surge and sway	310
surge forecasting	3, 25, 65
suspended load	114
Thames estuary	15
third order solutions	169, 173
three-dimensional models	40, 54, 65
threshold of motion	221
tidal current ellipse	10, 12, 13
tidal power	90
tides	10, 15
tolerances	244
transient response	210
turbine characteristics	89
turbine efficiency	99
two-dimensional models	15
two-dimensional spectra	143
Umbilicals	324, 337
Vocoidal wave theory	163
Wave breaking	137, 154, 160
wave climate	221
wave diffraction	140, 161
wave drift effect	305
wave energy	138, 157
wave forces on vessels	309
wave growth	165
wave impact	202
wave orthogonals	139
wave rays (see ray methods)	139
wave reflection	140, 161
wave refraction	140, 141, 146, 151
wave spectrum	140, 313
wave velocity	111
waves	60, 61
waves with viscosity	176, 179
weather window	238
Well's turbine	189
wind forces on vessels	308, 320

wind gusting	314
wind shear stress	179
wind stress	15, 71, 74